WORLD AIR POWER GUIDE

WORLD AIR POWER GUIDE

David Wragg

Pen & Sword
AVIATION

First published in Great Britain in 2010 by
Pen & Sword Aviation
An imprint of
Pen & Sword Books Ltd
47 Church Street
Barnsley
South Yorkshire
S70 2AS

ISBN 978 1 84884 282 3

A CIP catalogue record for this book is
available from the British Library.

Typeset by Acredula

Printed and bound in England
By CPI

Pen & Sword Books Ltd incorporates the imprints of Pen & Sword Aviation,
Pen & Sword Family History, Pen & Sword Maritime, Pen & Sword Military,
Wharncliffe Local History, Pen & Sword Select, Pen & Sword Military Classics,
Leo Cooper, Remember When, Seaforth Publishing and Frontline Publishing.

For a complete list of Pen & Sword titles please contact
PEN & SWORD BOOKS LIMITED
47 Church Street, Barnsley, South Yorkshire, S70 2AS, England
E-mail: enquiries@pen-and-sword.co.uk
Website: www.pen-and-sword.co.uk

Contents

Acknowledgements

In writing any book of reference, the author is always indebted to those who provide assistance, and especially those who provide good photographs. I am very grateful for their help to those in the many air forces and air arms, and of course those in the aircraft manufacturers and their archivists, as well as the service attachés and advisers in the embassies and high commissions who have put me in touch with their headquarters.

David Wragg

Edinburgh, August 2010

INTRODUCTION

In any book such as this, the contents are a snapshot of the situation at a particular time, although it has also been the intention to provide a brief history whenever possible and relevant for the air forces and air arms covered in this book. History is important, for the last century saw two global conflicts and many other more localised, but often serious, wars and campaigns. It is easy to forget those wars that don't directly affect us, but the lesson of, for example, the Falklands Campaign or the war between Iraq and Iran, or those between Pakistan and India, or between the Arab states and Israel, is that war happens all too easily. It is as well to be prepared. Deterrence is so often associated with massive nuclear retaliation that people forget that conventional forces have their part to play: well-balanced, well-equipped and well-trained armed forces are a deterrent against disagreements flaring up into war.

Yet these lessons have been forgotten by politicians and their advisers, as well as by the mass of the general population. One can understand this to some extent in those countries where the day-to-day struggle to survive dominates all else, but in the well-educated and affluent democracies, such negligence is inexcusable. In the United Kingdom, just 0.28 per cent of the population is in the armed forces, and the situation is even worse in other countries, not only in Western Europe but in Canada and Australia as well. In the United States the figure is 0.5 per cent. This is the cost of having taken the so-called 'peace dividend' on the collapse of the Warsaw Pact and the break-up of the former Soviet Union. Too many have forgotten that a dividend is a return on an investment. However, there has been little investment in many of the armed forces covered in this survey. The dividend, if one is deserved at all, has been taken in advance before any benefits have accrued. The terrorist attacks on New York and Washington showed that the world is, if anything, more unstable and uncertain today than during the second half of the twentieth century. The threats are more varied. Democratic governments delude

1

themselves and fail to provide leadership by burying their heads in the sand, claiming that there will be as much as ten years' notice of an emerging threat. No such period of warning existed before hijacked airliners, as devastating as any conventional bomb or missile, slammed into the twin towers of the World Trade Center and the Pentagon. Governments anxious to remain in office accord higher priorities to expenditure on social welfare than to defence, but without adequate defence, democracy itself is threatened and effectively taken hostage. There are other means of providing for education and health, but only governments can provide defence!

Twice during the twentieth century, democracy was saved by the timely intervention of the United States. The new century has started with what has to be regarded as a *Pax Americana*. The lesson learnt should have been greater preparedness on the part of the democracies, but the United States had to take the lead in the Gulf War and in action over the former Yugoslavia. The question is not one of American leadership, since that would probably be a fact of life given the massive economic and military power of the United States. The problem is that the USA has had to provide more than the sum of its European allies in terms of effort, time and time again. Despite posturing by the European Union, too few European countries have the capability or inclination for operations outside of their own territory. Aircraft and personnel are grudgingly deployed in penny packets. New command structures, duplicating those of NATO, divert scarce funds and skilled and experienced staff officers. NATO is the longest lasting and most successful alliance in the whole of history, but it is undermined at great risk by European politics. In the late 1930s, the Dutch, Belgian and French governments knew, as did the British, that in the event of a major war, British intervention would be essential. Yet there were no joint exercises, no attempts at collaboration. NATO has provided all of this, and a command structure as well, and no less important, it formally tied the USA and Canada into protecting Europe. Given current rates of defence expenditure, in terms of a percentage of GNP often half what they were a decade or so ago, lessons have been forgotten.

The British like to believe that they always 'punch above their weight', but there are limits. Only two out of three aircraft carriers are operational. The size of the carriers in service is less than half that of the ships operated in the 1950s through to the early 1970s. In 2006, the Royal Navy lost its Sea Harriers with their Blue Vixen radar, leaving the Fleet without an air defence fighter until 2016. How is it that the United States, with just five times the UK's population, can afford eleven large aircraft carriers, each more than four times the size of a British carrier, and keep a twelfth carrier in reserve? The French now maintain just one carrier, and would be embarrassed if the ship met with an unfortunate accident of the kind all too frequent at sea. Yet the French, most of all among the Europeans, like to pretend that they can do without the United States. The malaise is global. New Zealand might not face an obvious threat today, or even tomorrow, but she has scrapped her combat squadrons, and has had to hire preserved aircraft to provide air defence training for her small navy. Meanwhile, her nearest neighbour,

Australia, is at last doing something about power projection, with two helicopter carriers entering service.

History repeats itself, but never exactly. The British idea of ten years to prepare in the 1930s seems to have returned. That was a politicians' and civil servants' ten years, not that of an airman, sailor or soldier. Given the lengthy time taken to develop modern combat aircraft and get them into service, ten years is not enough. As events in the United States in September 2001 have shown, it does not even take three or four years for a major new threat to emerge. Given the situation in the states of the former USSR, where the populations are suffering a hangover from the old regime without feeling the benefits of a free society and market economy, a reversal of the new freedoms could have popular appeal. Where would that leave the democracies?

GLOSSARY

AAM – air-to-air missile
AEW – airborne early warning
AEWC – airborne early warning and command/control
ALCM – air-launched cruise missile
ASM – air-to-surface missile
ASuV – anti-surface vessel
ASW – anti-submarine warfare
COIN – counter insurgency
CSAR – combat search and rescue
EEZ – exclusive economic zone
ELINT – electronic intelligence
FAC – forward air control
JSOW – joint stand-off weapon
MANPAD – man portable air defence (usually missiles)
MAP – Military Aid Programme
MCM – mine countermeasures
MLU – mid-life update
MR – maritime-reconnaissance
SAR – search and rescue
SIGINT – signals intelligence
UAV – unmanned aerial vehicle

NOTES

Sterling/US dollar conversions are at US $1.5644 to £1.00.
Manpower figures are from *THE MILITARY BALANCE 2010*, The International Institute for Strategic Studies.

NATO DESIGNATIONS FOR RUSSIAN AIRCRAFT AND MISSILES

The secrecy with which many aircraft were developed during the Cold War meant that there could be an interval between an aircraft being detected and its design bureau and designation becoming known to the West. To enable identification to be made in intelligence reports, aircraft were given designations beginning with 'F' for fighters; 'B' for bombers; 'C' for transports; 'H' for helicopters; 'M' for maritime-reconnaissance, and, perhaps confusingly, 'M' for trainers. Air-to-air (AAM) missiles were designated with names beginning with 'A', Air-to-surface (ASM) with 'K', and surface-to-air (SAM), with 'S'. A 'U' suffix indicated a training variant of a combat aircraft. This gave designations, often with suffixes such as 'A', 'B', etc for uprated variants of an aircraft, or missile, with notable examples including:

An-2: *Colt*
An-12: *Cub*
An-14: *Cold*
An-22: *Cock*
An-24: *Coke*
An-26: *Curl*
An-32: *Cline*
Be-6: *Madge*
Be-10: *Mallow*
Be-12: *Mail*

Il-14: *Crate*
Il-18: *Coot*
Il-22: *Coot*
Il-28: *Beagle*
Il-28U: *Mascot*
Il-76: *Candid*
Ka-25: *Hormone*
Ka-27/28/32: *Helix*
Li-2: *Cab*
Mi-4: *Hound*
Mi-6: *Hook*
Mi-8/-17: *Hip*
Mi-10: *Harke*
Mi-24/-35: *Hind*
Mi-26: *Halo*
MiG-15: *Fagot*
MiG-15UTI: *Midget*
MiG-17: *Fresco*
MiG-19: *Farmer*
MiG-21: *Fishbed*
MiG-23/27: *Flogger*
MiG-25: *Foxbat*
MiG-29: *Fulcrum*
MiG-31: *Foxhound*
Mya-4: *Bison*
Su-7/17/22: *Fitter*
Su-9: *Fishpot*
Su-11: *Flagon*
Su-24: *Fencer*
Su-25: *Frogfoot*
Su-27/30/33: *Flanker*
Tu-14: *Bosun*
Tu-16: *Badger*
Tu-20/95/142: *Bear*
Tu-22: *Blinder*
Tu-104: *Camel*
Tu-114: *Cleat*
Tu-134: *Crusty*
Tu-154: *Careless*
Yak-11: *Moose*
Yak-40: *Codling*

Missiles included:

AA-2: *Atoll*
AA-3: *Anab*
AA-6: *Acrid*
AA-7: *Apex*
AA-8: *Aphid*
AA-10: *Alamo*
AA-11: *Archer*
AS-4: *Kitchen*
AS-7: *Kerry*
AS-10: *Karen*
AS-11: *Kilter*
AS-12: *Kegler*
AS-13: *Kingbolt*
AS-14: *Kedge*
AS-15: *Kent*
AS-17: *Krypton*
AS-18: *Kazoo*
SA-2: *Guideline*
SA-3: *Goa*
SA-4: *Ganef*
SA-5: *Gammon*
SA-6: *Gainful*
SA-8: *Gecko*
SA-9: *Gaskin*
SA-11: *Gadfly*
SA-12: *Gladiator/Giant*
SA-13: *Gopher*

National entries listing

Afghanistan
Albania
Algeria
Angola
Antigua and Barbuda
Argentina
Armenia
Australia
Austria
Azerbaijan
Bahamas
Bahrain
Bangladesh
Barbados
Belarus
Belgium
Belize
Benin
Bhutan
Bolivia
Bosnia-Herzegovina
Botswana
Brazil
Brunei
Bulgaria
Burkina Faso
Burundi
Cambodia
Cameroon
Canada
Cape Verde
Central African Republic
Chad
Chile
Chinese People's Republic
Colombia
Comoros Islands
Congo
Democratic Republic of Congo (Zaire)
Costa Rica
Cote d'Ivoire

Croatia
Cuba
Cyprus
Czech Republic
Denmark
Djibouti
Dominican Republic
Ecuador
Egypt
El Salvador
Equatorial Guinea
Eritrea
Estonia
Ethiopia
Fiji
Finland
France
Gabon
Gambia
Georgia
Germany
Ghana
Greece
Guatemala
Guinea
Guinea-Bissau
Guyana
Haiti
Honduras
Hungary
Iceland
India
Indonesia
Iran
Iraq
Ireland
Israel
Italy
Jamaica
Japan
Jordan
Kazakhstan

Kenya
Korea, North
Korea, South
Kuwait
Kyrgyzstan
Laos
Latvia
Lebanon
Lesotho
Liberia
Libya
Lithuania
Luxembourg
Macedonia
Madagascar
Malawi
Malaysia
Maldives
Mali
Malta
Mauritania
Mauritius
Mexico
Moldova
Mongolia
Montenegro
Morocco
Mozambique
Myanmar
Namibia
NATO
Nepal
Netherlands
New Zealand
Nicaragua
Niger
Nigeria
Norway
Oman
Pakistan
Palestine
Panama
Papua New Guinea
Paraguay
Peru
Philippines

Poland
Portugal
Qatar
Romania
Russia
Rwanda
Saudi Arabia
Senegal
Serbia
Seychelles
Sierra Leone
Singapore
Slovakia
Slovenia
Somali Republic
South Africa
Spain
Sri Lanka
Sudan
Surinam
Swaziland
Sweden
Switzerland
Syria
Taiwan
Tajikistan
Tanzania
Thailand
Togo
Tonga
Trinidad & Tobago
Tunisia
Turkey
Turkmenistan
Uganda
Ukraine
United Arab Emirates
United Kingdom
United States of America
Uruguay
Uzbekistan
Venezuela
Vietnam
Yemen
Zambia
Zimbabwe

AFGHANISTAN

- Population: 28.4 million
- Land Area: 250,000 square miles (647,697 sq.km.)
- GDP: $14.8bn (£9.6bn), per capita $534 (£346)
- Defence Exp: $180m (£115m)
- Service Personnel: 93,800

AFGHAN NATIONAL ARMY AIR CORPS

Founded: 1937

The Afghan National Army Air Corps is the successor to the Afghan Air Force which ceased to exist with the breakdown of central government during the closing years of the twentieth century, when some two-thirds of the country was controlled by the Islamic fundamentalist Taliban. The nation's military air power was divided between the Taliban and the government-supported National Islamic Movement. Some aircraft were also in the possession of the anti-Taliban Northern Alliance, including former Iraqi aircraft flown to safety in Iran during the Gulf War. The new regime, installed with United Nations support following the US-led police action against the Taliban and the al-Qa'eda terrorists, is preparing a new army with its own air power and which is heavily dependent on outside aid.

Internal dissent is not new in Afghanistan. An army air arm had been formed in 1924, using two Bristol F.2B fighters flown by German pilots. Afghan pilots were trained by the Soviet Union, which donated a squadron of R-2 reconnaissance aircraft. This fledgling air arm disappeared when the aircraft were destroyed during a civil war in 1928–29. A Royal Afghan Air Force was formed in 1937, receiving eight Hawker Hart bombers, sixteen Meridionali Ro 37 reconnaissance aircraft and eight Breda Ba 25 trainers in 1938. British and Italian instructors established a flying school, and Afghan pilots were trained by the RAF in India. Twenty Hawker Hinds were supplied in 1939, with some remaining in service until 1957, while five out of twelve ex-RAF Avro Ansons delivered in 1948 were still in service in 1968.

A 1955 agreement with the USSR led to MiG-17 fighters, Ilyushin Il-28 bombers and Il-14 transports, Antonov An-2 transports and SM-1 (Mi-1) being delivered in 1957 with MiG-15UTI and Yakovlev Yak-18 trainers. Russian instructors arrived, and airfields were built while Afghan personnel were trained in the USSR and India.

MiG-21 interceptors replaced the MiG-17s during the early 1970s, while Mi-4 helicopters and Il-18 transports followed. A republic was declared in 1973, but in December 1979, Afghanistan was invaded by Soviet troops. The Afghan Air Force suffered considerably during the intense fighting following the invasion. Rebel activity forced the withdrawal of Soviet forces during the late 1980s, but Russian arms deliveries later resumed with MiG-23 and Sukhoi Su-7, Su-17 and Su-22 combat aircraft and Mi-8, Mi-17 and Mi-25 helicopters, with Aero L-39 Albatross trainers and ground-attack aircraft. Many of these aircraft have since been destroyed. Iran is believed to have helped the Northern Alliance, part of the National Islamic Movement, with aircraft flown there from Iraq during the Gulf War. Although Pakistan has been a supporter of the Taliban regime, no military aircraft were ever supplied. US military action during late 2001 and early 2002 met little Afghan resistance.

The new ANAAC has about 3,000 personnel and still has many ex-Soviet types in service, including twenty-six Mil Mi-8 and M-17 transport helicopters and fifteen Mi-24 and Mi-35 combat helicopters, while there are nine Bell UH-1H utility machines. There are no operational fixed-wing combat aircraft, but there are four Lockheed Martin/Alenia C-27 Spartan, two Antonov An-26 and five An-32 transports, with two Aero L-39s. As many as eighteen additional C-27 Spartan transports are believed to be on order. There are also believed to be several hundred SA-2, SA-3, Stinger, SAM-7 and SAM-14 missile launchers available.

ALBANIA

- Population: 3.6 million
- Land Area: 11,097 square miles (28,741 sq.km.)
- GDP: $13.4bn (£8.6bn), per capita $3,694 (£2,361)
- Defence Exp: $254m (£162m)
- Service Personnel: 14,295 active

ALBANIAN JOINT FORCES AIR ELEMENT

Founded: 1947

An Albanian Air Corps was founded in 1914, but the Austrian Army seized the aircraft on the outbreak of World War I. Financial difficulties hindered attempts to form an air arm between the wars, and in April 1939, Italian forces annexed the country. At the end of World War II Soviet forces occupied Albania, leading to the formation of the Albanian People's Army Air Force in 1947, with Soviet instructors and key personnel accompanied by a gift of twelve obsolete Yakovlev Yak-3 fighters and some Polikarpov Po-3 biplane trainers. A founder member of the Warsaw Pact, in 1955, Albania received MiG-15 fighters and MiG-15UTI trainers. Relations with the Soviet Union cooled in favour of closer links with Communist China, and by the early 1970s, Shenyang F-6 and F-4 fighters were introduced, although Soviet An-2 and Il-14 transports and Mi-1 and Mi-4 helicopters remained until replaced by Chinese equivalents.

Some Western equipment has been introduced in recent years, while NATO based elements of its Operation Allied Force in Albania during 1999. In 2008,

11

Albania signed the protocols to become a member of NATO, but these have still to be ratified. As Europe's poorest country, Albania finds aircraft and fuel expensive so flying hours are just fifteen hours per annum, too low for operational effectiveness. Although sometimes referred to as the Albanian Air Force, it is an air element within the Joint Forces Command. It currently operates only helicopters with seven Agusta Bell AB206s and three AB205s, as well as an AW109 and five Eurocopter Bo-105s. There are believed to be a small number of SA-2 batteries.

ALGERIA

- Population: 34.2 million
- Land Area: 919,590 square miles (2,381,741 sq.km.)
- GDP: $161bn (£103bn), per capita $4,714 (£3,013)
- Defence Exp: $5.3bn (£3.9bn)
- Service Personnel: 147,000 active, plus more than 150,000 reserves

ALGERIAN AIR FORCE

Founded: 1962

Algeria became independent of France in 1962, and almost immediately started to create an air force out of the National Liberation Army, with assistance from Egypt, the USSR and Czechoslovakia. Egyptian pilots and technicians were seconded to fly the first aircraft, five ex-Egyptian MiG-15 fighters. Initially, it was known as the *Force Aérienne Algérienne,* but now has an Arabic title. More sophisticated equipment was introduced during its first decade. By the early 1970s, it had a total of 140 interceptors in ten squadrons, including MiG-17s and MiG-21s, as well as thirty Ilyushin Il-28 jet bombers in two squadrons, and a transport squadron of An-12s and Il-18s. Western equipment was also introduced, including twenty-eight Potez Magister armed trainers and twenty SA330 Puma helicopters, the latter operating alongside thirty Mi-4s.

Russia has remained the main source of equipment for the AAF, but aircraft are also obtained second-hand from other sources, while South Africa has also become a major supplier, upgrading Mi-24 armed helicopters. It is possible that Roivalk armed helicopters will be obtained in the near future. Recent years have seen American, French and Dutch transport and communications aircraft introduced.

There are 14,000 personnel. Flying hours average 150 annually. There are five fighter, three fighter-bomber and two reconnaissance squadrons operating thirty-five MiG-29s, which finally replaced more than fifty MiG-21bisMF/UMs, twenty-nine MiG-23BN/MS/Us, fourteen MiG-25/R/Us, refurbished in the Ukraine during 2000–01, thirteen Su-24 *Fencer* and twenty-two Su-22 *Fitter,* and up to thirty-five Su-30s, using AA-2, AA-6 and AA-7 AAM missiles. The combat aircraft are supported by four Il-78 tankers. There are thirty-two Aero L-39ZA/C Albatros armed trainers, some of which will be replaced by sixteen Yak-130s. Two MR squadrons use six Beech Super King Air B-200Ts, with more believed to be on order, while six 1900s provide SIGINT. Helicopters include thirty-three Mi-24 *Hind* attack helicopters, upgraded in South Africa to MkIII standard, four Mi-6s and sixty-four Mi-8/Mi-17s, as well as nine AS355F Ecureuils, two Bell 412s and three Ka-27s. Six AW101s are on order as well as four Super Lynx 130s. Transport aircraft include seventeen C-130H Hercules, nine Il-76MD/TDs, and a VIP unit with two F27-400M Friendships, three Gulfstream IIIs and two Falcons. Training aircraft include six T-34C Turbo Mentors, forty Algerian-assembled Zlin 142s, and twenty-eight Mi-2 *Hoplite* helicopters. There are three SAM regiments with SA-3, SA-6 and SA-8.

Future acquisitions may include maritime surveillance aircraft for EEZ duties, most probably from Russia.

ANGOLA

- Population: 12.8 million
- Land Area: 481,226 square miles (1,246,369 sq.km.)
- GDP: $77.7bn (£49.7bn), per capita $6,074 (£3,883)
- Defence Exp: $2.77bn (£1.77bn)
- Service Personnel: 107,000 active

ANGOLAN PEOPLE'S AIR FORCE/*FORÇA AÉREA POPULAIRE DE ANGOLA*

Founded: 1975

The Angolan People's Air Force was formed in 1975 following independence from Portugal. Strong support from Cuba included MiG-21 fighters and Mi-8 helicopters. The country was racked by civil war between different guerrilla groups, the strongest of which was UNITA, Union for the Total Independence of Angola, until a peace agreement was reached in 1995. The armed forces were reduced in strength following the peace deal, although the APAF received MiG-23s, Mi-17s and Mi-24 *Hind* shortly before hostilities restarted in 1998, and these were later joined by MiG-23s and Su-22s from Belarus. Eight ex-RAF Hercules C-130Ks are on order from Lockheed, while there have been reports of additional aircraft, Mi-17s, MiG-29s and Su-27s, being obtained from Eastern Europe, as well as a number of Mi-35s supplied by Russia in 2001.

The threat posed by UNITA has now virtually diminished, which should affect future procurement programmes and the size of the APAF, currently at 6,000 personnel. Fighter and strike aircraft include twenty-two MiG-23s, fifteen Su-22M4/Us and fifteen Su-25s, seven Su-27s, and twenty-four MiG-21MF/bis fighters. Training and counter-insurgency operations are handled by four Pilatus PC-9s and twenty-two PC-7s plus fourteen EMB-312 Tucanos and six Aero L-29 Delfins, as well as three Cessna 172s. Attack helicopters include fifteen Mi-25/35s, eight missile-armed SA342s, twenty-five IAR-316s, five Bell 212s and eighty-two Mi-8/Mi-17 transport helicopters. MR is provided by a Fokker F-27 and five C212MPA Aviocars, with another five for transport operating alongside eight Hercules C-130Ks, seven Il-76s, twenty-three An-/2426s, eleven An-32s, five An-72/74s, two VIP Boeing 707s and twelve BN2T Islanders, a Dornier 28, a Gulfstream III and a King Air 200. HOT anti-tank missiles and AT-2 ASM are used, with AA-2 AAM.

The Angolan People's Army uses six AS565 helicopters.

ANTIGUA AND BARBUDA

- Population: 85,632
- Land Area: 170 square miles (449 sq.km.)
- GDP: $1.2bn (£0.77bn), per capita $14,022 (£8,963)
- Defence Exp: $21m (£13m)
- Service Personnel: 170 active, plus 75 reserves

Antigua and its dependency Barbuda maintains a small army and navy, but has yet to acquire aircraft for either service. There is a US Strategic Command detection and tracking station on Antigua.

ARGENTINA

- Population: 40.9 million
- Land Area: 1,084,120 square miles (2,807,857 sq.km.)
- GDP: $277bn (£177bn), per capita $6,758 (£4,319)
- Defence Exp: $2.2bn (£1.4bn)
- Service Personnel: 73,100 active

ARGENTINE AIR FORCE/*FUERZA AÉREA ARGENTINA*

Founded: 1944

Argentine military aviation dates from September 1912, when a military aviation school was established using Farman, Bleriot and Morane aircraft. Aircraft

14

appeared on manoeuvres during 1914. Additional aircraft were difficult to obtain during World War I because the Argentine was neutral. It was not until 1919 that an Italian aviation mission provided six Ansaldo SVA Primo fighters, four Caproni Ca33 tri-motor bombers, two Fiat R2 and two Ansaldo reconnaissance aircraft, and two Savoia trainers, to establish a Military Aviation Service. In 1920, a French military aviation mission visited. A Military Aviation Factory was established in 1927, building 100 Avro 504R trainers, followed by forty Bristol F2Bs and forty Dewoitine D21C-1 fighters. This substantial increase enabled the MAS to organise into two fighter, two bomber and two reconnaissance groups, reorganising again into three regiments in 1938.

Nationally-designed aircraft appeared during the mid-1930s, the Ae MB-1 light bomber, Ae T-1 liaison aircraft, and Ae MO-1 training and observation aircraft. Licence-built aircraft followed in 1937, with 200 Curtiss Hawk 75-O fighters and 500 Focke-Wulf Fw44J Stieglitz trainers, while thirty-five Martin 139-W medium bombers were bought from the USA. External sources virtually dried up again during World War II, apart from some elderly DC-2 transports received in 1944, but the Military Aviation Factory, renamed the *Instituto Aerotécnico*, produced 100 IAe DL-22 trainers.

The MAS became a separate service, the *Fuerza Aérea Argentina* in 1944. Re-equipment had to await the end of the war, when 100 Fiat G55 fighters, twenty Avro Lancaster and Lincoln bombers, Douglas C-47 and C-54 transports, and thirty Beech AT-11 Kansan trainers arrived. The *FAA* was the first export customer for the Gloster Meteor jet fighter, followed by fifty de Havilland Dove, thirty Vickers Valetta and thirty Bristol 170 transports, 200 Percival Prentice and thirty Fiat G46 trainers. Locally-produced aircraft included 100 IAe Calquin light bombers and 150 DL-22 observation aircraft.

Argentina became a member of the Organisation of American States in 1948.

After an interval of some years, re-equipment restarted in 1957, with ninety Beech T-34 Mentor and forty-eight Morane-Saulnier MS760 Paris trainers built under licence. In 1959, North American F-86F Sabre fighter-bombers were introduced, followed later by more than forty Douglas A-4 Skyhawk fighter-bombers, de Havilland Canada Beaver and Twin Otter transports, and the Argentine Dinfia Guarani and Huanquero transport and general-purpose aircraft. In the 1970s, these were joined by Dassault Mirage III fighter-bombers, eighty Dinfia Pucara COIN aircraft, Lockheed C-130E Hercules and Fokker F-27M Troopships, as well as Sikorsky S-55, Bell UH-1H Iroquois and 47G Sioux and Hughes 269-HM helicopters. Mirage 5 fighter-bombers and Israeli-built Daggers followed.

For many years, the Argentine had claimed the British colonies of the Falkland Islands, some 400 miles east of Argentina, and its dependency, South Georgia, 1,500 miles away. On 2 April 1982, its forces invaded the Falklands and South Georgia, defended respectively by just seventy and twenty-two British marines. Initial landings were from the sea, but once the airfield at Port Stanley was secured, Hercules and F-27 transports, plus chartered civilian F-27s, landed troops. FAA and *Comando de Aviación Naval* aircraft, including Skyhawks, Mirages and Étendards attacked a British naval task force, the leading ships of which set sail on 5 April. A number of Pucaras were based at Stanley, attacked by British naval aircraft. Operating from the mainland, Argentine air force and naval aircraft succeeded in sinking two British destroyers, two frigates and a container ship converted to carry most of the task force's troop-carrying helicopters. Argentine pilots showed considerable skill and courage, with the conflict lost by the poor performance of the army, despite frequently outnumbering the attacking British forces.

Afterwards, the Argentine dictatorship was overthrown, leading to a gradual reduction in the size of the armed forces. Relatively little new equipment has been received, although Skyhawks and Mirages have been upgraded, but the poor economic outlook must cast doubt on spending plans for the immediate future.

The *FAA* has 14,600 personnel, a reduction of a quarter over the past thirty years. Six squadrons operate in the fighter and strike role, with two having twenty-four uprated Dagger/Neshers, and another having six Mirage 5Ps while a fourth squadron has six Mirage III/EAs. Another two squadrons operate twenty-three upgraded A-4AR/OA-4AR Skyhawks, and two operate thirty-four IA58 Pucaras on tactical operations. There are six air transport squadrons: one with four Boeing 707 tanker/transports; two have eight C-130B/H/L-100-30 Hercules, including two KC-130H tankers; the remaining three squadrons include one with four F-27s and four Saab 340s, one with four F-28s and one with six Twin Otters. Three IA-50s are operated on communications duties. A Boeing 707, four Learjet 35As and three IA50s in a single squadron cover reconnaissance operations. Helicopters include four Bell 212s, nine UH-1Hs and seventeen MD-500s, as well as two Chinooks. A presidential flight operates a Boeing 757, a Sabreliner 75A and an S-70 Black Hawk. Training is based on four TA-4Rs, thirty T-34C Turbo Mentors, twenty-nine EMB-312 Tucanos, three MD500s and twenty IA63 Pampas, manufactured by Lockheed Martin Argentina. ASM-2 Martin Pescadors are carried by *FAA* aircraft, while AAM include AIM-9B Sidewinder, R-530, R-550 and Shafrir. As many as thirty Pucaras may be in storage.

ARGENTINE NAVAL AVIATION/*COMANDO DE AVIACIÓN NAVAL*

Founded: 1919

An Argentine Naval Aviation Service was founded in 1919, encouraged by the arrival of an Italian mission. The first aircraft were two Macchi M7, two M9 and two Lohner L-3 flying boats. HS2L and F5L flying boats followed, while later Dornier Wal and Supermarine Southampton flying boats were operated. Training aircraft were obtained from Vickers, Avro and Savoia-Marchetti. Rationalisation followed by major re-equipment in 1937 took the form of Douglas DB-8A-2 bombers, Vought V65F and V142 Corsair reconnaissance-bombers, Consolidated

P2Y-3 flying boats, and, for operation from cruisers, Grumman J2F-2 and Douglas Dolphin amphibians.

Few changes took place until 1956, when ten F4U Corsair fighter-bombers arrived, followed in 1957 by Lockheed Neptunes for MR, Consolidated PBY-5A Catalina amphibians and Martin PBM-5 Mariner flying boats. Grumman F6F-5 Hellcat fighter-bombers, JRF Goose and J2F-2 amphibians came later, with North American T-6 Texan, Vultee 13T-13 and Beech AT-11 trainers. The fighter-bombers were for the aircraft carrier *Independencia*, formerly HMS *Warrior*, which arrived in 1958. The Royal Netherlands Navy carrier *Karel Doorman*, another former British carrier, was bought in 1969 and renamed *Veinticinco de Mayo*. *Independencia* was withdrawn in 1970.

During the 1970s, *Veinticinco de Mayo* operated Grumman F-9B Panther fighter-bombers and six Grumman S-2A Tracker ASW aircraft. The Neptunes and Catalinas remained, while other aircraft included twenty-four Aermacchi MB326K armed trainers, Grumman TF-95 Cougar, Beech C-45 and North American T-28 Fennecs in the training role, plus Short Skyvan, DHC-6 Twin Otter, Douglas C-47 and C-54 transports. Helicopters included Sikorsky UH-19s and SH-34Gs, and Bell 47 Sioux, while the arrival of two new Type 42 guided missile destroyers from the UK also introduced Westland Lynx to the Argentine. Aboard the carrier, Douglas A-4 Skyhawk and Dassault Super Étendard aircraft replaced the strike aircraft at the end of the decade.

During the invasion of the Falklands, naval aircraft had to operate from shore bases because the *Veinticinco de Mayo* was having her catapult serviced. A naval Super Étendard fired the Exocet missile that sank the British Type 42 destroyer *Sheffield*. Operations were limited by a shortage of Exocet missiles, in course of delivery at the start of the campaign. An Alouette III helicopter covering the invasion of South Georgia was forced down by fire from the defending marines.

The carrier was placed in reserve in 1997 and scrapped in 1999, but there have been improvements in capability elsewhere, with the introduction of six P-3B Orion long-range MR, operating in one squadron with the three remaining Trackers, modernised with turboprops. Naval aircrew are reported to have kept their carrier-landing skills honed on the Brazilian carrier using eleven Super Étendards. Six Beech King Air/Super King Airs provide additional MR as well as VIP transport, while other transports include two F-28s and a PC-6. Helicopters include twelve Sikorsky SH-3D/H Sea Kings operated from four *Almirante Brown* (German *MEKO 360*) destroyers and an icebreaker, with four AS555 Fennecs and three Alouette IIIs aboard three *Drummond* and six *Espora* (*MEKO-140*) frigates. Training uses MB326, MB339 and T-34C Turbo Mentor aircraft, as well as AS555 helicopters. Miscellaneous aircraft include a Queen Air and PC-6B Turbo Porter communications aircraft, and A109A utility helicopters. One Type 42 destroyer remains, converted to a fast troop transport.

The air arm accounts for a tenth of the navy's 20,000 personnel. Supporting the navy is a coastguard service, the *Prefectura Naval*, with eleven aircraft including

C212-300 Aviocar transports, Puma, Panther and Dauphin helicopters for transport and SAR. It may be merged into the navy in the near future.

ARGENTINE ARMY AVIATION/*COMANDO DE AVIACIÓN DEL EJÉRCITO*

Founded: 1959

An air branch was formed in the Argentine Army in 1959 for AOP, liaison and light transport duties. At first, Cessna 182J, 310 and Skymaster aircraft were operated, joined by Piper Apache and Aztecs, Bell 206 JetRanger helicopters, three DHC-6 Twin Otters and three C-47s. It now operates small and medium helicopters, including five A-109s, forty-three ex-US UH-1Hs and five Hiller UH-12s, as well as three Antarctic-based AS332B Super Pumas, three SA315Bs, two SA330 Pumas and twenty AS532 Cougar transports. Fixed-wing aircraft include a C212-300, three G-222s, two DHC-6 Twin Otters and two Merlin IVs, a photo-survey Queen Air and Citation 1 and 9 OV-1D Mohawks for surveillance.

ARMENIA

- Population: 3 million
- Land Area: 11,490 square miles (30,384 sq.km.)
- GDP: $11.9bn (£7.6bn), per capita $4,014 (£2,566)
- Defence Exp: $376m (£240m)
- Service Personnel: 46,684 active

ARMENIAN AIR FORCE

Founded: 1990

One of the many new states spun off in the break-up of the Soviet Union, the Armenian Air Force has some 2,900 personnel and is part of this land-locked state's army. As elsewhere in the former Soviet Union, initial equipment largely consisted of Soviet material *in situ* at the time of the collapse of central control. Locked for much of its post-independence history in a struggle with neighbouring Azerbaijan, equipment has been relatively poor and in limited quantities, although this may now be changing following a defence pact with Russia. A fighter squadron operates a MiG-25 and fourteen Su-25s. Another squadron operates twenty-four Mi-24P/K/RKR attack helicopters, with ten Mi-8MT and two Mi-9 transports as well as two Mi-2s on utility duties. Fixed-wing transport is provided by two Il-76s. Training uses four Aero L-39s.

Operational effectiveness is likely to be poor.

AUSTRALIA

- Population: 21.3 million
- Land Area: 2,967,909 square miles (7,682,300 sq.km.)
- GDP: $1,119bn (£724bn), per capita $56,632 (£36,200)
- Defence Exp: $24.2bn (£15.5bn)
- Service Personnel: 54,747 active, plus 19,915 reserves

The most prosperous and stable democracy in the southern hemisphere, Australia has long been suspected of relying too heavily on first the United Kingdom and then more recently on the United States for defence. The Second World War, when the UK was unable to keep its promise to send a strong fleet to defend Singapore, should have been sufficient warning. Despite this, over the last sixty years the country has seen its defence capability reduced, including the loss of its aircraft carriers.

This is now changing. A new government which it was feared would reduce defence expenditure has given the country's armed forces a 6.4 per cent increase in the budget for 2008–2009, promised a 3 per cent increase in real terms until at least 2017, and also earmarked part of the tax windfall resulting from Chinese demand for Australian minerals for a special defence reserve fund. In 2008–2009 the money from this fund, A$1 bn, will cover the costs of the country's involvement in Afghanistan, Iraq, East Timor and the Solomon Islands.

Over the next few years, Australia will once again be able to project power beyond the range of its aircraft through the addition of two 27,000-tonne amphibious landing ships being built in Spain, the addition of twenty-four Boeing Super Hornet combat aircraft and following them up to 100 Lockheed Martin F-35s, while the army will have grown from six to eight battalions. In the longer term, a third amphibious assault ship, a new submarine force with additional vessels and more surface vessels for the RAN, and the possibility of STOVL versions of the F-35 are all possibilities, but despite the budget increases, will it be enough?

ROYAL AUSTRALIAN AIR FORCE

Founded: 1921

Australian military aviation originated with the formation of the Australian Flying Corps in 1913, with two Royal Aircraft Factory BE2a and two Deperdussin aircraft. One of the BE2as and a Farman seaplane accompanied an Australian contingent to German New Guinea shortly after the outbreak of World War I in 1914. A small force was attached to the Royal Flying Corps in Mesopotamia in 1915, fighting against Turkish forces. Further service with the RFC came in 1916 and 1917, with four squadrons serving in Egypt and the UK with FE2bs and SE5as. Post-war, the AFC was disbanded.

It was reformed in 1920, and in 1921 became the Australian Air Force, an autonomous service that acquired the 'Royal' prefix in June. It used Australia's share of the 'Imperial Gift' of war surplus aircraft, more than 100 aircraft including DH9s, SE5as, ten Sopwith fighters, twenty-six Avro 504K trainers and six Fairey IIID seaplanes. Two squadrons were formed in 1925 for army co-operation duties, partly manned by regulars and partly reservists of the Citizen Air Force, while a fleet co-operation flight operated six Supermarine Seagull MkIII amphibians. Supermarine Southampton flying boats, Australian-built de Havilland Cirrus Moth trainers, Bristol Bulldog fighters and twenty-eight Westland Wapiti general-purpose biplanes were added later. The Depression prevented further equipment purchases until 1934, when Hawker Demon fighters, Supermarine Walrus amphibians, Avro Cadet trainers and Anson reconnaissance aircraft were ordered. The last of these arrived in 1937, when a three-year expansion programme started. In addition to the de Havilland factory, Australian businessmen formed the Commonwealth Aircraft Corporation in 1936, building the North American NA-33 trainer, known in Australia as the Wirraway.

Only twelve squadrons out of a planned thirty-two were operational on the outbreak of World War II, with 164 aircraft and 3,500 personnel, including a new squadron of Lockheed Hudson bombers. A squadron of nine Short Sunderland flying boats was transferred to the RAF. Plans to contribute six squadrons to the RAF proved impractical, but many Australians served with the RAF, and eventually the RAF included thirteen Australian squadrons, including No3, the top scoring Allied fighter unit in the Mediterranean. The RAAF also participated in the Empire Air Training Scheme, training 280 pilots, 184 observer-navigators and 320 wireless-operator/air gunners each month, plus basic training of others. Wirraways and Tiger Moths were joined by 200 Australian-designed Commonwealth CA-2 Wacketts. Several hundred Ansons and Fairey Battles were sent from the UK.

After the Japanese invasion of the Dutch East Indies, the RAAF received a number of ex-Royal Netherlands Air Force aircraft, including Curtiss Helldivers and Vultee Vengeances, Republic P-43s, Douglas C-47s and Dornier Do24 flying boats. Wartime aircraft included de Havilland Mosquito, Tiger Moth and Dragon Rapide, Boulton-Paul Defiant, Bristol Beaufort, Consolidated Liberator, Curtiss A-40 Kittyhawk, Douglas Boston, Handley Page Hampden and Halifax, Hawker Hurricane, Lockheed Lightning and Ventura, North American Mitchell and Supermarine Spitfire. The 1942 Japanese naval aircraft raid on Darwin, which lacked fighter cover, led to the development and production of the Commonwealth CA-12 Boomerang, an effective fighter which later did sterling work as a ground-attack aircraft. RAAF units operated in every theatre of the war, while personnel strength peaked at 200,000 men and women.

Post-war, a maximum personnel strength of 15,000 was set. The RAAF participated in the British Commonwealth Occupation Force in Japan. Re-equipment started, with Commonwealth Aircraft building 100 North American P-51 Mustang fighters, while the Government Aircraft Factory built seventy-five Avro Lincoln heavy bombers, followed by eighty de Havilland Vampire jet fighters to replace the Mustangs, which joined the Wirraways in the Citizen Air Force squadrons. Douglas C-47s joined the Berlin Airlift. A Gloster Meteor F8 squadron in Japan was attached to the US Fifth Air Force for operations over Korea, while in Malaya, a squadron of Lincolns and one of C-47s operated with British forces against Communist bandits.

During the early 1950s, North American F-86 Sabres re-equipped the fighter squadrons, while the bomber squadrons received English Electric Canberra jet bombers and Vampire trainers arrived: all three types were built in Australia. The Sabres differed from the standard aircraft through using Rolls-Royce Avon engines. Lincolns were replaced on MR duties by Lockheed P2V-5 Neptunes. In 1955, Australia became a founder member of the South-East Asia Treaty Organisation.

The next decade saw continued upgrading of the RAAF's capabilities, with manufacture under licence of 100 Dassault Mirage IIIs and seventy-five Aermacchi MB326Hs, while Lockheed C-130 Hercules and de Havilland Canada DHC-4 Caribous were also obtained. In the early 1970s, the Canberras were replaced by leased McDonnell Douglas F-4E Phantoms as an interim aircraft before deliveries of General Dynamics F-111s, while Orions started to replace the Neptunes. Two squadrons of Bell UH-1 Iroquois were also introduced, and operated in support of Australian forces fighting alongside the US in Vietnam from 1966 until Australia decided to withdraw in 1970. A Mirage squadron was based at RAAF Butterworth in Malaysia for many years. During 2000, the RAAF assisted in the United Nations occupation of East Timor, overseeing the removal of Indonesian forces and providing air transport support.

The RAAF has 14,056 personnel, with average flying hours at 175, or 200 on the F-111. The main strike and reconnaissance capability is moving from the extensively upgraded F-111C/RF-111C/F-111G in two squadrons to twenty-four F/A-18F Super Hornets. Three squadrons with fifty-five F/A-18A Hornets provide fighter and tactical air cover, with sixteen F/A-18Bs in an OCU; while thirty-three BAe Hawk lead-in fighter trainers equip two tactical training squadrons where they replaced MB-326Hs in 2001. FAC is provided by three PC-9As. Two squadrons operate nineteen AP-3Cs on maritime-reconnaissance duties. Six Boeing 737 Wedgetail AEWC aircraft are entering service. A transport and tanker group operates two squadrons with twelve C-130Hs and twelve C-130Js; another squadron operates four C-17 Globemaster II transports while five Airbus KC-30B tankers are being delivered, replacing five Boeing 707 tanker/transports. A VIP squadron has three CL-604 Challengers and three King Air 350s. Training uses sixty-four PC-9s. Missiles include AIM-7 Sparrow, AIM-9 Sidewinder and AMRAAM, with AGM-84A and AGM-142 ASM.

Future equipment is likely to include Lockheed Martin JASSM stand-off weapons, and a number of UAV.

ROYAL AUSTRALIAN NAVY FLEET AIR ARM

Founded: 1948

While the Royal Navy made extensive use of Australian bases to support the British Pacific Fleet during the closing stages of World War II, the decision to form a Fleet Air Arm for the Royal Australian Navy was not taken until 1948. Prior to this, RAAF amphibious aircraft had operated from cruisers and the seaplane carrier *Albatross*. Initial equipment comprised two squadrons each of Hawker Sea Fury fighter-bombers and Fairey Firefly anti-submarine aircraft based ashore at Nowra, near Sydney, and aboard the light fleet carrier HMAS *Sydney* (formerly HMS *Terrible*). A second carrier, HMAS *Vengeance* was leased in 1953 pending delivery of HMAS *Melbourne* (HMS *Majestic*) in 1956.

Delivered in 1948, *Sydney* played an important part in the Korean War, joining the British carriers in operations off the coast. Later, the Sea Furies and Fireflies were replaced by de Havilland Sea Venom jet fighters and Fairey Gannet ASW aircraft, while the RAN also acquired Vampire jet trainers and Bristol Sycamore helicopters. After the arrival of *Melbourne*, *Sydney* became a training carrier and then a fast troop transport during Australia's involvement in the Vietnam War. *Melbourne*'s aircraft were updated with the addition of McDonnell Douglas A-4G Skyhawk fighter-bombers and Grumman S-2E Tracker anti-submarine aircraft, as well as twenty-seven Westland Wessex (S-58) ASW helicopters and nine Bell UH-1H Iroquois, while the Vampires were replaced by Aermacchi MB326H jet trainers in 1970–71. The RAN originally adopted the Royal Navy's 700/800 squadron numbering scheme before switching to the USN scheme.

Plans to replace *Melbourne* with the Royal Navy's *Invincible* and acquire Sea Harriers were abandoned after the UK decided to keep the ship after the Falklands campaign. Wessex helicopters had been replaced by Sea Kings ashore and aboard the carrier. *Melbourne*'s withdrawal in June 1982 marked the end of Australian fixed-wing naval aviation, leaving only helicopters, operated from eight MEKO 200 and four American-built frigates.

Australian naval aviation accounts for 990 of the service's 13,230 personnel, but this may increase in the future with the introduction of two Canberra-class amphibious assault ships, possibly with a third being ordered later. The RAN operates a single anti-submarine helicopter squadron equipped with sixteen S-70B2 Sea Hawks, complemented by six Sea King Mk50s in the utility role with six AS350B Ecureuils on training duties, while six NH90s are entering service. There is speculation that the Canberra-class assault ships may be modified to operate STOVL versions of the F-35.

AUSTRALIAN ARMY AVIATION

Founded: 1968

Established as a separate corps in 1968, Army Aviation consisted of fifty Bell 47G Sioux and a small number of Alouette III helicopters for AOP and liaison duties.

These were soon joined by twelve Boeing CH-47D Chinook heavy-lift helicopters, and by Bell OH-58A Kiowa light helicopters assembled in Australia. Between 1966 and 1970, army helicopters joined the Australian contingent in Vietnam, where, at its peak, a fifth of the country's small army was based. As part of the rationalisation of Australian defences, Iroquois helicopters were transferred from the RAAF to the Army to operate in the gunship and utility roles. Battlefield mobility has been enhanced by the addition of Sikorsky S-70A-9 Black Hawks, while Eurocopter AS350B Ecureuils have been introduced for training.

The Army operates twenty-two Tiger combat helicopters which replaced Bell 206 Kiowas, thirteen CH-47D Chinooks, thirty-four Black Hawks, fourteen OH-58s, and forty NH90s which have replaced Bell UH-1H utility helicopters. Training uses an AS350 and thirteen OH-58s. A number of UAV are in service.

AUSTRIA

- Population: 8.2 million
- Land Area: 32,366 square miles (83,898 sq.km.)
- GDP: $419bn (£268bn), per capita $51,083 (£32,653)
- Defence Exp: $3.1bn (£2bn)
- Service Personnel: 27,300 active, plus 195,000 reserves

AUSTRIAN AIR FORCE/*OSTERREICHISCHE LUFTSTREITKRAFTE*

Formed: 1955

A short-lived *Deutschosterreichische Fliegertruppe*, German-Austrian Flying Troop, was formed in 1918 on the break-up of the Austro-Hungarian Empire. After World War I, it fought successfully against Yugoslav forces during the Corinthian War of 1919, but was disbanded by the Allied Control Commission which prohibited Austria from maintaining a military air service, and also banned the country from using the title of *Deutschosterreichische*.

Austria regained full sovereignty in 1936, with plans for an air force with ten operational squadrons, including fighter and bomber units, plus training squadrons. Fiat CR20, CR30 and CR32 fighters, Junkers Ju86D bombers and Ju52/3M transports, Messerschmitt Bf108B Taifan and Focke-Wulf Fw58 Weihe communications aircraft, and Fw44 Stieglitz trainers were quickly acquired and the combat squadrons were fully operational by 1938, when Austria was incorporated into Nazi Germany. The air force was incorporated into the *Luftwaffe*, with the fighter squadrons giving notable service in Europe and North Africa as the Jagdgruppen I/135 and I/138. Post-war, Austria was briefly divided into four occupied zones by the Allies, and again banned from maintaining a military air arm.

Austria was re-established as a sovereign state in 1955, compelled to maintain neutrality under post-war agreements. It immediately created the *Osterreichische Luftstreitkrafte*, (OLk), Austrian Air Force, within the Army. The first aircraft were eight Yakovlev Yak-11 and Yak-18 trainers donated by the Soviet Union, and shortly afterwards, Austria purchased some Zlin trainers. In 1957, three de Havilland Vampire jet trainers were obtained, along with Fiat G46-5Bs, three Bell 47G Sioux helicopters, six Piper Super Cubs and two Cessna 182s for liaison and communications. By 1960, these aircraft had been joined by six Westland S-55 Whirlwind and six Alouette II helicopters. Fighter aircraft arrived in 1961, with Saab J-29F Tunnan fighter-bombers, with a second squadron formed in 1962, when thirty Potez Magister trainers also arrived. Over the next decade, forty Saab 105OE strike aircraft replaced the J-29s. The OLk also established a transport squadron with six DHC Beavers and two Short Skyvans, two Sikorsky S-65, fourteen Alouette III and nine Alouette II, twenty-two Agusta Bell 204B and twelve 206 helicopters, backed by Saab Safir trainers and Magisters. More potent aircraft followed, with twenty-four ex-Swedish Saab J-35OE Draken fighters acquired in 1988, with AIM-9P3 AAM, relegating the 105OEs to the training role. Five additional Drakens were acquired in 1999 as spares. To enhance training in the confined Austrian air space and encourage experienced fast jet pilots to stay, much Draken training took place in Sweden.

Today, the service has just 2,700 personnel and is part of a 'Joint Command' integrated armed service. It operates fifteen Typhoon fighters in one squadron while a second squadron operates up to twenty-two 105OEs as a training unit. Transport is provided by three ex-RAF C-130Ks and reconnaissance and liaison by eight PC-6s. Most aircraft are helicopters, with nine UH-60s, twenty-three AB212s and eleven OH-58s, as well as twenty-four AS316Bs. Training is conducted on sixteen PC-7s as well as the 105OE, with eleven OH-58B Kiowas.

AZERBAIJAN

- Population: 8.2 million
- Land Area: 33,430 square miles (86,661 sq.km.)
- GDP: $51bn (£32.6bn), per capita $6,190 (£3,956)
- Defence Exp: $1.5bn (£0.96bn)
- Service Personnel: 66,940 active, plus 300,000 reserves

AZERBAIJAN AIR FORCE

Formed: 1990

Created on the collapse of the Soviet Union in 1990, the Azerbaijan Air Force used former Soviet aircraft and aircraft purchased from other sources; an interesting variety but a logistics and maintenance nightmare. The AAF was involved in the fighting with neighbouring Armenia over the disputed territory of Nagorno-Karabakh until Russia mediated a cease-fire in 1994.

There are 7,900 personnel. Serviceability is believed to be extremely low. A squadron operates thirteen MiG-29s and five MiG-25s of diverse variants in fighter and reconnaissance duties, while there is a ground-attack squadron with five MiG-21s, eleven Su-24s and thirteen Su-25s. Transport aircraft include an An-12 and a Tupolev Tu-134A with three Yak-40s. A helicopter regiment has forty-nine Mi-24s and twenty-three Mi-8/17s. Training is provided on twelve Aero L-39s. There are more than 100 SA-2, SA-3 and SA-5 SAM batteries.

BAHAMAS

- Population: 307,552
- Land Area: 4,404 square miles (11,406 sq.km.)
- GDP: $7.5bn (£4.8bn), per capita $24,612 (£15,732)
- Defence Exp: $46m (£29m)
- Service Personnel: 860 active

ROYAL BAHAMIAN DEFENCE FORCE

Formed: 1973

Primarily a naval force, the Royal Bahamas Defence Force was created after independence from Britain in 1973. It operates two light aircraft, a Cessna 208 and a King Air 350, which replaced a Cessna 404 and 421.

BAHRAIN

- Population: 728,709
- Land Area: 231 square miles (570 sq.km.)
- GDP: $20bn (£12.8bn), per capita $27,446 (£17,544)
- Defence Exp: $697m (£446m)
- Service Personnel: 8,200 active

BAHRAIN AMIRI AIR FORCE

Formed: 1986

The small Gulf state of Bahrain maintains relatively powerful armed forces for its size, with modern equipment. The BAAF has 1,500 personnel, and has grown in strength since 1986, when it received its first Northrop F-5E fighter-bombers. Currently, it has one squadron with eight F-5Es and four F-5Fs; two fighter squadrons with thirty-two F-16C/Ds, with the older aircraft being upgraded to match the last ten aircraft delivered early in 2002; a helicopter squadron has twenty-four AH-1E Cobras, while there are also twelve AB-212s, three BO-105s and a UH-60L Black Hawk. Transport includes a Boeing 727, a Gulfstream II and a III, plus an RJ85. Training is on three T67M Fireflies and six Hawk Mk129s.

Separate from the BAAF, the Bahrain Navy has four SAR BO-105s.

BANGLADESH

- Population: 156 million
- Land Area: 55,126 square miles (142,766 sq.km.)
- GDP: $89.1bn (£56.9bn), per capita $571 (£365)
- Defence Exp: $1.21bn (£0.77bn)
- Service Personnel: 157,053 active

DEFENCE FORCE AIR WING

Formed: 1972

Until 1972, Bangladesh comprised East Pakistan, and participated in Pakistan's armed forces. Independence was seized after the war between India and Pakistan in 1971. Initially the air force, part of a unified tri-service defence force, operated aircraft abandoned on independence, although most Pakistani aircraft had been destroyed by Indian aerial attack. Since that time, aircraft purchases have mainly been from China, adopting Pakistan's earlier procurement policy, no doubt because the country finds Western equipment too expensive. More recently, Russian equipment has been introduced, while the transport element has been boosted through purchase of ex-USAF Hercules.

Currently, the DFAW has 14,000 personnel. Flying hours are fairly low, at around 100 to 120 per annum, doubtless due to the strain fuel purchases place on the frail economy. Fighter and ground-attack operations are handled by four squadrons with six MiG-29 *Fulcrum*, plus another two for training, ten FT-6s (MiG-19UTs), thirty-one Chengdu F-7MG/Ms, and eighteen Nanchang A-5C *Fantans*. AA-2 AAM missiles are used. Training is conducted on ten Shenyang PT-6s, eight L-39ZA Albatros and twelve Cessna T-37Bs. Transport is provided by four upgraded ex-USAF C-130B Hercules, three An-26s, three An-32s, seventeen Mi-17s and eleven Bell 212s, while two 206s are used for helicopter training.

The Bangladesh Army has three Bell 206 helicopters.

BARBADOS

- Population: 284,589
- Land Area: 166 square miles (438 sq.km.)
- GDP: $3.8bn (£2.4bn), per capita $13,353 (£8,535)
- Defence Exp: $32.5m (£20.8m)
- Service Personnel: 610 active, plus 430 reserves

Barbados has a small army and a small navy, but neither service yet operates aircraft.

BELARUS

- Population: 9.6 million
- Land Area: 80,134 square miles (209,790 sq.km.)
- GDP: $60bn (£38.3bn), per capita $6,227 (£3,980)
- Defence Exp: $611m (£390m)
- Service Personnel: 72,940 active, plus 289,500 reserves

MILITARY AIR FORCES

Formed: 1990

Formerly part of the USSR, Belarus became independent in 1990, equipped with aircraft left behind by departing Soviet forces or earmarked for locally-raised reserve forces. Reorganisation has led to the Army's aviation units being absorbed into the Military Air Forces, VVS. Attempts are being made at standardisation, while the VVS is perhaps unusual in having a surplus of Il-76 transport aircraft, many of which are in store and being sold. To raise revenue, some transports are operated in civil colours. Economic problems mean that flying hours are believed to be just twenty-eight, suggesting low operational efficiency.

The VVS has 18,170 personnel. There seems to have been some inconsistency in aircraft procurement in recent years, with MiG-29s being sold to Algeria and Peru, but further aircraft of this type ordered! Fighters and strike aircraft include forty MiG-29s, twenty-two Su-27, seventy Su-25 and thirty-five Su-24 strike aircraft. Helicopters include fifty-two Mi-24/K/Rs, supported by 108 Mi-8/-17s, thirteen Mi-26s and two Mi-6s. Transport is provided by six Il-76s and six An-26s and a VIP fleet of a Tu-134, a Tu-154 and a Yak-40. Training includes variants of the combat aircraft, plus five Aero L-39s.

BELGIUM

- Population: 10.4 million
- Land Area: 11, 778 square miles (30,445 sq.km.)
- GDP: $479bn (£306.2bn), per capita $46,000 (£29,404)
- Defence Exp: $4.23bn (£2.7bn)
- Service Personnel: 39,420 active, plus 100,500 reserves

BELGIAN AIR COMPONENT

Founded: 2002

Part of a unified defence force formed in 2002, which replaced the *Force Aérienne Belge* which itself dated from 1946 but could trace its origins to the creation of an army flying corps in 1911, using Farman HF3 and F20 biplanes manufactured under licence in Belgium. In 1912, the force used one of its aircraft for the first European trials of the new American Lewis gun. During its first two years, the flying corps was a section within the Army Balloon Company, but in 1913 it became the *Compagnie des Aviateurs* and its sixteen aircraft were organised into four reconnaissance squadrons.

On the outbreak of World War I, the strength was hastily raised through pressing privately owned aircraft into service. A new title, *Aviation Militaire*, was adopted in 1915, but a shortage of pilots and the lack of a training organisation inhibited expansion as Belgium was almost overrun by German forces, with only a small area in the vicinity of Ostend left unoccupied. Pilots were often trained in Britain at their own expense, but enough became available to form a naval air squadron for service in Belgian Central Africa. Aircraft included Royal Aircraft Factory BE2as, RE8s, Sopwith 1½-Strutters, Pups and Camels, DH9s, Farmans, Spads, Hanriots HD1s, Caudron GIIIs and GIVs, and Breguet Br14s.

Post-war reorganisation saw the *Aviation Militaire* with twenty-six squadrons in eight groups: 1st Group operated balloons; 2nd (Observation) Group operated Ansaldo A300s and DH4s; 3rd (Army Co-operation) Group operated DH4s and F2Bs; 4th and 5th (Fighter) Groups operated Nieuport 27C and Spad 13 fighters; 6th (Reconnaissance) Group operated DH4s and DH9s, and Breguet Br14s; while 7th (Technical) and 8th (Flying School) groups operated an assortment of aircraft, including Avro 504Ks and Fokker DVIIs. Many of the aircraft were licence-built in Belgium. This was followed by a further reorganisation in 1925, with the twenty-six squadrons formed into three wings, which became regiments in 1927. New aircraft types entered service, including Avia BH21 fighters and Breguet Br19 reconnaissance-bombers, Stampe et Vertongen RSV 32/100 and RS 26/180 and Morane-Saulnier MS230 trainers. The British Fairey concern opened a factory in 1931, producing Firefly and Fox fighters, while a Belgian manufacturer, Renard, produced the R31 reconnaissance aircraft.

During the mid-1930s, Gloster Gladiator fighters entered service, with Fokker FVII transports, Stampe SU-5 general-purpose biplanes and Koolhaven FK56 trainers. By the time World War II started in September 1939, Belgium was producing Hawker Hurricane and Fiat CR42 fighters and Fairey Battle light bombers, but most aircraft were obsolescent and destroyed on the ground when German forces invaded on 10 May 1940. Those that did get into the air fought against overwhelming odds. As Belgian resistance collapsed, the flying school squadrons escaped to French Morocco and other personnel managed to reach Britain from Belgium and from Africa, to be absorbed into the RAF which eventually had 1,200 Belgian personnel. Those serving in the Belgian Congo joined the South African Air Force, many seeing service in North Africa. Later, two Belgian fighter squadrons were formed within the RAF, scoring 161 confirmed victories, thirty-seven probables and sixty-one damaged enemy aircraft.

The *Force Aérienne Belge* was formed in 1946 as an autonomous air force. The initial structure included four day-fighter wings with Supermarine Spitfires and one night-fighter wing with de Havilland Mosquitoes. A transport wing operated Douglas C-47s, Avro Ansons, Airspeed Oxfords and de Havilland Dominies (Rapides), while the army co-operation wing operated Percival Proctors. The Spitfires were replaced by Gloster Meteors in 1949, and followed by Republic F-84 Thunderjets in 1951 when Belgium became a member of NATO, and eligible for US military aid, releasing Meteors to replace Mosquitoes, before being replaced in turn by Avro Canada CF-100 all-weather fighters. Longer haul transports, such as Douglas C-54s and Fairchild C-119F Packets linked Belgium with the Congo. Meteor and Lockheed T-33A trainers were also introduced, and Bristol Sycamore helicopters. Prior to 1955, a number of pilots were trained in the United States, while training facilities were also established in the Congo to take advantage of the better flying conditions.

In the early 1960s, during a civil war in the Congo, the Packets dropped Belgian paratroops to rescue civilians held hostage.

By the end of the decade, the *FAB* was operating Republic F-84F Thunderstreak fighter-bombers and RF-84F Thunderflash reconnaissance aircraft as well as licence-built Hawker Hunters. During the late 1960s and early 1970s, these aircraft were replaced by fifty Lockheed F-104G interceptors, with additional F-104s for strike duties. More than sixty Dassault Mirage 5-BRs operated in two reconnaissance squadrons with twenty-seven Mirage 5-BAs in a fighter-bomber squadron. Franco-German Transalls replaced the Packets, and twelve Lockheed C-130H Hercules entered service in 1973. Training used thirty-six SIAI-Marchetti SF260s, forty Potez Magisters and seventeen Mirage 5-BDs. Helicopters included ten Sikorsky S-58s, half of them produced under licence in France.

The Belgian armed forces were subjected to cutbacks and reductions in flying time even before the end of the Cold War, with *FAB* personnel falling from 20,500 in 1972 to 8,600 in 2001, and now to 7,203: the current figure is deceptive as the

Air Component includes activities that would otherwise be part of the Army Component and Navy Component.

Belgian army aviation dated from re-establishment of the Belgian Army after the end of World War II, initially operating Auster AOP6s which were replaced by Alouette II and III helicopters, and eventually thirty-eight IIs and forty-two IIIs were operated alongside twelve Dornier Do27s in four squadrons. The Do27s were replaced by ten BN2A Islanders. The army's contribution to the Air Component consists of twenty-eight armed Agusta A109HA helicopters.

The Belgian Navy acquired its first helicopters during the early 1960s with two Sikorsky S-58Cs, soon joined by three SA316s for operation from warships and which now belong to the Air Component.

Earlier, Belgium had joined several other European NATO members in a joint programme to acquire McDonnell Douglas F-16As, with the original seventy-two F-16AMs operated in six squadrons in the interceptor and attack role now reduced to fifty-six aircraft in four squadrons. These are equipped with Sidewinder and AMRAAM AAM, and Maverick ASM. An OCU has thirteen F-16BMs. A transport squadron has the eleven remaining C-130H Hercules, while a second squadron has an Airbus A310-200 and an A330, two Embraer ERJ-135s and two ERJ-145s. Seven Airbus A400Ms are on order and are expected to replace the Hercules. A squadron of three Sea King 48s provides SAR. Tactical support for ground forces is provided by twenty-five AW109 helicopters with ten NH90s now entering service. Training is on twenty-nine Alphajets, thirty-two SF-260s and six SA219 helicopters. UAV include eighteen B-Hunters.

Each successive modernisation reflects a further cut in numbers, and there are plans to further reduce the current size of the defence force. Successive cutbacks in budgets mean that Belgium is no longer interested in the Lockheed-Martin F-35, and its F-16s will have to remain in service until 2015. There are twenty-four Mistral SAM launchers.

BELIZE

- Population: 307,899
- Land Area: 8,867 square miles (22,965 sq.km.)
- GDP: $1bn (£0.64bn), per capita $3,319 (£2,121)
- Defence Exp: $19m (£12.1m)
- Service Personnel: 1,050 active, plus 700 reserves

BELIZE DEFENCE FORCE AIR WING

Formed: 1983

The Belize Defence Force consists entirely of a small army, with two BN2B Defenders for maritime-reconnaissance and transport, with a T67-200 Firefly and a Cessna 182 for training. Support is provided by the presence of a British Army contingent, while US Drug Enforcement Agency aircraft within the country are nominally under BDF control.

BENIN

- Population: 8.8 million
- Land Area: 44,649 square miles (115,640 sq.km.)
- GDP: $7.2bn (£4.6bn), per capita $820 (£524)
- Defence Exp: $79m (£50.4m)
- Service Personnel: 4,750 active

PEOPLE'S ARMED FORCES OF BENIN/*FORCE ARMÉES POPULAIRE BÉNIN*

Founded: 1958

As Dahomey, Benin became an independent state within the French community in 1958, with an embryonic air arm within the Army, the *Force Aérienne du Dahomey*. This was equipped with the standard package of communications aircraft allotted to former French African colonies of a Douglas C-47, three Max Holste 1521M Broussards (Bushrangers) and an Alouette II helicopter. The USA donated an Aero Commander 500B VIP transport. While the equipment has been updated since, economic problems have meant that two An-26 transports are no longer operated. Current equipment includes two Do128s and an HS748, although there are four AW109s and an AS350. There are 250 personnel.

BHUTAN

- Population: 1.5 million
- Land Area: 19,000 square miles (46,620 sq.km.)

ROYAL BHUTAN ARMY

This small Himalayan kingdom has a Do228 and two Mi-8 helicopters attached to its army for communications duties.

BOLIVIA

- Population: 9.8 million
- Land Area: 424,160 square miles (1,119,782 sq.km.)
- GDP: $18.6bn (£11.9bn), per capita $1,900 (£1,215)
- Defence Exp: $243m (£155m)
- Service Personnel: 46,100 active

BOLIVIAN AIR FORCE/*FUERZA AÉREA BOLIVIANA*

Founded: 1941

Bolivian Army officers received flying training in 1917, but the *Cuerpo de Aviación* was not formed until 1924. At first, a number of different aircraft were operated, including Breguet Br19A-2 and Fokker CV-C reconnaissance bombers, Caudron C97 and Morane-Saulnier MS139 trainers. Vickers 143 fighters and Vespa III AOP biplanes were added in 1929, and were later joined by Junkers W34 bombers, Curtiss Hawk 1A fighters and Falcon AOP aircraft, and Curtiss-Wright Osprey general-purpose aircraft, eventually totalling some sixty aircraft. This force was engaged in a war with neighbouring Paraguay during the late 1920s and early 1930s: Paraguay won.

In 1937, an Italian air mission started reorganisation, but war in Europe intervened before Italian equipment could be delivered, although three Junkers Ju86 transports were expropriated from a German-owned airline. In 1940, Curtiss-Wright 19R and CW-22 trainers arrived, followed by an American air mission the following year, which divided Bolivia into four air defence zones. The present title was adopted, although the *FAB* remained under army control. New equipment was provided by the USA, with Grumman OA-9 amphibians, Douglas C-47s, Stinson 105 Voyager and Interstate L-8 AOP aircraft, with Beech AT-11 and North American NA-16-3 trainers, before American entry into World War II cut off this source of aircraft.

In 1948, Bolivia joined the Organisation of American States and became eligible for American military aid, including Republic F-47D Thunderbolt fighter-bombers, North American B-25J Mitchell bombers and T-6 Texan trainers. The US provided eight Boeing B-17G Fortress bombers in 1958 for conversion into transports. Aircraft supplied over the next two decades included twelve North American F-51D Mustang fighter-bombers and three AT-6 armed trainers, as well as twenty North American T-6 and T-28 trainers, seven Cessna 185 liaison aircraft, eighteen C-47s and a C-54, and, more up-to-date, twelve Hughes 500M and three Hiller H-23 helicopters.

In 1973, the *FAB* received surplus Canadian Canadair T-33 and AT-33 armed-trainers, which were augmented by aircraft from other sources. The *FAB* is now largely a transport and communications force, operating three C-130Bs, three Arava 201s, three King Air 90/200s, plus a Beech 1900, a C-212, a Turbo Commander 690 and two Learjet 25s. Helicopters include five Mi-17s, eleven UH-1Hs supplied by the USA for anti-drug operations, as well as an AS316 and an AS332, and two SA350s. Training uses eleven Pilatus PC-7s, while six K-8s are on order. Currently, the *FAB* has 6,500 personnel; although up to two-thirds are conscripts.

Although land-locked, Bolivia has a small navy maintaining river and lake patrols, *Las Fuerzas Navales Bolivianos*, which uses a single Cessna 402. The Bolivian Army, or *Ejército Boliviano*, has a King Air and a CASA C212.

BOSNIA-HERZEGOVINA

- Population: 4.6 million
- Land Area: 19,736 square miles (51,199 sq.km.)
- GDP: $18.3bn (£11.7bn), per capita $3,957 (£2,529)
- Defence Exp: $281m (£180m)
- Service Personnel: 11,099 active

BOSNIA-HERZEGOVINA STATE FORCES AIR WING

Formed: 2003

The break-up of the Yugoslav Federation following the death of the former President Tito led to civil war in Bosnia-Herzegovina between Orthodox Christian Serbs and the Muslim community. Eventually, Croatian Roman Catholics sided with the Muslims against Serbian ambitions to maintain the federation. In 1995 the Bosnia-Herzegovina forces raised an army of some 30,000 personnel, with a small air arm initially operating ten Mi-8 helicopters, although after NATO intervention in 1995, the United States provided fifteen UH-1H helicopters. Starting in 2003, the armed forces were merged into a single force based on the army, but with an air wing or, officially, an air regiment.

Current equipment includes ten J-1 Jastreb and seven J-22 Orao armed trainers, while helicopters consist of eleven Mi-8/-17s, thirteen UH-1Hs and seven SA342s.

BOTSWANA

- Population: 2 million
- Land Area: 222,000 square miles (574,980 sq.km.)
- GDP: $12.1bn (£7.7bn), per capita $6,076 (£3,883)
- Defence Exp: $293m (£187m)
- Service Personnel: 9,000 active

BOTSWANA DEFENCE FORCE AIR ARM

After independence from Britain in 1966, Botswana established its own army, with an air transport and communications capability. In 1996, it acquired ten CF-5A (F-5A) fighters, and three CF-5B conversion trainers, augmented in the COIN role by seven Pilatus PC-7s (also used for training) and ten Britten-Norman BN2A Defenders. Transport is provided by three ex-USAF C-130B Hercules, two CASA CN235s and two C212-300 Aviocars, and six Bell 412 helicopters, plus a VIP Gulfstream IV. There are two Cessna 152s and ten AS350 Ecureuil helicopter trainers, and nine anti-poaching Cessna O-2 Skymasters. Ground forces are increasing from 8,500 to 10,000 personnel in the near future, and the Air Arm may also increase from its 500 personnel.

BRAZIL

- Population: 198.7 million
- Land Area: 3,287,195 square miles (8,512,035 sq.km.)
- GDP: $1.72tr (£1.1tr), per capita $8,670 (£5,542)
- Defence Exp: $29.7bn (£19bn)
- Service Personnel: 327,710 active, 1,340,000 trained reserves

BRAZILIAN AIR FORCE/*FORÇA AÉREA BRASILEIRA*

Formed: 1940

The Brazilian Alberto Santos-Dumont made the first flight in Europe in 1906, but it was not until 1913 that the Brazilian Navy established a seaplane school with an Italian Bossi seaplane. An army flying school formed shortly afterwards. A combined total of seventeen aircraft was operated by both services by 1914, including Farmans and Bleriots. As a supporter of the Allies during World War I, in 1917 Brazilian officers started training in the UK, which provided Brazil with seventy-nine aircraft, and also received combat experience in Italy.

Post-war, French assistance was provided in the creation of a Brazilian Army Air Service. By 1922, this was operating a squadron of Breguet Br14A-2s and one of Spad S7 reconnaissance aircraft, as well as a variety of training aircraft. The Brazilian Navy Air Service had three squadrons operating fourteen Savoia reconnaissance seaplanes and twelve Felixstowe F5L flying boats, while there were also twelve Avro 504 trainers and the same number of Curtiss trainers. The inter-war years saw the Army operate Boeing 256 and 267 and Curtiss Hawk 75A fighters, North American NA-44 and Vultee V-119B light bombers, and Vought O2U-1 Corsair, de Havilland Fox Moth, Gipsy Moth, Beech D-17A, Morane-Saulnier and Avro 630 aircraft on training and communications duties. Meanwhile, the Navy operated Boeing F4B-4 fighters, Curtiss-Wright Osprey general-purpose aircraft, and Fairey Gordon reconnaissance-seaplanes. There were also Brazilian-designed Muniz M-7 and M-9 trainers. Other aircraft were built under licence, including Focke-Wulf Fw44J Stieglitz and Fw58B Weihe trainers. Given the vast size of the country and the natural barriers to surface communications of the Amazonian rain forests, the Army also found itself heavily involved in aerial survey work and communications duties. In 1940, Brazil followed the growing trend towards autonomous air services by merging the two air arms into the *Força Aérea Brasileira*.

Brazil joined the Allies in 1942 after its shipping had been attacked by U-boats. Bases were put at the disposal of the Allies, and military assistance was received from the United States. The *FAB* started to receive aircraft in quantity, with 100 Curtiss P-40 Warhawk fighters, small numbers of Douglas B-18B and Lockheed A-28 bombers, twenty-five North American B-25 Mitchell bombers, 100 Vultee BT-15 Valiants, 200 licence-built Fairchild PT-19s, 125 North American AT-6s, ten Beech AT-7 and ten AT-11 trainers. Additional aircraft followed, and in 1944, a Brazilian squadron was sent to Italy with Republic F-47D Thunderbolts. Additional Thunderbolts entered service in 1945, along with twenty-nine Vultee A-35B Vengeance dive-bombers, twenty-five Douglas A-20 bombers, twenty-one Piper L-4 and forty Aeronca L-3 liaison aircraft, eight Beech C-45, eleven C-47, eight Lockheed C-60 and thirty-three Cessna UC-78 transports. At the end of the war, the *FAB* had around 1,000 aircraft.

Peace meant the inevitable reduction in strength, but when Brazil joined the Organisation of American States in 1948, American aid resumed. Additional Thunderbolt and Mitchell aircraft were delivered, plus Boeing B-17G Fortress and Lockheed PV-2 Neptune MR aircraft, and North American T-6 Texan trainers. The first jets, sixty Gloster Meteor F8 fighters and T7 trainers, came in 1954. In 1956, twelve Fairchild C-82 transports were delivered, while Brazilian production of the Fokker S11 Instructor trainer started. These aircraft were soon followed by fifty Lockheed F-80C Shooting Stars and fifty T-33A jet trainers, and Fairchild C-119 Packet transports. Additional deliveries included ten Lockheed C-130E Hercules, six HS748s, a Vickers Viscount and a number of Beech light transports.

Post-war, a separate naval air arm had been formed with an aircraft carrier, but in 1965, all aviation was once again merged into the *FAB*.

Steady modernisation of the *FAB* saw sixteen Dassault Mirage III fighters and fifteen Douglas A-4F Skyhawk fighter-bombers enter service in the late 1960s. These joined Douglas B-26K bombers and fifty-four Lockheed TF-33 armed jet trainers, the Neptune and ex-naval Grumman S-2A Tracker MR. During the 1970s, twenty-four DHC-5 Buffaloes as well as many surviving C-47, C-54 and C-119 transports augmented the Hercules. Helicopters included six Bell SH-1Ds, six UH-1Ds and seventeen 206 JetRangers as well as eighteen 47G/Js delivered earlier, and five Sikorsky UH-1Ds. A wide variety of training types included 150 Neiva IPD-6201 Universals and seven Potez Magisters, as well as surviving examples of the earlier North American, Beech and Fokker trainers which were replaced by 112 Aermacchi MB.326 trainers built in Brazil during the mid-1970s. Later, Brazil joined Italy in a collaborative ground-attack aircraft project, the AMX, also available as an operational trainer, and augmenting its earlier acquisition of Northrop F-5E Freedom fighters.

Despite having a growing indigenous aircraft industry with export sales of Tucano trainers, the cost of buying and maintaining aircraft for many years led to an ageing force and problems with serviceability with upgrading existing equipment instead of replacement. Brazil has borders with ten other states and a substantial offshore EEZ to police. In recent years some modernisation has taken place, along with re-organisation with army and naval air arms re-established, while the domestic industry has shown considerable ingenuity in adapting its own designs to meet requirements, including adapting the EMB145 regional airliner as an AEW platform.

Currently, the *FAB* has 70,710 personnel, only about 10 per cent of whom are conscripts, an increase of more than 40,000 in the past thirty years. The *FAB* has its own designation for many aircraft, but fighter defences are provided by a squadron with ten Mirage 2000Cs, with another three squadrons operating fifty-seven upgraded F-5E/F/Bs. Ground-attack operations are covered by forty AMXs, designated by the *FAB* as the A-1. Close support and COIN operations are covered by four squadrons with fifty AT-29 or Super Tucanos, a squadron of AT-26s (EMB326/MB326), and one with AT-27 Tucanos, with a squadron of four AMX/RA-1s and another with eight RT-26s in the reconnaissance role. AWACS is provided by a squadron of five R-99As (EMB145RSA), with another three in the reconnaissance role, while surveillance and radar calibration is provided by four Hawker 800XPs, which also undertake Amazonian inspection flights. ELINT is provided by three R-99s, three R-35As (Learjet 36s) and four EC-93s (HS 125s) as well as nine EC-95s (Bandeirantes). Maritime-reconnaissance has passed from the Trackers to nine P-3AMs, upgraded P-3 Orions.

Given the size of the country and the difficulties in surface communication, there is a substantial air transport capability. Strategic transport is provided by three squadrons with twenty C-130E/H Hercules. Nine squadrons have fifty-nine C-95A/B/C Bandeirantes, one has ten C-95As (ERJ-145s), and another has twelve C-105s (C-295Ms), while fifty C-212s are entering service. Other aircraft include an Airbus VC-1A (Airbus ACJ), twenty-three Cessna 208s and seventy-two

EMB110s (Bandeirantes). The tanker fleet currently includes two KC-130Hs and two KC-137s (Boeing 707-320Cs), but this will be standardised with the future delivery of twenty-three KC-390s.

Helicopters include thirty-eight UH-1Hs, while up to eighteen UH-60Ls, twelve Mi-35s and sixteen EC725s have entered service. In addition there are eight Eurocopter AS332s, six AS355s and eight Bell 206s. Training uses eight AMX-Ts, four F-5F/FMs and two Mirage 2000Bs, 113 EMB312s, fifty EMB314s and twenty-one EMB326s, and thirty-one AS350s. Missiles include AIM-9B Sidewinder, R-530, Magic 2 and MAA-1 Piranha.

BRAZILIAN NAVAL AIR ARM/*FORÇA AERONAVAL DA MARINHA DO BRASIL*

Formed with US assistance in the early 1950s, with three Bell 47J helicopters and later with two Westland Widgeon. The Royal Navy's light carrier HMS *Vengeance* was acquired in 1957 and renamed *Minas Gerais*, for which thirteen Grumman S-2A Tracker ASW aircraft were acquired with a few North American T-28C armed trainers. Other aircraft were also introduced, including fifteen Westland Whirlwind and five Wasp helicopters, while three Hughes 269As and nine 200s were acquired for training. The fixed-wing aircraft were transferred to the *FAB* in 1965, although carrier operations continued. Subsequently six Hughes 500 and four Sikorsky SH-3D Sea Kings were also bought.

In recent years, re-establishment of naval fixed-wing flying has taken place, with twenty ex-Kuwaiti A-4s (designated AF-1) and three TA-4 Skyhawks for operations from the carrier *Sao Paulo*, formerly the French Navy's *Foch*, which replaced the *Minas Gerais* in 2003. Acquisition of four ex-Royal Navy Type 22 frigates has introduced the Westland Super Lynx to the *Marinha*, with earlier Lynxes being upgraded and equipped with the Sea Skua ASM.

Currently, 1,387 of the 67,000 naval personnel are connected with naval aviation. In addition to the Skyhawks and twelve Super Lynxes, there are thirteen Sea Kings for ASW and SAR, sixteen Cougars for SAR and transport, and four S-70s, seven AS332s, eight AS355s and two Bell 206s, while training uses sixteen Bell 206s and eighteen AS350s.

BRAZILIAN ARMY AVIATION/*AVIAÇÃO DO EXÉRCITO BRASILEIRO*

Given its size, with a total of 120,000 personnel, the Brazilian Army has a relatively small air corps, and is still heavily dependent upon the *FAB* for support. It is structured into two units, both with a light transport and armed scout element. There are four Sikorsky S-70A Black Hawk helicopters, as well as nineteen Eurocopter AS550 Fennecs and thirty-six HB350-1 Esquilos (AS365 Dauphin), for attack and reconnaissance duties, and eight AS532 Cougars for transport.

BRUNEI

- Population: 388,190
- Land Area: 2,226 square miles (5,765 sq.km.)
- GDP: $14.5bn (£9.3bn), per capita $37,937 (£24,250)
- Defence Exp: $395m (£252m)
- Service Personnel: 7,000 active, plus 700 reserves

ROYAL BRUNEI AIR FORCE

Part of an integrated defence force, the RBAF supports the ground forces using 1,100 personnel. The only fixed wing aircraft are three CN-235s used for MR, four PC-7s and two SF-260W trainers. Two squadrons of helicopters include one with five Bo-105 helicopters armed with 81mm rockets, and one with ten Bell 212s and a SAR 214, four Sikorsky S-70As and a VIP S-70C.

BULGARIA

- Population: 7.2 million
- Land Area: 42,818 square miles (110, 911 sq.km.)
- GDP: $51.2bn (£32.7bn), per capita $7,108 (£4,543)
- Defence Exp: $1.11bn (£0.7bn)
- Service Personnel: 40,747 active, plus 302,500 reserves

BULGARIAN AIR DEFENCE FORCE

Formed: c.1948

Bulgarian military aviation dates from the formation of an Army Air Corps to fight in the Balkan War of 1912–13, using twelve Bleriot and Bristol monoplanes flown by foreign pilots against Turkish forces. This improvised force was disbanded once hostilities ended, to be revived in 1915 with German and Austro-Hungarian assistance once Bulgaria allied herself with them during World War I. The Central Powers provided both aircraft and pilots. Allied victory in 1918 meant that the air corps had to be disbanded a second time and the Treaty of Neuilly banned Bulgarian military aviation.

In 1937, Bulgaria denounced the Treaty and formed the Bulgarian Air Force within the Bulgarian Army. Polish aircraft were bought, including twenty-four PZL P-24-G fighters, forty-three PZL P-43 reconnaissance-bombers, Avia B534 fighter and Letov S328 reconnaissance aircraft, while Focke-Wulf Fw44 Stieglitz trainers were built under licence in Bulgaria. Bulgaria allied herself with the Axis powers in 1941, allowing German forces to use Bulgarian bases for the invasion of Greece. *Luftwaffe* instructors and advisers were seconded, and military aid included Messerschmitt Bf109E fighters, Junkers Ju86D and Ju87B and Dornier Do17 bombers, as well as Fw58 Weihe communications aircraft and Arado Ar96 trainers. In 1943, 100 Dewoitine D520s and another 150 Bf109Es were supplied so that the Bulgarians could replace *Luftwaffe* units transferred to the Russian front. Bulgaria never declared war on the USSR, but despite pleas of neutrality, the country was invaded at the end of 1944.

Post-war, Bulgaria found herself within the Soviet sphere of influence. Initially limited to an air force, BVVS, of no more than 5,000 men and ninety aircraft, Soviet assistance later allowed this to rise. Initially, obsolete Yakovlev Yak-9 fighters and Ilyushin Il-2 close support aircraft were provided, with a few bombers, transports and trainers, including Yakovlev Yak-18s. In 1953 jets arrived: twenty-four Yakovlev Yak-23 and sixty MiG-15 fighters, as well as Il-10 ground-attack aircraft.

A member of the Warsaw Pact, Bulgaria continued to receive Soviet assistance and the air force strength grew to 12,000 by 1970. More modern equipment arrived, initially MiG-17s, then MiG-19s and eventually MiG-21 fighters, with Ilyushin Il-28 reconnaissance-bombers and Il-12 and Il-14 transports. Mil Mi-4 helicopters were also supplied, while training used Yak-18s, Aero L-29 Delfins and MiG-15UTIs. Growth continued throughout the Cold War years with personnel reaching some 20,000, and older aircraft were steadily replaced, with deliveries of MiG-23, -25 and -29 interceptors, and Sukhoi Su-22 and Su-25 attack aircraft. Bombers disappeared in favour of a tactical fighter and ground-attack force able to operate in close support of ground forces aided by Mi-8 transport and Mi-24 attack helicopters.

The collapse of the USSR and of the Warsaw Pact have left Bulgaria struggling to maintain an increasingly elderly air force, cut off from cheap Soviet equipment and fuel, so that flying hours average between thirty and forty per year and serviceability of aircraft is poor. The number of operational bases has reduced to just five from a dozen, while personnel has halved over the past six years to 9,344. Fighters form two squadrons, one with twenty MiG-29As and one with twenty-six MiG-21s, while two ground-attack squadrons operate thirty-four Su-25K/UBKs, some of which have been upgraded to be compatible with Bulgaria's new alliance, NATO. Transport consists of three An-26s, five C-27Js, three L-410s and a PC-12. Helicopters include seven Mi-24s in the attack role, thirteen Mi-8/-17s and twelve AS532s. Training uses six Bell 206s and six PC-9s.

As is usual with former Warsaw Pact countries, there is no separate army air corps, but the small Bulgarian Navy operates four Mi-14 ASW helicopters and six AS565s are entering service

BURKINA FASO

- Population: 15.7 million
- Land Area: 105,839 square miles (274,123 sq.km.)
- GDP: $9.3bn (£5.9bn), per capita $606 (£387)
- Defence Exp: $123m (£78.6m)
- Service Personnel: 10,800 active

BURKINA FASO AIR FORCE/*FORCE AÉRIENNE DE BURKINA FASO*

Formerly Upper Volta, Burkina Faso formed an air transport and communications force within the Army. Burkina Faso became a Soviet client state during the 1980s, and at one time the *FABF* received MiG-17s and MiG-21s, which have long since ceased to be operational. The current air arm has just 600 personnel, and its only armed equipment is five SF-260W/WL armed trainers, while there are also three Mi-17 transport helicopters, and a Eurocopter SA316B and an AS350, as well as two Cessna 152s for training and liaison. Transport is provided by a VIP Boeing 727, a Nord 262, an HS748, a CN-235 and three Beech King Air 200s.

BURUNDI

- Population: 9.5 million
- Land Area: 10,747 square miles (27,834 sq.km.)
- GDP: $1.11bn (£0.7bn), per capita $121 (£77)
- Defence Exp: $82m (£52.4m)
- Service Personnel: 20,000 active

BURUNDI NATIONAL ARMY/*ARMÉE NATIONALE DU BURUNDI*

A small air wing with 200 personnel is operated within the Army. Serviceability of the aircraft is believed to be extremely poor given the country's parlous economic state and after a civil war in the mid and late 1990s. Combat aircraft include five SF260W Warrior armed trainers, two Mi-24 and four SA342L helicopters, while transport aircraft include two C-47s, four Mi-8s, which are believed to be non-operational, as well as three Alouette III helicopters for light transport and liaison and two Cessna 150L trainers.

CAMBODIA

- Population: 14.5 million
- Land Area: 71,000 square miles (181,300 sq.km.)
- GDP: $10.2bn (£6.5bn), per capita $706 (£451)
- Defence Exp: $222m (£141.9m)
- Service Personnel: 124,300 active

ROYAL CAMBODIAN AIR FORCE

With its neighbours, Vietnam and Laos, Cambodia had a chequered history throughout much of the twentieth century. It was part of French Indo-China until independence in 1949, becoming an independent kingdom within the French Union until complete independence in 1955. An air force, Royal Khmer Aviation, was planned in 1953, for police and communications operations. Its first aircraft were seven Fletcher FD-25A/13 Defender light attack aircraft, followed in 1955 by seven Morane-Saulnier MS733 Alcyon trainers. American military aid started in 1956, with eight Cessna L-19 Bird Dog AOP aircraft and Douglas C-47 and DHC-2 Beaver transports. As attention centred on South East Asia, aircraft started to arrive from both sides of the Iron Curtain. France provided thirty surplus Douglas A-1D Skyraiders, Potez Magister trainers and Alouette II light helicopters; the United States, North American T-28D armed-trainers, Curtis C-46 Commando transports, Sikorsky S-58 helicopters, and Cessna T-37B and North American T-6G trainers; the Warsaw Pact provided MiG-15 and MiG-17 fighter-bombers and An-2 transports.

Cambodia claimed to be neutral, but allowed North Vietnamese troops to use its territory for operations against South Vietnam. In 1970, a pro-Western government took over, creating a republic and RKA became the Cambodian Air Force. The following year, Viet Cong guerrillas destroyed all the MiGs on the ground. The country suffered a major civil war from 1970 to 1975, during which most of the air force's remaining aircraft were destroyed and in 1975 the country became isolated as international opinion turned against the Khmer Rouge regime. This was overthrown in 1979 by invading Vietnamese troops, and a new constitution was established in 1981, with the country known as Kampuchea until the restoration of the monarchy.

The 1,500-strong Royal Cambodian Air Force operates a small number of combat aircraft, with a squadron of fourteen MiG-21bis/UMs, some of which have been upgraded in Israel. There are also five L-39 armed-trainers. Transport consists of two An-26s, two Y-12s, a BN2 and a Cessna 421. Helicopters include twelve armed AH-1E/Ps, thirteen Mi-8/-17s and an AS355. Apart from the L-39s, training uses five MiG-21UMs. The size of the RCAF has dropped by a quarter over the past six years.

41

CAMEROON

- Population: 18.9 million
- Land Area: 183,000 square miles (475,500 sq.km.)
- GDP: $24.1bn (£15.4bn), per capita $1,279 (£817)
- Defence Exp: $306m (£196m)
- Service Personnel: 14,100 active

CAMEROON AIR FORCE/*L'ARMÉE DE L'AIR DU CAMEROUN*

Originally a French African colony, Cameroon has a small air force with around 350 personnel, although it is active since the country's coastal borders are disputed with its neighbours Nigeria and Equatorial Guinea. A composite squadron operates armed trainers, four Alphajets, five Potez CM-170 Magisters and six ex-South African MB-326 Impala, although two are used for training; four armed SA-342L Gazelle helicopters and two MR Dornier Do128 Skyservants. Transport aircraft include three Lockheed C-130Hs, four DHC-5D Buffaloes, a VIP Boeing 707 and Gulfstream III, an IAI-201 Arava and a Piper Aztec, a Bell 206, an SA319 Alouette III, and two AS332L Super Pumas. The army is believed to have control of four SA342 helicopters.

CANADA

- Population: 33.5 million
- Land Area: 3,851,809 square miles (9,976,185 sq.km.)
- GDP: $1.47tr (£0.94tr), per capita $44,104 (£28,192)
- Defence Exp: $20.14bn (£12.85bn)
- Service Personnel: 65,722 active, plus 33,967 reserves

Like Australia, Canada has taken the decision to expand its armed forces and the current objective is to raise the overall personnel figure to 70,000, although the size of the reserve forces will be halved. In the recent past, questions have been asked about the viability of the Canadian armed forces, but these problems seem as if they are to be addressed. This is important as the country has forces deployed in fifteen foreign countries, many of them on peace-keeping or security duties, but also playing their part in NATO command structures. Even so, the proposed expansion will still mean that only 0.21 per cent of the population is in the armed forces, a very low figure.

While the Royal Canadian Navy, Royal Canadian Air Force and Canadian Army were merged into a single Canadian Armed Forces in 1967, with Air, Maritime and Land Forces Commands, much of this structure has been dismantled and while there is an overall Canadian Forces with a Chief of the Defence Staff, and a Canada Command (CANADACOM) for all domestic operations, a Canadian Force Expeditionary Command (CEFCOM) for all overseas operations, a Canadian Special Operations Forces Command (CANSOCOM) and a Canadian Operational Support Command (CANOSCOM), there are references to the Canadian Air Force, Canadian Navy and Army, even officially, and these recruit independently and present an individual identity to the outside world.

CANADIAN ARMED FORCES/*FORCES ARMÉES CANADIENNES*

Founded: 1967

Officially Canada has a unified defence force, but in recent years many of the changes made when the Canadian Army, Royal Canadian Navy and Royal Canadian Air Force were amalgamated in 1967 have been reversed, so that Air Command is often referred to as the 'Canadian Air Force'.

Canadian military aviation dates from 1914 and the formation of a Canadian Aviation Corps with a single aircraft that was scrapped the following year after accompanying Canadian troops to France on the outbreak of World War I. The British Royal Flying Corps established a flying school in Canada and many Canadians flew with the RFC in Europe. Licence-production of Avro 504 and Curtiss JN-4 aircraft started in 1916 and by 1918 two Canadian squadrons were flying SE5a fighters and DH9A bombers. The Royal Canadian Naval Air Service was formed in 1918, with twelve Curtiss HS2L flying boats and eight Sopwith seaplanes, but disbanded in 1919.

In 1920, the Canadian Air Force was formed within the Canadian Army with the British 'Imperial Gift' of war-surplus aircraft. Canada received eighty aircraft of Airco, Avro, Bristol, Curtiss and Sopwith manufacture. King George V bestowed the 'Royal' prefix in 1923. Initially, the new service was organised on a non-permanent basis to provide refresher training for personnel with wartime experience, but this proved unsatisfactory and the service was reorganised and re-established on a permanent basis in 1924. Many of the older flying boats were replaced in 1924 by eight Vickers Viking Vs, and in 1925, the first Canadian-built Vickers Vedettes joined the force. In the early years, as much as two-thirds of the budget was devoted to civil operations, including fire-watching, surveying and communications, but this started to change as the Depression took effect and a 20 per cent cut in strength occurred in 1932. New aircraft had been entering service in small numbers, although relatively few were combat types. Some Armstrong-

Whitworth Siskin III fighters were delivered, but other aircraft deliveries included Ford Trimotor transports, Armstrong-Whitworth Atlas liaison and AOP aircraft, de Havilland Gipsy Moth, Curtiss-Reid Rambler, Avro Avian, Tutor and Fleet trainers. Just before the cuts, some Fairchild 71-C seaplanes were introduced.

In 1935 an expansion programme began, and in 1936, most civil duties were transferred to the Department of Transport. New squadrons were formed, some with obsolete ex-RAF Westland Wapitis. New aircraft included Vickers Vancouver flying boats, followed by eighteen Supermarine Stranraer MR flying boats and twenty Northrop Delta transports. As the international situation deteriorated, re-equipment and expansion accelerated, with many aircraft being built in Canada, often by subsidiaries of British companies such as Avro and de Havilland. New aircraft included Grumman GE-23 fighters, Bristol Blenheim bombers and Blackburn Shark torpedo-bombers, Noorduyn Norseman transports and North American NA-16-3 trainers. As it expanded, the RCAF was organised into three commands, Eastern, Western and Training, and became a separate service in 1938. A British air mission visited Canada, to arrange Canadian production of aircraft for the RAF, and prepare for Canadian participation in the Empire Air Training Scheme, producing pilots for the RAF and other air forces.

The outbreak of war found the RCAF with 270 aircraft in eight regular and twelve reserve squadrons, but only a relatively small number were modern, including twenty Hawker Hurricane fighters and ten Fairey Battle light bombers. Airspeed Oxford and Avro Anson trainers were other new acquisitions. One RCAF Hurricane squadron served with distinction in the Battle of Britain. By 1941, six RCAF squadrons were based in the UK. During the war, the RCAF served mainly in Europe and in protecting shipping in the North Atlantic, but a number of squadrons operated from its other coastline, over the Pacific, and in 1942, one was posted to Ceylon (Sri Lanka). Allied governments decorated some 8,000 RCAF personnel, and more than 17,000 died in action. Wartime aircraft included Hawker Hurricane and Typhoon, Supermarine Spitfire, Curtiss Kittyhawk and Tomahawk, Bell Airacobra, de Havilland Mosquito, Bristol Beaufighter and North American Mustang fighters, night-fighters and fighter-bombers, Fairey Battle light bombers and Albacore torpedo-bombers, Vickers Wellington, Handley Page Hampden and Halifax, Avro Manchester, Lancaster and Lincoln, Bristol Blenheim and Consolidated Liberator bombers. Northrop Nomad, Lockheed Hudson and Ventura, and Douglas Digby MR aircraft were joined in this role by Consolidated PBY-5A Catalina amphibians and Saunders-Roe Lerwick flying boats. Transport aircraft included Lockheed Lodestars and Douglas C-47s, while trainers included de Havilland Tiger Moths, Fleet Forts and Finches, Fairchild Cornells, Boeing-Stearman PT-27s and Cessna T-50s.

At the end of the war, RCAF strength was cut sharply from ninety squadrons to eight regular and fifteen reserve squadrons, although limited expansion was also planned. The RCAF also gained its first peacetime overseas commitment, joining the occupation forces in Germany, then remaining as part of NATO's defences throughout the Cold War with a peak of twelve fighter squadrons based in the former West Germany during the 1950s.

The first post-war aircraft were North American B-25 Mitchell light bombers and Beech C-45F Expeditor light transports, plus extra Mustangs and Lodestars. Long-range transport came with Canadair DC-4M North Stars – Canadian-built C-54s. The first jets, de Havilland Vampires, arrived in 1948. These were followed by Avro Canada CF-100 Canuk all-weather fighters and Canadair built F-86E Sabres under licence. Lockheed T-33A trainers were also built in Canada. Canada became a founder member of NATO in 1949.

Post-war, Lancaster and Lincoln heavy bombers were used for MR, but these were replaced first by Lockheed P2V-7 Neptunes and then by the Canadair Argus, a variant of the Yukon transport developed from the Bristol Britannia. Canadair also built the CC-109 Cosmopolitan; licence-built turboprop Convair Metropolitans to replace the C-47s, and some fifty Fairchild C-119 Packet transports also entered RCAF service. Two de Havilland Comet 1A jet transports were bought, and later adapted for radar-calibration. Other aircraft included DHC-1 Chipmunk trainers, Vertol H-21A, Sikorsky S-51, S-55 and S-58, and Bell 47G helicopters. De Havilland Canada became the world's largest builder of short take-off aircraft, providing the RCAF and other air forces with DHC-2 Beavers, DHC-3 Otters, DHC-4 Caribou and DHC-5 Buffalo transports during the 1950s and 1960s.

On many occasions, Canada provided transport and communications support for UN forces, starting initially in the Suez Canal zone after the Anglo-French withdrawal following the Suez fiasco of 1956.

Canadian-built Lockheed CF-104 Starfighters replaced the Sabres in Germany during the 1960s, initially with six interceptor and two reconnaissance squadrons, but by the end of the decade, these had been reduced to two squadrons as part of a long series of defence cuts. The CF-100s were replaced by licence-built McDonnell CF-101 Voodoo interceptors during the early 1960s, but were replaced later by uprated US-built versions. Canadair-built Northrop CF-5As were introduced during 1969 and 1970, although not all of these saw operational service.

Meanwhile, the Royal Canadian Navy had created an air arm, building on its wartime experience when its personnel had manned two escort carriers and four Royal Navy air squadrons. Post-war, the squadrons transferred to the RCN, and a light fleet carrier, HMCS *Warrior* was obtained from the UK in 1946, operating two squadrons, one with Supermarine Seafire fighters and one with Fairey Firefly ASW aircraft. In 1948, *Warrior* went into reserve and was replaced by HMCS *Magnificent*, on loan from the RN, while Hawker Sea Furies replaced the Seafires and 100 Grumman TBM-3E Avengers, mostly shore-based, replaced the Fireflies. In 1957, *Magnificent* was returned to the RN and the former HMS *Powerful* was commissioned as HMCS *Bonaventure*. Until her withdrawal in 1968, this ship operated McDonnell F2H-3 Banshee fighters, Grumman S2F-1 Tracker anti-submarine aircraft and Sikorsky S-55 helicopters. The S-55s and some Bell HTL-4s were also operated from some other Canadian naval vessels.

Canadian Army aviation restarted in 1946 with Auster AOP aircraft, replaced in 1954 by Cessna L-19A Bird Dogs, Bell 47 Sioux and Sikorsky S-51 helicopters.

In 1967, Canada's centenary year, the three services were formed into a unified defence force, with ranks modelled on army lines rather than those of the British armed forces. The new CAF was organised into commands: Headquarters, Training, Material, Mobile, Maritime, Air Defence, Air Transport and CAF Germany. Initially, the CAF had a personnel strength of 90,000, but this reduced to 59,100 at one time. At first, Air Defence Command had sixty-six McDonnell Douglas F-101 Voodoo interceptors in three squadrons, as well as Bomarc B SAM squadrons. Mobile Command operated twenty-six Canadair CF-5A fighter-bombers in two squadrons, fifty Bell CUH-1N Iroquois and seventy-four OH-58A Kiowa helicopters, and fifteen DHC-5 Buffalo transports. Air Transport Command operated five Boeing 707-320C tanker/transports, twenty-three Lockheed C-130Es in two squadrons, a Canadair CC-106 Yukon squadron and a Cosmopolitan squadron, with a handful of Caribou and C-47 transports. Six squadrons of DHC-3 Otters were in reserve. CAF Germany operated many of these aircraft on rotation, plus forty CF-104 Starfighters in two squadrons. Maritime Command brought together the RCAF's four MR squadrons and the RCN squadrons, with a squadron of Trackers and one of Sikorsky CHSS-3 Sea King ASW helicopters: the last mentioned were usually based aboard the nine destroyers then able to operate helicopters. Training Command replaced its Chipmunks with Beech Musketeers, Canadair CL-4 Tutors, CF-5Bs, Beech C-45 Expeditor and Douglas C-47 aircraft, and Hiller UH-12 helicopters. Eight DHC-6 Twin Otters were delivered during the early 1970s for rescue duties.

The Argus was replaced in the early 1970s with thirty Lockheed P-3B Orions, known in Canada as the CP-140 Aurora. Later, a purchase of CF-188A Hornets (F/A-18) standardised interceptor and strike force aircraft.

Given the size of the country and its many sparsely populated areas, and the need to patrol two oceans (a third coastline, the Arctic, is frozen for most of the year), transport and MR loom large in Canadian defence planning. The end of the Cold War has meant that its Arctic regions are no longer a 'front line' between East and West, but in common with many other countries, it has meant that defence planning has become more difficult and threats more varied. Canadian involvement with UN operations has also meant that a long reach has become essential. Continued reductions in defence expenditure have seen decisions deferred, including new medium-lift and SAR helicopters, although these have now been resolved. Upgrades of the Aurora and Hornet have been accompanied by a steady reduction in force sizes, with the official line being that 'less is better', on the basis that new or upgraded equipment is better than a greater quantity of older equipment.

Once again, Canada is taking a leading role in training for other air forces. In 1997, the Canadian government approved the NATO Flying Training in Canada

Programme, NFTC, operated by Bombardier, which now has most of the Canadian airframe industry, using at least twenty-six CT-155 (BAE Hawk 115) jet trainers and twenty-four CT-156 Beech Harvard II turboprop trainers.

Increasingly referred to not as the Canadian Armed Forces but as the Canadian Land Force, Canadian Air Force and Canadian Navy, the Air Force or Air Command has 19,922 personnel, while another 2,344 primary reservists are integrated into its force structure. Today, the CAF operates seventy-two CF-188AB Hornets in four squadrons, using AIM-7 Sparrow and AIM-9 missiles, while thirty-one CF-188Bs are used for training and conversion. Three MR squadrons operate twenty CP-140 Auroras. Five transport squadrons operate seventeen CC-130Js which have replaced most of twenty-two CC-130E/Hs, and one with five KCC-130H Hercules tankers, while long-range transport is provided by a squadron of five CC-150 Polaris (A310) airliners capable of conversion into freight, while one can also be converted for VIP use. At the other end of the scale, six CC-115s (Buffalo) provide light transport and SAR, six CC-144 Canadair Challenger jets provide transport and VIP while four CC-138 Twin Otters remain for additional SAR and light transport. The CH-113 Labrador SAR helicopters have now been replaced by fourteen CH-149 Cormorants (Merlin), but replacement of the twenty-eight CH-124 Sea King helicopters in three ASW squadrons usually from frigates and destroyers will be by the S-92, while eighty-eight CH-146 Griffon (Iroquois) helicopters, of which thirty are armed, provide light transport for ground forces and additional SAR cover. Training is provided on a wide variety of types, but includes twenty-seven CT-114 Tutors, some of which are in the aerobatic team. Training also uses eleven Grob 120As, two CT-145 Super King Airs and thirteen CH-139 JetRangers, as well as the NFTC's twenty-six CT-156s (T-6A) and CT-155 Hawks.

Heron and Skyhawk UAVs are operated, mainly by Land Command although there is a Heron unit within the Air Command or Air Force.

CAPE VERDE

- Population: 429,474
- Land Area: 1,580 square miles (4,040 sq.km.)
- GDP: $1.72bn (£1.1bn), per capita $4,017 (£2,665)
- Defence Exp: $8.8m (£5.6m)
- Service Personnel: 1,200 active

CAPE VERDE AIR FORCE/*FORÇA AÉREA CABOVERDAINE*

A former Portuguese colony in the Atlantic off the West Coast of Africa, Cape Verde maintains a small air force, with 100 personnel. Two An-26 transports are owned, although serviceability might be low, but the main role is the provision of an MR Do228-201 and an EMB110 in support of the coastguard's three patrol craft.

CENTRAL AFRICAN REPUBLIC

- Population: 4.5 million
- Land Area: 234,000 square miles (616,429 sq.km.)
- GDP: $2.1bn (£1.3bn), per capita $476 (£304)
- Defence Exp: $22m (£14.1m)
- Service Personnel: 3,150 active

CENTRAL AFRICAN SQUADRON/*ESCADRILLE CENTRAFRICAINE*

Formerly part of French Equatorial Africa, the Central African Republic became independent in 1960, after which it received military aid from France. Originally created as the Central African Air Force, the small air arm initially had the more or less standard French departing gift of a single C-47, three Max Holste 1521M Broussards (Bushrangers) and an Alouette II helicopter. Severe financial difficulties have resulted in reductions since and the reduction in status to an air arm of the Army with around 150 personnel. It has a number of aircraft but availability is uncertain, including a C-130, a Cessna 337 and a Falcon 20, while helicopters include an Mi-8, an AS350 Ecureuil and an Alouette II.

CHAD

- Population: 10.3 million
- Land Area: 488,000 square miles (1,282,050 sq.km.)
- GDP: $8.1bn (£5.2bn), per capita $806 (£515)
- Defence Exp: $151m (£96.5m)
- Service Personnel: 25,350 active

CHAD NATIONAL FLIGHT/*ESCADRILLE TCHADIENNE*

Formerly the most northerly part of French Equatorial Africa, Chad became independent in 1960, after which it received military aid from France, including the standard French package of a single C-47, three Max Holste 1521M Broussards (Bushrangers) and an Alouette II helicopter for what was known originally as the Chad Air Squadron. Since 1970, the country has been involved in numerous border disputes with its neighbour Libya, acquiring a number of Pilatus PC-7 Turbo Trainers via France in 1985, of which two are still operational. It managed to capture four SF260W armed-trainers from Libyan forces and put these into service, although probably only one is still operational. In recent years it has acquired six Su-25s, two An-26s and a C-130H. Other aircraft operated by the ET, which has 350 personnel, are a Reims-Cessna FTB337, two Mi-24s and five Mi-17s, seven AS550s and three Alouette III helicopters, a PC-6B Turbo Porter on communications duties and a PC-9 trainer.

CHILE

- Population: 16.6 million
- Land Area: 286,397 square miles (738,494 sq.km.)
- GDP: $160bn (£102.3bn), per capita $9,621 (£6,149)
- Defence Exp: $5.23bn (£3.3bn)
- Service Personnel: 60,560 active, plus 40,000 reserves

CHILEAN AIR FORCE/*FUERZA AÉREA DE CHILE*

Founded: 1930

Chilean military aviation started with the foundation of a flying school in 1913, followed shortly afterwards by the establishment of the Chilean Military Aviation Service. Initially three Bleriot aircraft were operated, followed by three more, as well as three Sanchez-Besa, a Deperdussin and a Voisin. By 1915, there were ten aircraft in two squadrons. In 1917, six Bristol M1Cs were introduced, one of which made the first flight across the Andes in 1918. In 1919, the Naval Aviation Service was established to operate seaplanes, and an aircraft factory was also founded. The Military Aviation Service continued to grow steadily but slowly, buying war-surplus DH4 bombers in 1921.

The two air arms were merged to form the *Fuerza Aérea de Chile* in 1930. Early *FAC* aircraft included Vickers Wibault and Curtiss Hawk III fighters, Junkers R-34 and Dornier bombers, Vickers Vixen and Curtiss Fox general-purpose aircraft, Dornier Wal flying boats and Fairey IIIF seaplanes, as well as Loening C2 and Sikorsky S.38 amphibians, de Havilland Gipsy Moth and Avro 504 trainers. Focke-Wulf Fw44, Avro 626 and Nardi FN305 trainers, and Arado Ar95 general-purpose aircraft later joined these. An American aviation mission reorganised the *FAC* in 1941, providing Curtiss P-40 and Republic F-47 fighters, Douglas B-24 and North American B-25 bombers, Sikorsky OS2U-3 seaplanes and Consolidated PBY-5A amphibians, as well as Fairchild PT-19, North American T-6 and Vultee BT-13 trainers. There were few new aircraft during World War II, but afterwards additional F-47s were delivered. During the 1950s, Beech T-45 Mentor and D18S trainers, and Twin Bonanza communications aircraft entered service. Many of the aircraft bought during this period were transport aircraft, including DHC-2 Beaver and DHC-3 Twin Otters. The first jet aircraft, de Havilland Vampire T55 trainers, did not arrive until the end of the decade, and were joined by Douglas B-26 Invader bombers and fifty Chilean-built Chincol trainers.

Chile became a founder member of the Organisation of American States in 1948, but never became entirely dependent on US aircraft.

Jet fighter-bombers were delivered during the 1960s: twenty Lockheed F-80C Shooting Stars, replaced in 1969 and 1970 by Hawker Hunter FGA9s, with some Hunter T7 trainers. A number of second-line aircraft were also obtained, including fourteen Grumman HU-16B Albatross MR amphibians, twenty Beech C-45s and twenty-five Douglas C-47s and four DC-6 transports, as well as Bell 47 and UH-1D Iroquois, Hiller UH-12E and Sikorsky H-19 helicopters. A change of government in 1970 made Chile's relationships with Western countries difficult, which continued after a military dictatorship took over before an eventual return to democracy. During this period, the *FAC* managed to update itself with Dassault Mirage 5 and Northrop F-5 fighter-bombers, as well as a few Lockheed Hercules. The *FAC* joined the Coalition Forces fighting for the liberation of Kuwait during the Gulf War of 1990–91.

Today, the *FAC* has 7,760 personnel, its strength having reduced by some 40 per cent since 2001 while flying hours are low at 100 per year. Economic difficulties have delayed modernisation, although the main aircraft types have been upgraded. Mirage 5s have been upgraded locally as the Pantera, or Mirage 50 interceptor/reconnaissance aircraft, of which there are fifteen. The Hunters were replaced in 1995–96 by ex-Belgian Air Force Mirage 5s upgraded before delivery and operated as the Elkan, with twenty 5BAs and five 5BDs. The Mirages were replaced by twenty-six F-16s, with one squadron operating Block 50s and F-16D, and another F-16AM/BM, while a third fighter squadron has thirteen F-5Es and three F-5Fs, upgraded in Israel. Other combat aircraft include thirteen A-3Bs and nine A-36s (C-101) armed jet trainers. A single Israeli-modified Phalcon AEW aircraft supports these aircraft. Reconnaissance support depends on a King Air A-100, three DHC-6-300s and two Learjet 35As. Maritime-reconnaissance, transport and ELINT duties are covered by five Beech 99As. Transport aircraft include a Boeing 767 and a 737, which replaced four 707s, three C-130B/Hs and a Boeing 707 tanker. Smaller transports include twelve DHC-6 Twin Otters, used mainly in the Antarctic, as well as four C-212 Aviocars. A Citation I, two Learjet 35s, a Gulfstream IV and two Beech 100/200s provide VIP transport.

Helicopters include fourteen UH-1Hs, eight Bell 412s, an S-70, and up to twelve Mi-17s are supposed to be on order. Training uses variants of the F-5 and F-16, as well as nineteen T-35A/B Pillans, nine C-101BBs and two Bell 206s, while twelve EMB314 Tucanos have just entered service. The main missiles are

AS-11 and AS-12 for ASM duties, Sidewinder, Shafrir and Python III AAM, and some MATRA Mitral and Mygalle SAM launchers.

CHILEAN NAVAL AVIATION/*COMMANDANCIA DE AVIACIÓN NAVAL DE LA ARMADA DE CHILE*

Chilean naval aviation was absorbed into the new air force in 1930, but reappeared post-war, initially with Bell 47G Sioux helicopters and Beech T-34 Mentors used on communications duties. Added impetus was given to naval aviation with the arrival of two new Leander-class frigates in 1973, with Westland Wasp helicopters. In recent years, the significance of naval aviation has grown with two additional Leanders acquired from the Royal Navy and the arrival of four ex-British County-class guided missile destroyers. Today, eight destroyers and frigates are operated.

Today, the CAN has some 600 personnel out of the naval total of 16,500. It has an MR squadron operating four P-3As with three CN-295MPAs in course of delivery which may replace some of the five EMB110s. Transport is provided by three C-212s. Armed helicopters included five AS532 Cougars deployed at sea equipped with torpedoes or Exocet anti-shipping missiles, six SA365s, four SA365s, three Bell 206s and five Bo105s. Training uses a squadron with eight Pilatus PC-7s and another with two Bell 206s.

CHILEAN ARMY AVIATION/*COMANDO DE AVIACIÓN DEL EJÉRCITO DE CHILE*

The Chilean Army originally used aircraft primarily for communications and support duties, but in recent years has developed a more substantial combat capability. The mainstay of the force is twenty MD530F armed helicopters, as well as two AS332 Super Pumas, two AS350 Ecureuils, seven SA330 Pumas, and an SA315 Alouette II, a Bell 206 and a UH-1H. Transport is provided by six C-212 Aviocars, three CN235s and three Cessna 208s and a Citation I. Training is provided on Enstrom 280FX helicopters and six Cessna R172 Hawks.

CHINESE PEOPLE'S REPUBLIC

• Population: 1,339 million
• Land Area: 3,768,000 square miles (9,595,961 sq.km.)
• GDP: $4.86tr (£3.1tr), per capita $3,634 (£2,322)
• Defence Exp: $70.3bn (£44.9bn)
• Service Personnel: 2,285,000 active, plus an estimated 510,000 active reserves

Once considered the 'sleeping giant', after the Communist takeover, China became Asia's most heavily-armed nation, initially using Soviet equipment but after relations soured with the USSR, the country steadily became self-sufficient in defence equipment production and today has export customers that include Pakistan. At first, Chinese-built aircraft were copies of Soviet equipment, but an indigenous design capability has emerged. In recent years, very strong economic growth has also provided a platform for investment in the armed forces to the extent that China is now seen as the next superpower and poses a threat to the stability of the Pacific Region. In addition to border disputes with India, the country also influences North Korea, which may emerge as a 'client state', and has been undermining Nepal.

PEOPLE'S LIBERATION ARMY AIR FORCE

Founded: 1949

For most of the twentieth century China was split by warring factions, with a formal division occurring in 1949, when Communist forces gained control of the mainland and established the People's Republic of China, leaving the Nationalist Chinese to the island of Formosa.

Before the overthrow of the Emperor in 1911 and the declaration of a republic, one or two aircraft had already made an appearance in China. Nevertheless, it was left to a number of localised air arms, sponsored by opposing warlords, to introduce military aviation to the country. Yuan Shih-k'ai at Shanghai introduced twelve Caudron GIII and GIV aircraft in 1914, while in northern China, Tuan Chi-Jui formed a small air arm and joined the Allies in 1917. Official Chinese Army and Navy air arms also came into existence around this time.

British and American assistance brought the Chinese Aviation Service into existence in 1919, merging the Army and Navy air arms. In 1920, the CAS received sixty Avro 504K trainers, followed by forty Handley-Page O/400 bombers and Vickers Vimy transports. Morane-Saulnier trainers followed, with Breguet Br14B-2 and Ansaldo A300 bombers. These aircraft were soon operating against insurgents throughout China. A Manchurian Air Arm also operated O/400s and Br14B-2 bombers, as well as Potez XXV bombers and Caudron C59 and Schreck FBA flying boats.

Britain refused a request for assistance by the Central Government after Japan invaded Manchuria in 1931. An American aviation mission supplied Boeing 218 and Curtiss Hawk fighters and a six-squadron Central Government Air Force was operational by 1934. The USA also supplied Northrop Gamma, Vought V65 Corsair and Douglas O-38 bomber and attack aircraft, as well as Fleet trainers. In 1935, an Italian mission brought Fiat CR30 and CR32 and Breda Ba27 fighters, Fiat BR3, Caproni Ca111, Savoia-Marchetti SM7 and Heinkel He111A bombers, and Breda Ba25 trainers. These were later joined by Blackburn Lincock fighters and Vickers Vespa VI. Meanwhile, the provincial air arms survived, including the

Kwantung Air Force, mainly operating Russian aircraft, and the Kwangai Air Force, mainly operating British aircraft, and which allied itself with the Central Government in 1936.

Renewed attacks by the Japanese started in 1937, with the invaders having the advantage in numbers, equipment and training. USSR support for the CGAF included Polikarpov I-15 and I-16 fighters and Tupolev SB-2 bombers, with Russian pilots manning six squadrons. The Chinese gained air supremacy until the Japanese moved their latest aircraft into the area, and occupied all of central China as well as Manchuria. As Japanese authority was consolidated, some of the occupied provinces were allowed to operate air arms under Japanese control using Japanese equipment. In Manchuria, colonised by Japan as Manchouko, the Manchoukuoan Air Force operated Nakajima fighters and Kawasaki bombers, and many of its personnel were absorbed into the Japanese forces. After the fall of Nanking, the Cochin Chinese Air Force operated Nakajima Ki34 transports and Tachikawa Ki9 trainers.

Despite these setbacks, the CGAF remained in the air, and was supported by an American Volunteer Corps, formed in 1941 with ninety Curtiss P-40B fighters and known as the Flying Tigers (post-war, veterans formed the airline of the same name). After the United States entered World War II, this force was absorbed into the USAAF and the Central Government received renewed US aid. US aircraft supplied to the CGAF included 129 Vultee 48C, 377 Curtiss P-40B/E, 108 Republic P-43, fifteen Lockheed P-38 Lightning and fifty North American F-51 Mustang fighters; 130 North American B-25 Mitchell bombers; twenty-eight Curtiss C-46 Commando, eighty Douglas C-47 and some C-53 transports; twenty North American AT-6, eight Beech AT-7, fifteen Cessna AT-17, 150 Boeing-Stearman PT-17 Kaydet, 135 Fairchild PT-19, seventy Ryan PT-27 and thirty Vultee BT-13 trainers.

Post-war, the CGAF was reorganised and restyled as the Chinese Air Force. Surviving wartime aircraft were augmented by additional Mustangs, Lightnings and Mitchells, plus Republic F-47 Thunderbolt fighter-bombers and Consolidated B-24 Liberator heavy bombers, and 250 ex-RCAF de Havilland Mosquito fighter-bombers were purchased. Nevertheless, the rapid advance of Communist forces forced the nationalists to withdraw to the offshore island of Formosa in 1949.

Soviet aid for the Chinese Communists had not started until 1945 as the USSR had maintained a policy of neutrality against Japan until the last days of the war. The Air Force of the People's Liberation Army used captured CGAF and then CAF aircraft, including Mustangs, Mitchells, C-46s and C-47s. Soviet aid provided Yakovlev Yak-9 and Lavochin La-11 fighters, while the USSR established a flying school for the Air Force of The People's Liberation Army in Manchuria in 1948.

Preoccupied with consolidating its victory, the new regime did not play a part in the Korean War at the outset in 1950, but China's first jets, MiG-15 fighters supplied in 1951, eventually saw action over Korea.

The immediate post-Korean War period saw the two major Communist powers as allies, with the USSR supplying technology and equipment. Soviet aid included additional MiG-15s, and then its successors, the MiG-17, MiG-19 and MiG-21; Tupolev Tu-2 and Tu-4 (B-29 copy) bombers, and Ilyushin Il-28 jet bombers; Lisunov Li-2 (C-47 copy), Antonov An-2, and Ilyushin Il-12 and Il-14 transports; Mil Mi-1 and Mi-4 helicopters; and Yakovlev Yak-12 and Yak-18, and MiG-15UTI and Il-28U trainers. China started to build these aircraft in addition to Soviet supplies, a policy that accelerated as relations between the two powers cooled rapidly over ideological differences and competition to take the lead in the Communist world. China's ability to produce replacements made the rift possible. Under Chinese manufacture, the MiG-15 series became the Shenyang F-2, the -17 the F-4, the -19 the F-6 and the -21 the F-8. Copying Soviet practice, aircraft operated by the Civil Air Bureau, or CAAC, were also available for military service. Later, China also started to produce transport aircraft and helicopters, and more recently has developed its own designs across the whole range of military aircraft.

Relatively little activity was undertaken by the AFPLA during the Vietnam War, with North Vietnamese forces relying largely on Soviet assistance, due as much to Vietnamese suspicions of China as rivalry between China and the USSR. Air power could contribute little, other than reconnaissance, to the Chinese annexation of Tibet due to the shortage of landing fields in the inhospitable terrain. Nor did air power feature greatly in the border clashes between Russia and China during the 1960s and 1970s. The break-up of the Soviet Union and Russia's abandonment of Communism has, paradoxically, led to an improvement in relations, with China obtaining equipment from Russia and the Ukraine. A new source of equipment and expertise, especially in upgrading equipment, is now Israel. China's own economy has also shown considerable growth in recent years, making the cost of maintaining the world's largest military air arm affordable. Nevertheless, despite its 300,000–330,000 personnel, the PLAAF, as it is now known, and much modernisation, much of its equipment is still obsolete. Nevertheless, flying hours have increased to around 150 per annum for fast jet pilots, while transport pilots fly in excess of 200 hours. In recent years defence expenditure has been increased, although personnel numbers are being reduced.

The PLAAF is divided into seven military air regions, administered from four air army headquarters. There are twenty-nine air divisions, each with up to four air regiments, usually of ten to fifteen aircraft. The main concentration of strength lies in those air regions in the south-east.

There are four air regiments operating H-6E/F/H (Tu-16 *Badger*) with some able to carry the YJ-63 cruise missile, while a fifth regiment is nuclear-capable. There are still 250 H-5 (Il-28 *Beagle*) bombers in service. PLAAF FGA strength has been boosted by deliveries from Russia of around 200 Sukhoi Su-27/-30s, plus 200 licence-built as the J-11, which equip up to fourteen regiments. There are also 300 J-8/-10 Chinese-designed interceptors and up to 400 J-7s (MiG-21) of different marks. In the FGA role, there are also 500 Nanchang Q-5s (MiG-19

Fantan), and some J-6A/B/C developments of the MiG-19 fighters, but the J-5 (MiG-17) has finally disappeared. Reconnaissance is provided by a force of 100 Chinese-designed JZ-6s (MiG-19) and two regiments of JZ-8s, and a regiment of Y-8Ds provides ECM. AEW and AWACS is provided by a regiment with KJ-2000s (Il-76). Transport aircraft include two Boeing 737s and two CL-601 Challengers for VIP use, eighty-three An-26s and up to forty-six Il-76s, as well as two Tu-154s, plus 300 Y-5s (An-2), thirty Y-7s (An-24), fifteen Y-11s and eight Y-12s. Survey work uses eight An-30s with two Y-12s. The helicopter force includes 350 Z-5s and Z-6s (both Mi-4). Training uses more than 200 Hondgu K-8s and 500 JJ-5s, as well as fifty Y-7s, up to 1,500 CJ-6 trainers and 150 Shenyang JJ-6s. A variety of UAVs are operated, including the Chang Hong, Firebee and Harpy. Missiles include *Adder*, *Alamo*, *Archer* and the Chinese-produced PL-12, PL-8 AAM, as well as *Kedge*, *Kyrpton*, *Kazoo* and Chinese YJ-63, YJ-88 and YJ-91 ASM.

THE PEOPLE'S NAVY

Officially known as The People's Liberation Army Navy, its aviation activities are primarily land-based, apart from a small number of helicopters aboard its more modern destroyers and landing ships. Nevertheless, this may change in the future. In 1998, a Chinese business concern acquired the ex-Soviet aircraft carrier *Varyag* from the Ukraine, ostensibly as an entertainment venue, but the vessel must have provided valuable insights to the APN, given Chinese ambitions for superpower status. As with army aviation, equipment is a mixture of Russian and French, including licence-built French helicopters, although at present, helicopters are a small part of this force. Roughly 26,000 personnel are engaged in naval aviation out of the service's total 255,000 personnel.

The PLA Navy is structured into three fleets, the *Beihai* or North Sea Fleet; the *Donghai* or East Sea Fleet and the *Nahai* or South Sea Fleet. Its air component seems to be more dated than that of the PLAAF.

Substantial shore-based forces include two bomber regiments with H-6D/Ms (Tu-16 *Badger*) and one of H-6D, while fighters include a regiment of Su-30Mk2, one with J-7E (MiG-21), two with J-8I/F, and one J-7II, one with JH-7A and Q-5C (MiG-19) and three with JH-7A. Long-range MR is provided by a regiment of SH-5s and one with Y-8s (An-12). ASW helicopters include eleven Kamov Ka-28 *Helix*, twenty-one Z-9s (SA365 Dauphin), seven Z-8s (SA321 Super Frelon), with the last two types also having a transport role, with at least six Z-5s (Mi-4) and eight Mi-8s. Fixed-wing transport aircraft include ten Y-7s (An-24) and forty Y-5s (An-2 as well as some Yak-42s).

PEOPLE'S LIBERATION ARMY AVIATION CORPS

Founded: 1988

The PLAAC was formed in 1988 by the transfer of utility and transport helicopters from the PLAAF. There is as yet only a limited combat capability available with the small number of Gazelles. The PLAAC seems to be looking to the West for its requirements, even building the Eurocopter SA365 Dauphin under licence.

The PLAAC has eight SA342L-1 Gazelle helicopters, used for training, but can also use HOT anti-tank missiles, seven SA321 Super Frelons for SAR, fifty-three AS350 Ecureuils, ninety-five Mi-171s, three Mi-6 *Hook*, twenty-eight S-70C2 Black Hawks, ninety Mi-8/-17s, four Mi-26 transports and twenty-five Harbin Z-9s (SA365 Dauphin) and twenty Changhe Z-11 (AS352) utility helicopters. Six AS332 Super Pumas are used for VIP transport. A wide variety of UAV are used, including the hand-launched ASN-15 and the larger ASN-104, ASN-105, ASN-206, W-50, WZ-5 and WZ-6.

COLOMBIA

- Population: 43.7 million
- Land Area: 462,000 square miles (1,139,592 sq.km.)
- GDP: $267bn (£171bn), per capita $6,115 (£3,908)
- Defence Exp: $6.51bn (£4.16bn)
- Service Personnel: 285,220 active, plus 61,900 reserves

COLOMBIAN AIR FORCE/*FUERZA AÉREA COLOMBIANA*

Founded: 1943

Military aviation did not start in Colombia until after World War I, with a flying school founded in 1922 with a single Caudron GIIIA biplane, while a naval flying boat flight was also formed. At first, both air arms remained small. The Navy operated a handful of Seversky SEV-3MWW amphibians, while the Army operated Curtiss Hawk fighters and Falcon reconnaissance aircraft; Bellanca 77-140 bombers; Curtiss-Wright CT-32 Condor, Ford 4-AT-E and Junkers W33 and W34 transports. North American NA-16-3, Consolidated PT-11 and Curtiss Fledgling were used for training.

In 1943, an American mission proposed a merger to form an autonomous air force, the *Fuerza Aérea Colombiana*. Aircraft were in short supply during the war years, but in 1948, Colombia joined the Organisation of American States and American military aid followed. Initially, this included a squadron each of Republic F-47 Thunderbolt fighter-bombers; Boeing B-17G Fortress and North American B-25 Mitchell bombers; Convair PBY-5A Catalina amphibians and Douglas C-47 transports; Boeing-Stearman PT-17 Kaydet and North American T-6 Texan trainers. These were followed by DHC-2 Beaver transports in 1951. In 1954, the first jets, six Lockheed T-33A trainers, arrived plus Beech T-34 Mentors to replace the other earlier trainers. Operational jets, Canadair CL-13 B Sabre Mk6 fighters, arrived in 1956. Many of these aircraft remained in service for some considerable time, with deliveries of additional North American Sabres and Douglas B-26 Invader bombers. Growing emphasis on COIN operations saw Cessna T-37C and T-41D armed-trainers enter service. Douglas C-54 and Lockheed Hercules were obtained, with Otters for the utility role and Aero Commander 680s for communications. Helicopters included twenty Bell 47G/D/J Sioux, and smaller numbers of UH-1D Iroquois and Kaman HH-43B Huskie, as well as twelve Hughes OH-6A and six 269s, with four Hiller UH-23s for training.

During the 1980s and 1990s, the *FAC* gained stronger front-line combat capability with the arrival of Mirage 5 and then IAI Kfir fighter-bombers, later upgraded to Mirage 50 and Kfir C7 standard. COIN remained important, often supported by US aid, to assist in operations against drug producers and traffickers.

FAC personnel strength has grown over the past fifty years from 6,000 to 13,134 and is now structured in six regional commands, replacing an earlier system of air combat, tactical air support, military air transport and training commands. Combat aircraft include a squadron with eleven Mirage 50s, of which probably only six are operational, and another with fifteen Kfirs, augmented by a squadron of ten A-37/OA-37s, a squadron with seven AC-47s, one with seven OV-10s, a squadron with twelve Tucanos and two with a total of twenty-five Super Tucanos. In the reconnaissance role are six Schweizer SA2-37As, two Super King Airs and three Aero Commanders, as well as five Cessna 208s and five Citation Ultras. Transport aircraft include a Boeing 727 and Arava 201, but the backbone of the force is a squadron with seven C-130B/Hs and another with four C-212s, seven CN235s, four C-295s, an Arava and an ATR42. Helicopters are vitally important against drug trafficking, with three Bell 205s, twelve 212s, two 412s, and twenty-six UH-1H helicopters, plus sixteen Sikorsky UH-60A/Ls, ten MD-500ME/Ds and three MD-530Fs. The *FAC* has a number of Piper, Beech and Cessna light aircraft – some seized from drug traffickers – for communications. Training Command has two Mirage 5CODs, twenty T-27 Tucanos, three T-34Ms, thirteen T-37s (A-37) and eight T-41s (Cessna 172), with two UH-1Bs, four UH-1Hs and eleven Bell 206s. Sidewinder and R-530 AAM are deployed. UAV include Scan Eagle.

COLOMBIAN NAVY AIR ARM/*AVIACIÓN ARMADA DE COLOMBIA*

The Colombian Navy has just 146 of the service's 30,729 personnel and uses aircraft mainly for communications and liaison, but has acquired an MR capability with five CN235MPAs and a Piper Navajo. Other aircraft include a C-212 transport, a Navajo and four Cherokees for communications duties with six UH-1Ns, two AS555SN Fennecs, two Bo-105CBs and a BK117, with the last three types able to be operated from four Almirante Padilla-class frigates.

COLOMBIAN ARMY/*EJÉRCITO DE COLOMBIANA*

The Colombian Army's aviation brigade has expanded rapidly, largely with US aid for the campaign against drugs. The Colombian government earlier provided eighteen Mi-17s, but there are also forty-six UH-60Ls and fifty-two UH-1H/Ns, as well as a CV-580, an An-32, two C-212s, and two King Air 90/200s and two Turbo Commanders, as well as an assortment of light aircraft, mainly seized from drug traffickers, including four Piper Senecas and four Cherokees. Training uses five UTVA-75s.

COMOROS ISLANDS

- Population: 400,000
- Land Area: 838 square miles (2,170 sq.km.)

COMOROS DEFENCE FORCE

Since independence from France, a small defence force has been established in this small group of islands off the coast of East Africa. An AS350B Ecureuil helicopter is believed to be operated.

CONGO

- Population: 4 million
- Land Area: 132,000 square miles (341,880 sq.km.)
- GDP: $11.6bn (£7.4bn), per capita $2,880 (£1,841)
- Defence Exp: $112m (£71.6m)
- Service Personnel: 10,000 active

CONGO AIR FORCE/*FORCE AÉRIENNE CONGOLAISE*

Founded: 1958

The former French Congo, the southernmost part of French Equatorial Africa, reached independence in 1958, and was provided with the standard military aid package of a Douglas C-47, three Max Holste 1521M Broussards (Bushrangers) and a Sud Alouette II helicopter. It soon became the People's Democratic Republic of the Congo, and fell within the Soviet sphere of influence, receiving MiG-17s and MiG-15UTI trainers. These were later joined by sixteen MiG-21s in 1986, plus An-24 transports.

Today, it has 1,200 personnel and serviceability of its aircraft is believed to be low due to a shortage of spares and the difficulty of keeping aircraft which are little used in the open in a hot and very humid climate. Twelve of the MiG-21s are believed to have survived, with eight MiG-17Fs and MiG-15s, but are not operational. VIP transport is provided by a Boeing 727, while transports include five An-24s, an An-26 and an N-2501 Noratlas. Helicopters include six Mi-35s, three Mi-8s and an Mi-24, which are in store, an SA365 Dauphin, an Alouette II and an Alouette III.

DEMOCRATIC REPUBLIC OF CONGO

- Population: 68.7 million
- Land Area: 895,000 square miles (2,345,457 sq.km.)
- GDP: $11.3bn (£7.2bn), per capita $170 (£109)
- Defence Exp: $168m (£107.4m)
- Service Personnel: between 140,000 and 150,000 active

CONGO AIR FORCE

Formed: 1961

The former Belgian Congo became independent in 1961 and almost immediately descended into civil war as the province of Katanga, rich in mineral resources, attempted to break away. Known initially as the *Force Aérienne Congolaise*, the air force was formed with light aircraft to counter aircraft operated by the rebel forces. After the civil war ended, the government and rebel air arms were merged, and the latter's foreign pilots provided an initial pool of experience. An Italian air mission soon replaced Belgian advisers and instructors. In 1969, seventeen MB326GB armed-trainers were delivered, augmenting up to twenty North American T-6 Harvards and T-28 Trojan armed-trainers. Douglas C-47 and DC-4, and DHC-4 Caribou transports were also acquired, as well as light aircraft including de Havilland Dove and Beech 18. Six Alouette III helicopters were joined in 1971 by the first of nine SA330 Pumas. Training took place on Piaggio 148 trainers.

The country renamed itself Zaire and acquired more potent aircraft in the form of Dassault Mirage Vs, as well as Lockheed C-130Hs. Equipment purchases were relatively easy, given the country's wealth. Nevertheless, civil war once again struck, with the overthrow of the central government in 1997, with foreign forces fighting both on the side of the new government, with troops from Angola, Namibia and up to 12,000 from Zimbabwe, and on the side of the rebels, including troops from Burundi, Rwanda and Uganda. Little use seems to have been made of the air force in this internal conflict, which has been ended by the presence of troops from fifty different nations under UN auspices.

There are reported to be 2,548 personnel in the CAF, but it is suspected that the spares and support organisation has broken down, leaving many aircraft unserviceable. There are believed to be thirteen MB326s, six Mirage 5Ms, two MiG-23s and two Su-25s, with two DHC-5 Buffaloes and an An-12. Helicopters include six Mi-24/35s, eleven SA330 Pumas and an SE3160, but none of the thirty-five Mi-8s are believed to be operational. Any training takes place on five SF260MZs and twelve Cessna FRA150s.

COSTA RICA

- Population: 4.3 million
- Land Area: 19,653 square miles (50,909 sq.km.)
- GDP: $28.6bn (£18.3bn), per capita $6,720 (£4,295)
- Defence Exp: $180m (£115m)
- Service Personnel: 9,800 paramilitary active

CIVIL GUARD AIR SURVEILLANCE UNIT/*GUARDIA CIVIL SECCIÓN AÉREA*

Although Costa Rica maintained a small air force until the end of a civil war in 1948, it has since depended on paramilitary security forces, with a small air surveillance unit with 400 personnel and unarmed observation and transport aircraft. The largest aircraft is a DHC-4 Caribou. Light aircraft include two Piper Navajos, a Seneca, four Cessna U206Gs, two Cessna T210s and two MD500E helicopters.

COTE D'IVOIRE

- Population: 20.6 million
- Land Area: 124,510 square miles (322,481 sq.km.)
- GDP: $24.4bn (£15.6bn), per capita $1,183 (£756)
- Defence Exp: $360m (£230m)
- Service Personnel: 17,050 active, plus 10,000 reserves

IVORY COAST AIR FORCE/*FORCE AÉRIENNE DE CÔTE D'IVOIRE*

Formed: 1960

Formerly part French West Africa, the Ivory Coast became independent in 1960, receiving the standard independence military aid package of a C-47, three Max Holste 1521M Broussards and an Alouette II helicopter. It soon embarked on a limited expansion programme, acquiring another two C-47s and two more Broussards, an additional Alouette II and three Alouette IIIs. It received a gift of an Aero Commander 500 from the US. Two SA330 Puma helicopters were bought later, followed by a Fokker F-27M Troopship and three Cessna 337s in 1971.

In the 1990s, the *FACI* acquired a limited combat ability with five Alphajet armed-trainers. The biggest expansion of the *FACI*, which has 700 personnel, was in VIP transport, for which it at one time had an airliner and two executive jets as well as two helicopters. Today it has two MiG-23s and two Su-25UBKs, as well as four Mi-24s, of which probably only one is operational, plus three SA330 Pumas and an SA316, as well as seven Alphajets for training.

CROATIA

- Population: 4.5 million
- Land Area: 21,824 square miles (56,524 sq.km.)
- GDP: $67.4bn (£43.1bn), per capita $15,016 (£9,599)
- Defence Exp: $1.02bn (£0.65bn)
- Service Personnel: 18,600 active

CROATIAN AIR FORCE

Formed: 1991

Formerly a Yugoslav constituent republic, Croatia declared its independence in 1991, initially using aircraft that had been part of the Yugoslav Air Force. It has 3,500 personnel and flying hours are believed to be just fifty annually. Financial problems have made re-equipment difficult, with an upgrade of the MIG-21s in Israel rejected as too expensive, and an offer of fourteen ex-USAF F-16A/Bs rejected for the same reason. Mainstays of the force remain two squadrons with a total of twelve MiG-21bis/UMs. Transport uses two An-32s, with twenty-three Mi-8/-17 transport helicopters, as well as eight Bell 206Bs. Four CL415 fire-bombing amphibians are also operated. Training uses twenty PC-9s and a Zlin Z242L.

CUBA

- Population: 11.5 million
- Land Area: 44,206 square miles (114,494 sq.km.)
- GDP: $61.1bn (£39.1bn), per capita $5,336 (£3,410)
- Defence Exp: $2.29bn (£1.46bn)
- Service Personnel: 49,000 active, plus 39,000 ready reserves

ANTI-AIRCRAFT DEFENCE & REVOLUTIONARY AIR FORCE/*DEFENSA ANTI-AÉREA Y FUERZA AÉREA REVOLUCIONARIA*

Founded: 1959

A Cuban Army air arm was planned in 1915, but it was not until 1917 that the first aircraft, six Curtiss JN-4D trainers, arrived. There do not seem to have been any further acquisitions until 1923, when six DH4B bombers and a number of Vought VO-2 AOP aircraft entered service with the Aviation Corps. Most of these aircraft were lost in a hurricane in 1926, leaving the Aviation Corps to rebuild. In 1934, it was reorganised into Army Aviation and Naval Aviation; by this time it was operating Waco D-7 general-purpose biplanes, Bellanca Aircruiser and Howard transports, with Stearman A73 and Curtiss-Wright 19R-2 trainers.

Cuba offered bases to the Allies during World War II, and in return received Grumman G21 Goose amphibians, Aeronca L3 AOP aircraft and trainers, including Boeing-Stearman PT-13 and PT-17 Kaydet, and North American T-6 trainers. When Cuba became a member of the Organisation of American States in 1948, further US military aid followed, including North American F-51D Mustang fighters and B-25J Mitchell bombers and Douglas C-47 transports. In 1955, a reorganisation resulted in the entire force once again passing into army control, becoming the Cuban Army Air Force, *Fuerza Aérea Ejército de Cuba*. That same year the first jets, Lockheed T-33A trainers, entered service, followed later by ex-Royal Navy Sea Furies. Three DHC-2 Beaver transports entered service in 1957, followed by two Westland Whirlwind (S-55) helicopters in 1958.

From 1956 to 1959 Cuba endured a revolution, with fierce fighting with Communist forces under Fidel Castro, whose victory early in 1959 saw both a change of government and a political reorientation towards the USSR. The title of *Fuerza Aérea Revolucionaria* was adopted and Soviet military advisers flowed into Cuba, although the US prevented the island from becoming a base for ballistic missiles. Less successful was a US-backed attempt at counter-revolution by Cuban exiles, with the invaders being attacked by *FAR* fighters.

Soviet equipment and cheap fuel meant a rapid increase in size, although initially the equipment provided was obsolescent, such as MiG-15 fighter-bombers. Later MiG-17s and MiG-19s arrived followed by MiG-21 interceptors during the early 1970s. By this time, the *FAR* had 12,000 personnel, operating fifty MiG-21s, forty MiG-19s, seventy-five MiG-17s and twenty MiG-15s. Transport aircraft were also provided; the total of fifty including Antonov An-2s and An-24s, as well as Ilyushin Il-14s. Helicopters included twenty-five Mil Mi-4s and thirty Mi-1s, with thirty MiG-15UTIs and Zlin 226 trainers.

New aircraft during the 1980s included MiG-23 and, later MiG-29 attack and fighter aircraft, as well as Mi-24 *Hind* attack helicopters. The collapse of the USSR led to the withdrawal of advisers and an end to the supply of cheap aircraft and, more important, cheap fuel. It created a shortage of spares, aggravated by Cuba's weak economy.

Today, the *FAR* has 8,000 personnel, with an average of just fifty flying hours annually. Analysts differ over the number of aircraft that are no longer airworthy, but it could be as high as 80 per cent. Three fighter/ground attack squadrons operate a total of thirty-one MiG-29A/UB/UMs and four MiG-21MLs, with around 180 MiG-21s and MiG-23s in storage. ASW Mi-14s are no longer operational. Two helicopter squadrons operate eight Mi-17s and four Mi-35s, with many more in store. Transport aircraft include three VIP Yak-40s, as well as three An-24s. Training aircraft now include seven MiG-21UMs, seven L-39Cs and five Zlin Z-142s, with many more aircraft in storage. Missiles include AS-7 ASM, AA-2, AA-7, AA-8, AA-10 and AA-11, while there are thirteen sites with SA-2 and SA-3 SAM.

Some sources indicate that the Cuban Navy has an Mi-14.

CYPRUS

- Population: 1.1 million
- Land Area: 3,572 square miles (9,251 sq.km.)
- GDP: $24.9bn (£15.9bn), per capita $23,296 (£14,891)
- Defence Exp: $562m (£359m)
- Service Personnel: 10,000

CYPRUS AIR FORCE

Formed: 1973

An air wing was created in the Cyprus National Guard following the Turkish invasion and subsequent partitioning of the island in 1973. Officially, all units are classified as non-active under an agreement reached between the two Cypriot authorities at Vienna. The National Guard Air Wing was renamed the Cyprus Air Force in 1996, when it adopted the national military markings of Greece. Originally it had approximately 140 personnel, operating four SA342L-1 Gazelle helicopters capable of firing HOT anti-tank missiles, plus three 206L3 LongRanger and two MD500 helicopters; a BN2B Maritime Defender and two PC-9s for patrol and pilot training. Over the past six years it has also added twelve Mi-35P combat helicopters and two UH-1Hs. The Greek Cypriot police also have two Bell 412 helicopters and a BN2T.

In Turkish Northern Cyprus, an army of 5,000 personnel has no aircraft of its own, but Turkish Army Cessna U-17s and Bell UH-1D Iroquois are at their disposal, stationed permanently.

CZECH REPUBLIC

- Population: 10.2 million
- Land Area: 30,461 square miles (78,864 sq.km.)
- GDP: $205bn (£131bn), per capita $20,122 (£12,862)
- Defence Exp: $3.19bn (£2.04bn)
- Service Personnel: 17,932 active

CZECH AIR FORCE & AIR DEFENCE/*CESK LETECTVO A PROTIVZBUSNA OBRANA*

Founded: 1995

Formerly the larger and more prosperous part of the former republic of Czechoslovakia, the Czech Republic split with the other main element, Slovakia, in 1995.

Czechoslovakia itself had only dated from 1918 on the federation of Bohemia, Moravia, Slovakia, Ruthenia and part of Silesia on the break-up of the Austro-Hungarian Empire. On its creation, a Czechoslovak Army Air Force was formed almost immediately, based on air units that had operated with the Czech Legions that had existed in France and Russia during World War I. War-surplus aircraft were pressed into service with assistance from France. The Czechs were given licences to produce French aircraft, but were soon producing their own designs, including Aero A-18 fighters, Smolik Sm-1 and Sm-2 bombers and Letov S-10 trainers. Within a decade, the CAAF had 400 aircraft in twenty-five squadrons. The main types included Avia BH21 and Letov S-20 fighters; Aero A-24 and Letov S-16 bombers; Aero A-11 and A-12 AOP aircraft; Avia BH10 and BH11, and Letov S-10 and S-18 trainers.

During a period when aircraft became obsolescent far more quickly than today, this force changed significantly over the next decade. Avia BH33 and BH34 fighters entered service, joined by Potez 63 fighter-bombers, and Fokker FVII and FIX tri-motor bombers. Other newcomers were Aero A-30 and A-100 reconnaissance aircraft and A-32 AOP aircraft. Towards the end of the 1930s, the

Tupolev SB-2 bomber was built under licence as the Aero B-17, while the CAAF also received Avia 135 fighters and A-300 bombers. The Munich Agreement of 1938 bought a year's breathing space for Britain and France before the start of World War II, but at the cost of Czechoslovakia ceding a third of her territory to Germany. In 1939, Germany occupied the rest of the country, dismantled the Czechoslovak state, and made Slovakia a separate state. No resistance was offered to the invader's overwhelming force, but Czech personnel escaped to join the French and Polish air forces, and after the fall of France, many flew with the Royal Air Force, which had three Czech fighter squadrons and a bomber squadron. Other Czechs fought with the Soviet forces. The Germans eventually formed a Slovak Air Force to fight alongside the *Luftwaffe*, flying Messerschmitt Bf109G fighters.

The country was reunited at the end of World War II, with a new Czechoslovak Air Force using abandoned *Luftwaffe* and Slovak Air Force equipment, as well as the aircraft flown by the former RAF squadrons, including Liberator bombers. Small numbers of de Havilland Mosquito fighter-bombers, Lavochin La-7 fighters and Petlyakov Pe-2 bombers soon joined these aircraft.

A successful Communist *coup d'état* in 1948 saw Czechoslovakia forced into the Soviet sphere of influence, and the CAF purged of ex-RAF elements as Russian advisers moved in and Soviet forces were stationed in the country. The country became a member of the new Warsaw Pact when this was established in 1955, but even before this, the USSR provided aircraft including Ilyushin Il-10 ground-attack aircraft and Li-2 (C-47) and Ilyushin Il-12 transports. The first jets, 200 MiG-15 fighters, entered service in 1951, at about the same time as Antonov An-2 transports and Zlin 226 trainers. Eventually, all of the MiG series of fighters and interceptors entered service up to and including the MiG-21 accompanied by Sukhoi Su-7B ground-attack aircraft and Ilyushin Il-28 bombers. Constant updating of the CAF came to an abrupt end in 1968, when the Soviet armed forces intervened in Czechoslovak internal politics after support for a less authoritarian regime was perceived as undermining the cohesion of the Warsaw Pact. There was a gap of some years before Czechoslovak forces received further equipment in the form of Sukhoi Su-22M4s and then Su-25BKs, as well as Mil Mi-24 attack helicopters.

Post-war strength peaked in the early 1970s at around 18,000 personnel, by which time it had 150 MiG-21s and 100 MiG-19s, as well as 150 Su-7Bs, alongside older aircraft, including eighty each of MiG-17s and MiG-15s. The break-up of the USSR and the collapse of the Warsaw Pact saw Czechoslovakia return to its Western links. The CAF underwent further change following Slovakia's vote for independence, ending the federation, in 1995. In 1999, the Czech Republic became a member of NATO, and is receiving NATO assistance in making its aircraft compatible with NATO standards. Modernisation has included a squadron of twelve JAS39C Gripen fighters, while there are twenty L-159A armed-trainers, with another two Gripen and four L-159s for conversion training.

The CAF has 4,938 personnel, down by two-thirds over the past six years. Transport aircraft are in two squadrons with two Airbus A-319CJs, four C-295s, five An-26s and six L-410s. Helicopters include two squadrons with twenty-four Mi-24/-35 attack helicopters, and thirty-two Mi-8/Mi-17s for transport, as well as eight WZL P-3s. Training uses ten L-29 Delfins, as well as eight Zlin 142s in addition to the conversion trainers already mentioned. Combat aircraft are now equipped with AIM-9M Sidewinder and AIM-120 AMRAAM missiles.

The number of Gripens should have been twenty-four, but financial considerations prompted a reduction and this will be a problem for the foreseeable future.

DENMARK

- Population: 5.5 million
- Land Area: 16,611 square miles (42,192 sq.km.)
- GDP: $337bn (£215bn), per capita $61,286 (£39,175)
- Defence Exp: $4.58bn (£2.93bn)
- Service Personnel: 17,514 active, plus joint personnel of 9,017, but which figure includes civilians, plus a Home Guard or reserve force of around 100,000 personnel

ROYAL DANISH AIR FORCE/*KONGELIGE DANSKE FLYVEVBEN*

Founded: 1950

Although the RDAF dates from a merger of army and naval aviation in 1950, Danish military aviation dates from 1912, when the two services received their first aircraft: the Royal Danish Navy received a Henri Farman and the Army received a Danish-designed B&S monoplane. In 1913, the RDN received two Donnet-Leveque flying boats, while the Army gained a Henri Farman and a Maurice Farman in 1914. Danish neutrality through World War I meant that the services were dependent on aircraft built at the Royal Army Arsenal. Post-war, the RDN obtained ex-German aircraft and some Avro 504 trainers purchased in 1920. The Army bought LVG BIII, Fokker C1 and Potez XV aircraft.

Reorganisation in 1923 saw the creation of a Naval Flying Corps and an Army Flying Corps. The Naval Flying Corps was further reorganised in 1926 with the formation of two squadrons, No1 Luftflotille, with Hansa-Brandenburg reconnaissance seaplanes replaced in 1928 by Heinkel He8s; and No2 Luftflotille with Hawker Danecocks (licence-built Woodcocks), replaced by Hawker Nimrods in 1935. The inter-war aircraft of the NFC also included Hawker Dantorp torpedo-bombers and de Havilland Gipsy Moth and Avro 621 trainers. An ex-Lufthansa Dornier Wal flying boat joined three He8s on survey work in Greenland, of considerable use to the wartime Allies when they established a chain of bases to ferry aircraft across the Atlantic. In 1932, further reorganisation created an Army Aviation Corps, *Havaens Flyvertropper*, with five squadrons or *escadrille*. Bristol Bulldog fighters equipped No1 Eskadrille until replaced by Gloster Gladiators in 1935; Fokker CV reconnaissance aircraft equipped No2 Eskadrille, and No3 when its Fokker CI were replaced in 1934, as well as No5 when it formed in 1935. Defence cuts meant that No4 was never formed.

Denmark hoped to remain neutral during World War II, but the country was overrun by German forces in spring, 1940. Having lost most of their aircraft on the ground, a number of pilots managed to reach the UK to join the RAF.

Post-war, the two air arms were re-established, initially sharing six Percival Proctor and some liaison aircraft. A joint flying school was formed in 1946, and a joint air staff in 1947, in preparation for the creation of an autonomous air force. In the meantime, many new aircraft were delivered, including Supermarine Spitfire IX fighters and Sea Otter amphibians, and North American AT-6 Harvard trainers. The first jets, Gloster Meteor fighters, arrived in 1949. When formed in October 1950, the new RDAF had five squadrons: Eskadrille No721 operated Convair PBY-5A Catalinas, Sea Otters and Airspeed Oxfords; No722 with Spitfires and Oxfords; No723 had Meteor F4s until these were replaced by Meteor NF11 night-fighters in 1952; No724 had Meteor F8s; while No725 had Spitfires. New aircraft in 1950 included twenty-seven DHC-1 Chipmunk basic trainers. This force was soon increased to eight squadrons, as Denmark became a founder member of the North Atlantic Treaty Organisation, NATO. In 1951–53, the USA supplied 200 Republic F-84E/G Thunderjets and a quantity of Lockheed T-33A jet trainers, initially replacing 725's Spitfires and then equipping new squadrons 726–730. Other deliveries at this time included Douglas C-47 transports and Bell 47D helicopters.

This much enhanced force was updated during the late 1950s with thirty Hawker Hunter Mk51 fighters and ten Republic RF-84F Thunderflash reconnaissance-fighters, followed later by North American F-100D Super Sabres and, during the early 1960s, Lockheed F-104G Starfighters. These acquisitions were accompanied by ex-Canadian Fairey Fireflies for target-towing, Pembroke C52 communications aircraft and Sikorsky S-19 helicopters. The Catalinas soldiered on until replaced by Sikorsky S-61A helicopters in 1970. By this time, the RDAF had 11,000 personnel. Squadrons included two with F-104G interceptors; two with F-100D Super Sabre and one with Saab F35 Draken fighter-bombers; a Draken RF35 reconnaissance-fighter squadron and a Hunter ground-attack squadron. Most squadrons had sixteen aircraft, but the Draken squadrons had twenty-three each. Transport aircraft included eight Douglas C-47s and five C-54s, while an SAR squadron had eight S-61As. Training was conducted on Chipmunks and T-33s, with conversion trainers for the fast jet combat aircraft. During the 1970s, naval and army aviation reappeared, largely due to the growing role for the helicopter, but at this stage the RDAF still controlled twelve Hughes 500M helicopters and KZ VII Larks used by the Army, and eight Alouette IIIs for the RDN.

Even before the end of the Cold War, Denmark cut defence expenditure, despite occupying an important strategic position at the exit from the Baltic to the North Atlantic, and also having the remote Faeroes and the vast expanses of Greenland to police. Collaboration on procurement with other NATO European countries enabled it to replace its Starfighters and Drakens with McDonnell Douglas F-16 Fighting Falcons, but combat squadron strength was cut to just three. These aircraft have received an MLU. Despite being a member of the Nordic Standard Helicopter Programme, Denmark broke ranks with the other members and opted for the EH101 to replace the S-61s. In common with a number of other NATO members, Denmark announced in 1999 that its defence posture would move from that of the Cold War to the provision and support of expeditionary troops within international operations, and for this bought four C-130J Hercules II.

Currently, the RDAF has just 3,466 personnel, down by a quarter over the past seven years; flying hours have been cut from 180 to 165 per annum. Two fighter squadrons operate forty-eight F-16s, with AIM-9 Sidewinder and AIM-120A AMRAAM AAM, and AGM-12 Bullpup ASM, which will eventually be replaced by forty-eight F-35As. A transport squadron operates four Hercules and three CL604 Challengers (two for MR and one VIP), that replaced Gulfstream IIIs. The SAR squadron operates helicopters with fourteen EH101s, although at least five are used for troop transport. Flying training uses twenty-eight SAAB T-17s. SAM defences include the Hawk missile.

ROYAL DANISH NAVAL FLYING SERVICE/*SOVAERNETS FLYVET JENESTE*

Having replaced the eight Alouette IIIs with which Danish naval aviation was re-established with eight Lynx helicopters, the RDNFS is responsible for fisheries' protection both off the Danish coast and its offshore territories of the Faeroes and Greenland. Up to four helicopters are embarked at any one time, usually on support vessels, as the RDN does not have frigates capable of carrying helicopters. The Lynxes were updated during the early 1990s and are expected to remain in service until 2015.

ARMY FLYING SERVICE/*HAERENS FLYVET JENESTE*

The Danish Army returned to aviation with twelve MD500Ms originally supported by the RDAF, later augmented by twelve AS550 Fennecs.

DJIBOUTI

- Population: 724,622
- Land Area: 8,958 square miles (23,200 sq.km.)
- GDP: $1.04bn (£0.66bn), per capita $1,473 (£941)
- Defence Exp: $13m (£8.3m)
- Service Personnel: 10,450 active

DJIBOUTI AIR FORCE/*FORCE AERIENNE DJIBOUTIENNE*

Formed: 1977

Djibouti gained independence from France in 1977. The country received the standard package of three Max Holste Broussards (Bushranger) communications aircraft, a C-47 and an Alouette II helicopter while a significant French presence has remained. The region is important strategically, at the narrow southern end of the Red Sea, and unstable politically. The small air arm has just 250 personnel and its main role is transport and communications, although it is supposed to have at least one Mi-24 attack helicopter. Other helicopters include three Mi-8s and an AS355 Ecureuil, while other aircraft include one An-28, an L-410, a Cessna U-206G and a 208 Caravan I, as well as an EMB314 Super Tucano trainer.

DOMINICAN REPUBLIC

- Population: 9.7 million
- Area: 18,699 square miles (48,430 sq.km.)
- GDP: $44.1bn (£28.2bn), per capita $4,574 (£2,923)
- Defence Exp: $318m (£203m)
- Service Personnel: 49,910 active

DOMINICAN AIR FORCE/*FUERZA AÉREA DOMINICANA*

The *Aviacion Militar Dominicana* was formed by the Army during the early 1940s, with some Aeronca L-3 AOP and American training aircraft, including Boeing-Stearman PT-17 Kaydets, Vultee BT-13s and North American AT-6s. It was not until 1948 that an expansion plan was authorised, buying obsolete Bristol Beaufighters as well as slightly more modern de Havilland Mosquito and Republic F-47D Thunderbolt fighter-bombers and a small number of Boeing B-17G Fortress

69

bombers. By the early 1950s, these aircraft were replaced by a squadron of twenty North American F-51D Mustang fighter-bombers, later augmented by the *AMD's* first jets, twenty de Havilland Vampire fighter-bombers, as well as seven Douglas B-26 Invader bombers. This period also saw the arrival of two Convair PBY-5A Catalina amphibians; six Douglas C-47s and six Curtiss C-46 Commandos, three DHC-2 Beavers and three Cessna 170s for transport and communications aided by Dominica's membership of the Organisation of American States. Helicopters arrived, initially two Bell 47s and two Sikorsky H-19s, followed by an Alouette III and two Alouette IIs. Training used North American T-6 Harvard and Beech T-11 aircraft. In the early 1970s, seven Hughes OH-6A helicopters were obtained.

Army control ended with the creation of the Dominican Air Force, but COIN and support for ground forces have become more important, plus SAR. Currently, the *FAD* has 5,500 personnel, a 50 per cent increase over thirty years, but flying hours remain at just sixty. A squadron of eight EMB314 Super Tucanos has replaced six Cessna A-37 Dragonfly armed jet trainers and two O-2 Super Skymasters. Five Ipanema crop-dusting aircraft are used on anti-drug operations. Transport is provided by two CASA C212-400 transports, a Cessna 206 and a Navajo, while helicopters include a VIP Bell 430, nine UH-1Hs and eleven OH-58s. Training uses eight T-35B Pillans, acquired from Chile, five T-41D Mescaleros, and a few T-6s.

The UH-1H and OH-58 are operated for the Army, which has its own R-22 and R-44 light helicopters.

ECUADOR

- Population: 14.6 million
- Land Area: 104,505 square miles (265,443 sq.km.)
- GDP: $52.3bn (£33.4bn), per capita $3,587 (£2,292)
- Defence Exp: $1.1bn (£0.7bn)
- Service Personnel: 57,983 active, plus 118,000 reserves

ECUADORIAN AIR FORCE/*FUERZA AÉREA ECUATORIANA*

Formed: 1935

In 1920 an Italian aviation mission visited Ecuador and the *Cuerpo de Aviadores Militares* was formed within the Army with Ansaldo, Aviatik and Savoia aircraft. It remained primarily a training and liaison operation even after 1935, when the present title was adopted and a Curtiss-Wright Osprey AOP aircraft purchased with some of the same manufacturer's 16E trainers. Meridionali reconnaissance aircraft were introduced in 1938. Fighter aircraft followed a visit by an American aviation mission in 1941, which supplied Seversky P-35s as well as Fairchild, Ryan PT-20 and North American NA-16 trainers. The first transport aircraft were a German-owned airline's Junkers Ju52/3 transports. There were no further aircraft deliveries while World War II continued.

Ecuador became eligible for American military aid on joining the Organisation of American States in 1948. This included Republic F-47D Thunderbolt fighter-bombers, Convair PBY-5A amphibians, Douglas C-47 and Beech C-45 transports, and North American T-6 trainers. The first jets arrived in 1954, twelve Gloster Meteor FR9 fighters and six English Electric Canberra B6 bombers. The US provided the first helicopters in 1960; Bell 47s accompanied by twelve Lockheed F-80C Shooting Star interceptors and two Douglas DC-6B transports, Lockheed T-33A, North American T-28 and Cessna T-41A trainers. By the early 1970s, personnel strength had grown to some 3,500. During the early 1960s, Cessna A-37B Dragonfly armed jet trainers, Hiller FH-1100 helicopters, and a Short Skyvan light transport were introduced, with Cessna 180 AOP aircraft operated for the Army.

Border disputes with Ecuador's neighbours led to further expansion, despite a reduction in US military support in an attempt to reduce tension. Sepecat Jaguar strike aircraft and BAe Strikemaster armed jet trainers entered service during the 1970s, followed by Mirage F1 interceptors and Israeli Kfir attack aircraft. The *FAE*'s operation of two national airlines, TAME, Transportes Aereos Militares Ecuatorianos, and Ecuatoriana introduced Boeing and Fokker airliners. Separate naval and army air arms came into existence during the 1970s.

Mirage, Kfir and Jaguar aircraft have been extensively upgraded in recent years, with attrition deliveries of Kfirs to maintain operational strength. Currently the *FAE*, despite the relatively small personnel strength of 4,200, has five squadrons of combat aircraft: one with up to twenty-one A-37s; one with nine Kfir CE/TC-2s; one with eleven Mirage F-1JE/1JBs and a 50EV; while two squadrons have twenty-four EMB314 Super Tucanos. Attempts to buy used A-4M Skyhawks from the US and MiG-29s from Russia have failed. Transport aircraft are mainly operated through the airlines, but the *FAE* has a Military Air Transport Group with five Lockheed C-130B/Hs, two Boeing 727s and three DHC-6s. TAME has six Boeing 727s, a DHC-6, an F-28 and three HS748s, while Ecuatoriana has three Boeing 707-320s, a DC-10-30 and two Airbus A-310 airliners. There are two North American Sabreliners for communications and VIP work. Helicopters include a VIP 212, eight 206Bs, and four Alouette IIIs with nine Dhruv ALHs. Training uses thirteen Beech T-34C Turbo Mentors and nine Bell 206s, as well as three Mirage conversion trainers and twenty Cessna 150s. Missiles include R-550 Magic, Super 530, Shafrir, Python 3 and Python 4, all AAM.

ECUADORIAN NAVAL AVIATION/*AVIACIÓN NAVAL ECUATORIANA*

The *ANE* developed for transport and communications, while funding difficulties prevented it acquiring naval helicopters in 1999 for operation from two ex-British

Leander-class frigates. Aviation accounts for 375 of the service's 7,283 personnel. It now operates two CASA CN235Ms and three Beech 200/300s for offshore surveillance, six Bell 206A/Bs and two 230 helicopters. Training aircraft include two T-34C Turbo Mentors and four T-35B Pillans.

ECUADORIAN ARMY AIR SERVICE/*SERVICIO AÉREO DEL EJÉRCITO ECUATORIANA*

The Ecuadorian Army operates a light aircraft on survey and liaison duties, and attack and transport helicopters. Armed helicopters are eighteen SA342K/L Gazelles, with three Super Pumas, three Pumas and five Mi-17s for transport, three AS350B Ecureuils on communications with two SA315B Lamas. Transport aircraft include a Citation II for reconnaissance, a DHC-5D Buffalo, four IAI-201 Aravas, a PC-6B Turbo Porter, two C-212s and two CN235Ms. Training uses three T-41Ds (Cessna 172), two CJ-6s (Yak-18) and three MX-7-235s.

EGYPT

- Population: 78.97 million
- Land Area: 386,198 square miles (999,740 sq.km.)
- GDP: $186bn (£118.9bn), per capita $2,370 (£1,514)
- Defence Exp: $3.27bn (£2.1bn)
- Service Personnel: 468,500 active, plus 479,000 reserves

ARAB REPUBLIC OF EGYPT AIR FORCE

Formed: 1954

Egyptian military aviation dates from the formation of the Egyptian Army Air Force in 1932, equipped with five de Havilland Gipsy Moths for training, surveying and army co-operation. During its first five years the EAAF was largely operated by RAF personnel, but it received new equipment during this period, including ten Avro 626s in both 1933 and 1934, with a Westland Wessex in the latter year. Hawker Audaxes, Avro Ansons and Miles Magisters further expanded this force. The Anglo-Egyptian Treaty of 1936 made Egypt a sovereign state. In 1939, the link with the Army was broken, and the title of Royal Egyptian Air Force adopted. This coincided with another round of equipment purchases, with ex-RAF Gloster Gladiator and Hawker Hart fighters, some Bristol Blenheim bombers, and Westland Lysander and Percival Q-6 AOP aircraft, plus additional Ansons. These aircraft formed two fighter and three AOP squadrons. The outbreak of World War II found the REAF assisting the RAF in protecting the Suez Canal for the first two years, until Egypt declared itself neutral in 1941, so that the REAF saw no action throughout the war years. Despite neutrality, it received two squadrons of Hawker Hurricanes and one of Curtiss Tomahawk fighters.

Post-war, RAF assistance resumed allowing considerable expansion as war-surplus aircraft were supplied. These included Supermarine Spitfire VB and IX fighters; Handley Page Halifax, Short Stirling and Avro Lancaster heavy bombers (seldom used due to a shortage of suitable aircrew); Curtiss C-46 Commando, Douglas C-47 and de Havilland Dove transports; and North American T-6 Harvard and Miles Magister trainers. This assistance ended abruptly with an arms embargo after Egypt invaded Israel in May 1948, although the REAF had little success and the fighting ended in January 1949. Egypt next received de Havilland Vampire FB5 and Hawker Fury fighter-bombers, Gloster Meteor F4 and Macchi C205 fighters and Fiat G55B fighter-trainers and Meteor T7 trainers. A follow-on order for a further twenty Meteor F4s was banned by the British, but the REAF obtained thirty Vampire FB52s from Italy via Syria in 1953, and obtained additional Meteors and Vampires once the embargo was lifted.

An abrupt change in procurement followed the ending of the monarchy in 1954, with the 'Royal' prefix dropped from the title. Soviet equipment was ordered, initially via Czechoslovakia, in 1955, including MiG-15 fighters, Ilyushin Il-28 jet bombers, Il-14 transports, and MiG-15UTI and Yakovlev Yak-11 trainers. Most of these aircraft were destroyed on the ground by British and French air attacks during the Suez crisis of November 1956. The aircraft were soon replaced, with MiG-15s, MiG-17s and Il-28s supplied during 1957. Training was provided by Russian and Indian advisers in Egypt and in Czechoslovakia. Through the end of the decade and into the early 1960s, additional Soviet equipment was provided, with additional MiG-17s followed by MiG-19s and a small number of Tupolev Tu-16 heavy bombers and Antonov An-12B transports, some of which were shared with the then national airline, United Arab Airlines. Mil Mi-4 and Mi-6 helicopters were also supplied. Even though it was still receiving assistance itself, the EAF helped the new Indonesian Air Force, the AURI, and the Yemen Air Force, which at one time consisted mainly of Egyptian personnel. For a period, Egypt and Syria united in the short-lived United Arab Republic, and the Syrians also received Egyptian help. As the scale of support grew, the USSR provided a complex air defence system to cover both the Suez Canal and the Aswan Dam as well as Cairo and Alexandria. Egypt was the first country outside the Soviet Bloc and China to receive the MiG-21 interceptor. Having grown to thirty-one squadrons, most of the EAF's aircraft were destroyed on the ground during the Arab-Israeli War of June 1967, and the Suez Canal was closed.

Once again, the USSR replaced the EAF's heavy losses, providing SA-2 and SA-3 SAM missiles, and many of the aircrew for the aircraft. By the early 1970s, the EAF had grown again to 20,000 personnel. Its aircraft by this time included 150 MiG-21 interceptors; 105 Sukhoi Su-7B and 180 MiG-15 and MiG-17 fighter-bombers; twenty-eight Il-28 and fifteen Tu-16 bombers; forty Il-14 and twenty An-12 transports; forty Mi-4 and smaller numbers of Mi-6 and Mi-8 helicopters; and a total of 150 MiG-15UTI, L-29 and Yak-18 trainers. Later, MiG-23 interceptors were supplied, flown by Russian personnel, while the air defences were strengthened further with the addition of SA-6, -7 and -9 missiles.

Despite a change of leadership in Egypt in 1970, a further Egyptian and Syrian attack on Israel followed in October 1973. On this occasion, Israeli losses were far heavier, losing 200 aircraft as against just thirty in 1967. Nevertheless, as SAM missile stocks ran low and the Arab ground forces over-stretched their supply lines, the conflict once again moved against the Arabs.

An uneasy peace prevailed until diplomatic relations were established between Egypt and Israel, including the re-opening of the frontier in 1980. After purchasing Shenyang F-6s and Chengdu F-7As, this period also saw a move away from Soviet influence and renewed links with the West. Gradually, Soviet equipment was replaced by Western aircraft, including fifteen Boeing CH-47C Chinooks. Army and naval aviation also emerged, with the Army having almost 100 Gazelle light helicopters, some of which were loaned to the Navy, and thirty Commando (S-61) troop-carrying helicopters, while the Navy acquired ASW Sea Kings. The EAF acquired substantial numbers of Mirage 5s, but most of its aircraft were from the United States, with F-4 Phantoms being followed by more than 100 F-16s. Naval and army aviation returned to the EAF, acquiring suitable helicopters, including AH-64A Apache attack helicopters for the ground support role and ex-USN Super Seasprites for the Egyptian Navy. More recently, the Chinese-Pakistan K-8 Karakorum trainer has been introduced.

The EAF has 30,000 personnel, with another 80,000 in a separate Air Defence Command based on AAA and SAM. There are 165 F-16A/C in seven squadrons and a squadron with fifteen Mirage 2000EM interceptors. Older aircraft include sixty-three Mirage 5DE/E2/SDRs in four squadrons, while thirty-four F-4E Phantoms equip two squadrons. Seven squadrons operate 119 MiG-21s and J7s, and two have forty-four Shenyang F-6/FT-6 (MiG-19) attack aircraft. While most of the forty-five Alphajets are used for advanced jet training, thirteen of them can also be used in the attack role. Six attack helicopter squadrons operate thirty-six AH-64A Apaches and seventy-four SA342L/M Gazelles, forty-four of which are able to fire HOT missiles. The interceptors are supported by six Grumman E-2C Hawkeye AEW aircraft. There are nineteen CH-47C/D Chinook heavy-lift helicopters, and twenty-eight Westland Commandos for medium lift, although some are converted to the VIP and ELINT roles, as well as fifty Mi-8/-17s. Two Agusta AS-61s are used in the communications role, while there are two Sikorsky UH-60A/L Black Hawk VIP helicopters. Ten re-manufactured ex-USN Super Seasprites and five Sea Kings provide ASW cover for the Egyptian Navy, which also uses five Gazelles. Transport aircraft include twenty-two C-130H Hercules, with another two converted for ELINT. Three Boeing 707s and a Boeing 737, three Falcon 20s and six GULFSTREAM III/IVs provide VIP transport, as do two Beech 1900Cs, with another four used on ELINT duties. Five DHC-5 Buffaloes are in the transport role with another four for navigational training. Apart from the Alphajets and conversion versions of the combat aircraft, supported by forty-seven F-16B/6D twin-seat aircraft mainly used for conversion training, there are also thirty-nine L-

59Es, which are suitable for ground-attack, fifty-four EMB312 Tucanos, nineteen UH-12Es, 100 K-8 Karakorums and seventy-four Grob G115EGs. Missiles include a combination of Western and Russian AAM, *Atoll*, Sidewinder, Sparrow and Magic, and SAM, but ASM are predominantly Western, including Maverick, Harpoon, Hellfire and Exocet, as well as HOT anti-tank missiles. UAV include Redeye and Scarab.

EL SALVADOR

- Population: 7.2 million
- Land Area: 8,236 square miles (21,331 sq.km.)
- GDP: $22.2bn (£14.2bn), per capita $3,086 (£1,972)
- Defence Exp: $132m (£84.4m)
- Service Personnel: 15,500 active, plus 9,900 reserves

SALVADORIAN AIR FORCE/*FUERZA AÉREA SALVADURENA*

Latin America's smallest and most densely populated country, El Salvador established a Military Aviation Service in 1923 with five Aviatik trainers. Small numbers of Waco and Curtiss-Wright Osprey general-purpose aircraft and Fleet trainers followed these. The first combat aircraft, four Caproni ground-attackers, arrived in 1939. Post-war Beech AT-11, Fairchild PT-19, Boeing-Stearman and Vultee trainers were obtained. After joining the Organisation of American States in 1948, C-47 transports and North American T-6 trainers were provided, followed by six Vought F4U Corsair and six North American F-51D Mustang fighter-bombers for two squadrons, and Beech T-34 trainers.

Jet equipment arrived much later, Dassault Ouragan fighter-bombers and Magister armed jet trainers, followed by Cessna A-37B Dragonfly armed trainers and O-2A Super Skymasters. COIN operations became increasingly important during a civil war, during which the government forces relied heavily on US support, including MD500E and Bell UH-1 Iroquois helicopters.

Today, the *FAS* has 950 personnel. Ouragans and Magisters are grounded, while there are nine surviving A/OA-37B Dragonfly armed jet trainers and fourteen Super Skymasters in a single squadron. A squadron operates five MD500D/E and three UH-1M armed helicopters, with another operating fifteen UH-1Hs in the transport and SAR roles. Transport is also provided by three BT-67s and two C-47s, an Arava 201, a Merlin IIIB and a Gulfstream Commander. Training uses five ex-Chilean Air Force T-35Bs, as well as five Rallye 235Ss, six Hughes 269s and a T-41D.

EQUATORIAL GUINEA

- Population: 633,441
- Land Area: 10,830 square miles (28,051 sq.km.)
- GDP: $13.3bn (£8.5bn), per capita $21,067 (£13,466)
- Defence Exp: $11m (£7m)
- Service Personnel: 1,320 active

NATIONAL GUARD/*GUARDIA NACIONAL*

The National Guard of this former Spanish colony operates a small air arm with 100 personnel which has four Su-25s, while on transport and communications duties there are a VIP Falcon 900, an An-32, a Cessna 337, as well as three Mi-24s and two SA316 Alouette III helicopters.

The National Guard operates two Enstrom 480 and three Mi-24 helicopters.

ERITREA

- Population: 5.6 million
- Land Area: 45,754 square miles (118,503 sq.km.)
- GDP: $1.57bn (£1bn), per capita $285 (£182)
- Defence Exp: not known
- Service Personnel: c. 201,750 active, plus 120,000 reserves

ERITREAN AIR FORCE

Formed: 1994

After many years of internal unrest, Eritrea became independent from Ethiopia in 1993. Internal unrest continues, and border disputes with Ethiopia have resulted in conflict. The small air force was formed in 1994 with L-90 Redigo trainers, later joined in 1996 by Aermacchi MB339FD armed jet trainers. Late in 1998, ten MiG-29s were acquired, possibly from Moldova, and flown by Ukrainian mercenary aircrew, but half of these have been lost since. An Mi-35 helicopter was captured from the Ethiopian Air Force and pressed into service.

Currently, the EAF has around 350 personnel, less than half the number in 2001. It has the five surviving MiG-29s, three MiG-23s, three MiG-21s and four Su-27s as well as four MB339 armed-trainers. Transport is provided by three Harbin Y-12s, and an IAI-1125 Astra. There is an Mi-24 attack helicopter, and for transport three Mi-8/-17s. There are eight L-90 Redigo trainers still in use.

ESTONIA

- Population: 1.3 million
- Land Area: 17,410 square miles (45,610 sq.km.)
- GDP: $20.2bn (£12.3bn), per capita $15,528 (£9,925)
- Defence Exp: $382m (£244.2m)
- Service Personnel: 4,750 active, plus 25,000 reserves

ESTONIAN ARMY AVIATION

Founded: 1991

In common with the other small Baltic states, Estonia was occupied by the Soviet Union in 1940 having enjoyed independence only since 1920. After being occupied by German forces from 1941 to 1945, the country became a constituent republic of the USSR and did not achieve independence until 1991. The small army has a small air arm, with 300 personnel and equipped with two An-2 transports that were joined in 2000 by four Robinson R44 training helicopters, a gift from the USA.

There is also a Border Guard, the maritime element of which doubles up as a coastguard, which operates two L-410UVP light transports, a Cessna 172 and two Mi-8s, two AW-139s and a Schweizer 300C helicopter.

ETHIOPIA

- Population: 85.2 million
- Land Area: 432,403 square miles (1,096,900 sq.km.)
- GDP: $28.2bn (£18.1bn), per capita $331 (£212)
- Defence Exp: $317m (£203m)
- Service Personnel: 138,000 active

ETHIOPIAN AIR FORCE

Founded: 1975

Ethiopian military aviation dates from 1930 and the delivery of six Potez 25 bombers and three de Havilland Gipsy Moth trainers to the then Imperial Ethiopian Aviation. In 1935, Italy invaded, with the far stronger and better-equipped Regia Aeronautica destroying the IEA. British forces liberated Ethiopia in 1941, but no attempt was made to re-start military aviation until 1946, when the Imperial

Ethiopian Air Force was established with help from a Swedish aristocrat. Its first aircraft were ten de Havilland Tiger Moth trainers, soon replaced by thirty more modern Saab-91A Safirs and a handful of Cessna AT-17 Bobcat trainers. In 1948, thirty Saab-17A bombers were delivered, with a similar number delivered during the early 1950s, and eight Fairey Firefly fighters, Stinson L-5 AOP aircraft and Douglas C-47 transports.

American military aid commenced in 1960, providing the first jets, twelve North American F-86F Sabre fighters and a number of Lockheed T-33A armed-trainers. Re-equipment during the late 1960s included later versions of the Safir, Northrop F-5A fighters and de Havilland Dove and Ilyushin Il-14 transports. Canberra jet bombers were also acquired. By the early 1970s, the IEAF had 3,000 personnel, and was operating the F-5As in one squadron of eight aircraft, with another squadron operating the twelve Sabres, while a bomber squadron had six Canberras. COIN operations were becoming increasingly important and involved two squadrons, one with eight Saab-17As and another with six North American T-28s and three Lockheed T-33As. There were six C-47 and two C-54 transports, plus a solitary Il-14 and three Doves in a single squadron. Relatively few helicopters included three Alouette IIIs and an Agusta Bell 204B. Training used Saab-91C/D Safirs, North American T-28s and Lockheed T-33As. After the monarchy was overthrown in 1975, the present title was adopted. The USSR became the main supplier of equipment, with MiG-17 fighter-bombers, followed by MiG-21s and An-12s, and Mi-8 helicopters.

Increasingly, the EAF found itself involved in operations against Eritrean secessionists, resulting in independence for Eritrea in 1993. This was followed by war between the two states over disputed borders. After the break-up of the USSR, Russia continued to be the regime's main arms supplier, with MiG-23s and Su-27s as well as Mi-24 and Mi-35 attack helicopters and Mi-17 transport helicopters. The EAF suffered relatively heavy losses during the fighting in 1998, with Eritrea claiming to have shot down as many as seven MiG-23s. During the conflict, the EAF relied on Russian pilots to fly most, if not all, of its Su-27s.

Today, the EAF has some 3,000 personnel. It operates fourteen Su-27A/Us in one squadron, while there are also eight Su-25s, ten MiG-23BNs and fifteen MiG-21MFs. There are fifteen Mi-24/-35 attack helicopters. Transport is provided by three C-130B Hercules acquired in 1998, fifteen An-12s, an An-32 and two DHC-6s as well as twenty-four Mi-8/-17, with five Alouette IIIs on communications duties. A VIP Yak-40 is also operated. Training uses four SF260TPs and twelve L-39Z Albatros, as well as MiG and Sukhoi conversion trainers.

The Ethiopian Army operates two DHC-6 Twin Otters and twelve UH-1H helicopters.

FIJI

- Population: 944,720
- Land Area: 7,055 square miles (18,272 sq.km.)
- GDP: $2.7bn (£1.73bn), per capita $2,897 (£1,851)
- Defence Exp: $52m (£33.2m)
- Service Personnel: 3,500 active, plus 6,000 reserves

FIJI MILITARY FORCES AIR WING

Founded: 1989

The extensive group of islands that is Fiji gained its independence from Britain in 1970, and established a small army and navy. An air element was created within the army when France donated an AS355 Ecureuil II helicopter in 1989, although this machine and an SA365 Dauphin were both sold in 2000 and the Air Wing disbanded.

FINLAND

- Population: 5.3 million
- Land Area: 130,120 square miles (330,505 sq.km.)
- GDP: $258bn (£165bn), per capita $49,180 (£31,436)
- Defence Exp: $4.21bn (£2.7bn)
- Service Personnel: 22,600 active, plus reserves of 350,000

FINNISH AIR FORCE/*ILMAVOIMAT*

Founded: 1920

Finland seized her independence from Russia during the Russian Revolution, and in the war that followed the Finnish forces flew a Swedish-built Albatros reconnaissance-bomber which was soon joined by additional aircraft of the same make. Other aircraft included a Friedrichshafen seaplane, a flying boat of unknown origin, Nieuport 12s and 17C-1s and Morane Parasols. Most of the pilots were Swedish volunteers. The war ended in 1920, and the new air arm, known as the Finnish Flying Corps, became a separate service, the Finnish Air Force, *Ilmavoimat*. France provided support, including twenty Breguet Br14B-2 reconnaissance-bombers, twelve Georges-Levy flying boats, Caudron GIII, C59 and C60 trainers, while more than 100 Hansa-Brandenburg A22 seaplanes were built under licence. A British aviation mission visited in 1924, when a small number of Gourdou-Leseurre C1 and Martinsyde F4 Buzzard fighters entered service.

The USSR did not accept Finnish independence and demanded the use of Finnish bases. The continuing tension between the two countries ensured that a steady stream of new aircraft was introduced, including Blackburn Ripon II reconnaissance and torpedo seaplanes, Gloster Gamecock II and Bristol Bulldog fighters, de Havilland and Letov trainers, and licence-built Fokker DXXI fighters and CX reconnaissance aircraft. There were also the Finnish-designed Viima and Tuisku trainers. While Germany detained thirty-five Fiat G50 fighters en route to Finland before World War II broke out in 1939, eighteen Bristol Blenheim bombers were delivered from the UK. The Soviet Union invaded Finland on 30 October 1939, amassing vastly superior forces both on land and in the air. At first, the *Ilmavoimat* held its own against some 900 obsolescent Soviet aircraft, but the USSR transferred additional aircraft to the Finnish front and the numbers reached 2,000 by 1940. Despite deliveries of additional Blenheims, Gloster Gladiators, Hawker Hurricanes, Brewster 239s and Curtiss Hawk 75A and A-4 fighters, and Westland Lysander AOP aircraft, Finland was forced to cede territory to the USSR in 1940. The following year, when the Russo-German alliance collapsed, Finland allied herself with Germany, and received Morane-Saulnier MS206 and Messerschmitt Bf109G fighters, Junkers Ju88 and Dornier Do17 bombers, and Do22W seaplanes. As the war turned against Germany, Finland surrendered to Russia a second time in 1944, and was forced into war against Germany.

Post-war, the *Ilmavoimat* received sufficient surplus Bf109G fighters to standardise on the type with four squadrons. The 1947 Treaty of Paris restricted Finland to sixty combat aircraft and 3,000 personnel, among other measures enforcing Finnish neutrality during the Cold War. The first jets were not received until 1955, when de Havilland Vampire FB52s were delivered, as well as two Hunting Pembroke C53 communications aircraft. In 1958, Folland Gnat light fighters were introduced, and licence production of Potez Magister jet trainers started. The following decade saw high performance aircraft with twenty Mikoyan MiG-21 interceptors equipping two squadrons, while a third squadron operated nine Gnats and another operated Magister armed-trainers. Transport aircraft included ten Douglas C-47 Dakotas as well as the Pembrokes, DHC-2 Beavers, plus two Mi-8s, four Mi-4s and two SM-1s (Polish-built Mi-1), a Bell JetRanger and two Alouette II helicopters. Training was provided on thirty SAAB-91C Safirs, fifty-five Potez Magisters, and a small number of MiG-15UTI and MiG-21UTI trainers.

Saab Drakens eventually replaced the MiG-21s, while BAe Hawks were introduced to replace the Magisters, with basic training using L-70 Vinka aircraft. During the late 1990s, F/A-18 Hornets were introduced, with the last deliveries in August 2000, when the Drakens were retired. In 1997, the helicopter force was transferred to the Army. The *Ilmavoimat's* strength remains within the Paris Treaty limits.

Today, the *Ilmavoimat* has 2,750 personnel, although its wartime strength would be 35,000, mainly employed on AA duties. It has fifty-four F/A-18C Hornets in three squadrons, with another seven F/A-18Ds for conversion training. There are forty-nine Hawk 51/51As for training, on which nine L-90 Redigo and twenty-eight L-70 Vinka are also used. A Fokker F-27-100 provides ASW and another F27-400M transport, with two C-295s. Three Learjet 35s are used for VIP transport, ECM training and target-towing. Nine L-90 Redigos, seven Piper Arrows and six Chieftains are used on liaison and communications duties, mainly in support of the Army. AAM include AMRAAM and Sidewinder, while SAM include SA-11, SA-16 and SA-18, as well as Crotale.

FINNISH ARMY AVIATION/*MAAVOIMAT*

Founded: 1997

The *Ilmavoimat's* helicopter force, seven elderly Mi-8 transport helicopters plus a couple of MD500Ds for training, transferred to the Finnish Army in 1997. Since then, while the Mi-8 force has reduced to just three machines, they have been joined by fourteen MD500s and an initial twenty NH90s selected under the Nordic Standard Helicopter Programme. Ranger UAV are also used.

FRANCE

- Population: 64.4 million
- Land Area: 212,919 square miles (550,634 sq.km.)
- GDP: $2.87tr (£1.83tr), per capita $44,669 (£28,553)
- Defence Exp: $47.8bn (£30.6bn)
- Service Personnel: 235,595 (excluding 103,376 Gendarmerie for fair comparison with other countries) active, plus 30,000 reserves

While France has followed many other countries in reducing its armed forces following the collapse of the Soviet Union and, within it, the Warsaw Pact, it has not done so as drastically as many, especially the United Kingdom. One can draw a comparison with the French reaction to the end of the Cold War with the retreat from the colonies, for the French have maintained military links with their former possessions and there is a substantial French presence east of Suez, including a new naval base in the Gulf opened during 2009. Of all the European nations, it is now the French who have the global presence and who now have the strongest navy.

No less impressive is the way in which the country has maintained its defence industrial base. Impatient with progress on the Eurofighter project, the French pressed ahead with the Rafale. Export orders for the aircraft have been slow in

81

coming, but no doubt by the time this is published, the first export orders will have been received. This means that it may no longer be true that without export orders, the *Armée de l'Air* would not exist, as one French air force general said in the days when the Mirage was a major export success, but it is certainly true that equipping the French armed forces is not a drain on the balance of payments and that the taxes paid by workers in the defence equipment industry can offset the investment in that equipment. This self-sufficiency also means that French foreign policy is not compromised by concerns over whether or not a decision will lead to restrictions on arms sales to the country.

The French nuclear deterrent might not be as capable as that of the UK, but it is French and not dependent on American willingness to support an ally whose interests will, inevitably, differ from time to time. The French maintain that while they could not obliterate an enemy, such as the former Soviet Union, they could at least 'tear off an arm and a leg'. Who needs more?

Of course, only the superpowers can have everything or even almost everything. There are weaknesses in the French position. There is the lack of a second aircraft carrier, so if anything goes wrong with the nuclear-powered *Charles de Gaulle*, the *Marine Nationale* is without a strike carrier. On the other hand, the number of amphibious assault ships is impressive, and it seems likely that a third Mistral-class will be ordered. Another weakness is the absence of a heavy-lift helicopter now that the Super Frelon force is almost finished. Even France has had to buy AEW and AWACS aircraft from the United States. Much will also depend on the fate of the Airbus A400M, which is running late and over-budget, as collaborative projects are wont to do, making one wonder whether an order for C-17s would have been better value, especially if a mix of C-17s and C-130Js had been considered. Surely too, the Atlantique MR aircraft is also past its best and needs replacing?

That said, overall the French position is impressive and one that shows what can be achieved given the political will. The French armed forces have truly global scope and in 2009 a base was opened in the United Arab Emirates.

FRENCH AIR FORCE/*ARMÉE DE L'AIR*

Re-formed: 1943

French forces used a balloon for spotting at the Battle of Fleurus in 1794, but the origins of the present *Armée de l'Air* date from 1910. In that year, the French Army established the *Aviation Militaire*, with a Bleriot, two Farman and two Wright aircraft, some flown by naval officers preparing for the formation of a naval air arm, the *Service Aéronautique*, later that year. The following year saw no less than thirty aircraft and balloons operated. The *AM* was involved in early experiments in aerial photography and radio transmission. In 1912, the squadron, *l'escadrille*, was created. Later, the roundel marking was invented.

On the outbreak of World War I in 1914, the *AM* had twenty-one squadrons in France: five with Maurice Farmans, four each with Henri Farmans and Bleriots, two each with Voisins and Deperdussins, and one each with Breguets, Caudrons,

Nieuports and REPs. There were another four squadrons in the colonies. Initially, air power was used for reconnaissance, although bombing emerged as observers dropped 90mm shells fitted with fins over the side of aircraft, while aerial combat appeared in 1915. Morane-Saulnier Type C aircraft formed the *AM*'s first fighter squadrons. Rapid expansion under wartime conditions saw the *AM* suffer 60 per cent casualties, the highest of any Allied service, in the air or otherwise. The return of peace found it with 3,480 combat aircraft in 255 squadrons. These included 1,600 reconnaissance and spotting aircraft in 140 squadrons; 480 bombers in thirty-two squadrons, and 1,400 fighters in eighty-three squadrons. Aircraft included Caudron GIIIs and Maurice Farman AOPs; Caudron GIV, Dorand AR-1 and Voisin reconnaissance aircraft; Breguet-Michelin IVs, licence-built Caproni tri-motors and Breguet Br14 bombers; Morane-Saulnier, Spad S7C and S13C, Nieuport 17C-1 Bebe and Caudron RXI fighters.

Peace saw a reduction to 180 squadrons spread across France, Germany and in the colonies, including Algeria and Tunisia, and elsewhere in Africa. In 1920, a squadron was formed in French Indo-China with Breguet Br14 bombers. New aircraft started to appear, including Breguet Br16 bombers and Nieuport 29C fighters. Peacetime duties consisted mainly of police action in the colonies, but the uprising led by Abd el Krim that started in Morocco in 1925 required reinforcements and continued until 1934. New aircraft continued to enter service, and during the 1920s these included Lioré-Gourdou-Leseurre 32, Nieuport-Delage 62C, Spad 81, Wibault, Lioré-et-Olivier LeO 20, Amiot 122 and Bleriot 127M fighters; Potez 25 reconnaissance aircraft; Morane-Saulnier MS35, 130 and 138, Hanriot-Dupont 32 and Caudron C59 trainers. Nieuport-Delage 629C and Morane-Saulnier MS225C-1 fighters; Lioré-et-Olivier LeO 206 four-engined bombers and Potez 39 AOP aircraft followed during the 1930s.

Post-war, the AM had grown into the 'fifth arm' of the French Army, behind the infantry, cavalry, artillery and engineers. In 1928, it became an autonomous air force, the *Armée de l'Air*. It included control of six shore-based naval squadrons, including two with Dewoitine D500 fighters. The new service soon suffered neglect, as the uncertain French political situation aggravated the problems already experienced during the Depression, and the lower priority accorded defence expenditure in peacetime. In 1935, events in neighbouring Germany and Italy forced an urgent modernisation and expansion programme, with deliveries of Dewoitine D501 and D520 fighters, Morane-Saulnier MS406 and Bloch MB151 fighters, Farman F221, Bloch MB210 and Lioré-et-Olivier LeO 45 bombers, plus Potez and Bloch general-purpose aircraft. The programme soon fell behind, aggravated by the decision to nationalise the French aircraft industry in 1936 and 1937. Aircraft were ordered from the USA, but these could not be delivered in time.

On the outbreak of World War II, most obsolete aircraft were transferred to areas of low risk, although some were used for leaflet-dropping. There was a reluctance to mount offensive operations against Germany for fear of retaliation, but once the German invasion started, there were some successes in aerial combat. The Germans invaded the Low Countries in May 1940, bypassing the French defensive Maginot Line. The *Luftwaffe*'s overwhelming aerial superiority forced French and British forces into retreat, with France surrendering in June. Germany planned to disband the *AdlA* immediately, but the British attack on the French fleet at Mers el Kebir and Dakar led to hopes that Vichy France would support the Axis Powers militarily. Units in France were finally disbanded in 1942, after some units in North Africa defected to the Allies following the landings in Algeria and Morocco.

Earlier, many French airmen had managed to escape to the UK, and a Free French Air Force was formed within months. A few aircraft had been flown out of France, but for the most part, the FFAF flew Hawker Hurricane and Supermarine Spitfire fighters, Bristol Blenheim and Martin Maryland bombers, and Westland Lysander army co-operation aircraft. Operations started in 1942, and in 1943 a merger with former Vichy units allowed the *Armée de l'Air* to be re-established. New equipment during the final years of the war included North American F-51 Mustang, Republic F-47 Thunderbolt, Bell Airacobra, Lockheed Lightning and de Havilland Mosquito fighters and fighter-bombers; Douglas Dauntless, Lockheed Lodestar and Handley Page Halifax bombers, and Douglas C-47 transports.

Post-war saw French forces struggling to quell insurrections in French Indo-China as nationalists attempted to take over before the colonial power could re-establish itself. Many personnel arrived before their equipment, Supermarine Spitfires, so they flew ex-Japanese Nakajima Ki 43 fighters instead.

In France, the *AdlA* started to modernise and re-equip itself, standardising on fewer aircraft types to simplify maintenance and training. De Havilland Vampire F1 jet fighters were produced under licence, while the Dassault MD450 Ouragan entered production. As a founder member of the North Atlantic Treaty Organisation, NATO, the country was eligible for American military aid, initially consisting of Republic F-84F/G Thunderjet fighter-bombers and Lockheed T-33A jet trainers. These joined a number of other types in the late 1940s and early 1950s, including Grumman Bearcats, Gloster Meteor NF11 night-fighters, Morane Saulnier MS500s, Nord 1100s (a Bf108 development), Beech C-45s, Bell 47 Sioux and Sikorsky S-55 helicopters. The *AdlA* had 123,000 personnel in seventy-five squadrons. Gradually, as the French aircraft industry re-established itself, French designs – the Dassault Mystère fighter, Sud Vatour fighter-bomber, Nord Noratlas transport and the Morane-Saulnier MS733 Alcyon trainer – came to predominate. At the time of the Anglo-French invasion of Suez in 1956, Mystères and Thunderjets flew many of the operational sorties.

During the late 1950s and early 1960s modernisation continued, coupled with a determination to 'fly French'. Morane-Saulnier MS760 Paris and Potez Magister jet trainers replaced the Lockheed T-33As, while the Dassault Étendard IV fighter-bomber entered service, followed by the highly successful Dassault Mirage IIIC interceptor. The export success of this aircraft and the Mirage 5 development was credited by one French air force chief with making a modern air force affordable. Sud Alouette II and III helicopters also appeared during the 1950s. The mid-1960s saw the first appearance of the Mirage IVA bomber to carry the first generation of French atomic weapons and the keystone of the French *Force de Frappe*. As development costs soared, France turned to international collaborative projects, with the Franco-German C-160 Transall transport appearing towards the end of the 1960s to replace the Noratlas. By the early 1970s, only a few foreign aircraft remained. These included twelve Boeing KC-135F tankers; two Douglas Skyraider squadrons based overseas; three Tactical Air Force North American F-100D Super Sabre squadrons; and a few Douglas C-47s and C-54s, Sikorsky H-34 helicopters, and six special duties Canberra B6 jet bombers. Collaborative projects continued, with almost 200 Anglo-French Jaguar strike aircraft entering service during the early 1970s, followed by 130 Franco-German Alphajet trainers during the middle of the decade.

The North Atlantic Council was moved out of Paris to Brussels after France refused to co-operate in the alliance's command structure after the mid-1960s.

Steady contraction also came to affect the *AdlA*, with personnel falling from 105,000 in 1973 to 57,600 today. Decolonisation meant that overseas commitments fell, but serious difficulties were only encountered in French Indo-China and in Algeria. Despite its lukewarm attitude to NATO, France co-operated in both the Gulf War, as a member of the Coalition Forces established to liberate Kuwait after the Iraqi invasion in 1989, and in NATO operations over the former Yugoslavia during 1999, as well as more recently in the invasion of Afghanistan.

France still maintains substantial forces in its former colonies, often with strong *AdlA* support. This is in addition to UN peace-keeping commitments. Major changes taking place include the move to all-professional armed forces, and French commitment to the so-called European Rapid Reaction Force that has grown out of the Franco-German Eurocorps and can be expected to include its own tactical air power and air transport element for rapid deployment.

The Mirage series continued with the development of the first non-delta wing version, the Mirage F1, which was followed by the Mirage 2000 series, including the 2000N designed to deliver nuclear weapons and replace the Mirage IVA. The one major national aircraft project in recent years has been the Rafale, a fighter and fighter-bomber replacement for the Jaguar and much of the Mirage force, but due to budgetary constraints, this programme was stretched, and the first operational squadron was not functional until 2005. It remains to be seen just how many Rafale will eventually enter service, but it could exceed 225 if existing and projected orders are completed, but in the meantime, the Mirage 2000 remains in the front line.

Today, with 57,600 personnel, the *AdlA* is organised in a series of brigades: Combat; Air Mobility; Air Space Control and Security and Intervention, with the last-mentioned concerned with base security. There is also an Air Training Command and an Air Strategic Forces Command. Combat Brigade has two fighter squadrons with thirty-five Rafale F2-B/Cs, and another eight with 140 Mirage M-

2000B/C/5s and twenty-two Mirage F-1CTs. There are thirty-nine F-1CR Mirages in two reconnaissance squadrons. Air Strategic Forces Command has three squadrons with up to sixty Mirage M-2000Ns, supported by a C/KC-135 inflight refuelling squadron of fourteen aircraft. Air Mobility Brigade 2 leased A340-200s as well as its own three A310-300s and two VIP A-319s in one squadron, plus six squadrons with a total of fourteen C-130H/130-H-30s and forty-two C-160 Transalls, augmented by seven light squadrons for tactical transport, SAR and training, with twenty CN-235Ms, five DHC-6s, six VIP Falcon 50/900s, and a number of smaller aircraft including TBM-700s, plus fifteen C-160BG tankers. Helicopters include five squadrons with twenty-nine SA330 Pumas, seven SA332 Super Pumas, three AS532 Cougars and forty-two AS555 Fennecs. Air Space Control has a surveillance and control squadron with four E-3F Sentries and two ELINT C-160 Gabriels, as well as eight Crotale and Mistrale SAM missile squadrons. Air Training Command operates ninety-one Alphajets, twenty-five EMB312 Tucanos, twenty-eight EMB121 Xingus, twenty-five TB-30s and eighteen Grobs, but there are also conversion units for the aircraft of the two operational commands. AAM include Mica and Magic, with AS-30L, SCALP and Apache ASM.

NAVAL AIR ARM/ *AÉRONAUTIQUE NAVALE –* *L'AÉRONAVALE*

Formed: 1945

French naval aviation started just slightly later than that of the Army in 1910, when the *Service Aéronautique* was formed with a Henri Farman and a Voisin seaplane, soon joined by Breguets and Nieuports. Naval exercises during summer 1913 proved the value of the aircraft for reconnaissance, leading to conversion of the cruiser *La Foudre* to a seaplane tender.

During World War I, two German cargo vessels, *Ann Rickmers* and *Rabenfell*, were seized and used as seaplane tenders. French naval aircraft reconnoitred the Austrian fleet in the Adriatic, and flew reconnaissance missions against Turkish forces in the Middle East for the British Army. Before the end of the war, a platform was built over the forward gun turrets of the battleship *Paris* for take-off trials. *Service Aéronautique* strength increased to 1,260 aircraft during the war years.

In 1925, the *Service Aéronautique* was renamed the *Aéronautique Maritime*, and the French coastline was divided into six *Districts Maritimes*, later being reduced to four in 1929, among which the *AM*'s aircraft were divided. Six shore-based squadrons were lost to the *Armée de l'Air* in 1928, but overall the French Navy, or *Marine Nationale*, retained control of its own aircraft, and had already acquired its first aircraft carrier, the converted battleship *Bearn*, in 1925. The ship had earlier, in 1921, been used for further trials with a wooden platform laid over her forward gun turrets. In 1929, a new seaplane tender, the *Commandant Teste*, joined the

Bearn. A major role became support for ground forces fighting rebels in Morocco. Equipment included shore-based Farman Goliaths, dating from the war, as well as CAMS 37s and 55s, Latham 43s, Gourdou-Leseurre GL-32s and Dewoitine D1s, Levassaeur PL7Bs and PL10Rs, operating from the carrier, while Latécoère 29s and Gourdou-Leseurre GL-810s operated from the seaplane tender. Wibault 74s replaced the D1s, and these in turn were replaced by Dewoitine D373 fighters aboard the *Bearn.* The carrier's other aircraft remained unchanged until after the outbreak of World War II, when they were based ashore while she became an aircraft transport, ferrying aircraft from the USA. New seaplanes and flying boats did enter service during the 1930s, including Lioré-et-Olivier LeO 257s, Levasseur PL15s, Latécoère 298s and Lioré 210 seaplanes, as well as Breguet Bizerta (Short Calcutta), and Lioré 70 flying boats.

France's first two purpose-designed aircraft carriers were planned before the outbreak of war, but the first of these, *Joffre,* was only a quarter complete when Germany invaded, and the second, *Painleve,* wasn't started. *Joffre* would have been unusual, having a heavily offset flight deck to compensate for a large island, and two hangars despite being just 18,000 tons, indicating little protective armour.

Shore-based aircraft enjoyed some success against German and Italian forces before the fall of France. Some aircraft and their pilots escaped to Britain, but others were in North Africa and found themselves being attacked by the Royal Navy after the Vichy commander refused to surrender. After the Allied landings in North Africa, those *AM* aircrew who joined the Allies were assigned to the *Armée de l'Air.*

In January 1945, the French Navy was loaned an escort carrier, the former HMS *Biter,* as an interim measure before the light carrier HMS *Colossus* could be transferred later that year, renamed *Arromanches.* The *Aéronautique Navale,* often referred to as the *Aéronavale,* was born. A second light aircraft carrier followed in 1951, the Independence-class USS *Langley,* on loan as *La Fayette.* These two ships were to prove invaluable during the war in French Indo-China. A third ship, the loaned Independence-class USS *Belleau Wood, Bois Belleau* in French service, arrived in 1953 but too late for the war. Carrier-borne aircraft were needed as shore bases had become untenable as Communist forces advanced. Aircraft operated during this period were primarily of US origin. Aboard the carriers, Douglas Dauntless and Curtiss Helldiver dive-bombers operated alongside Grumman F6F Hellcat and Chance-Vought Corsair fighter-bombers, and Grumman TBM-3 Avenger anti-submarine aircraft. Ashore, there were Convair P4Y-2 Privateer and Avro Lancaster MR aircraft, although later replaced by Lockheed P2V-6/7 Neptune patrol aircraft, as well as Short Sunderland and captured Dornier Do24 flying-boats, and Grumman JRF-5 Goose and Consolidated PBY-5A Catalina amphibians. The first jets were Sud Aquilan (Sea Venom) fighters. Helicopters were also introduced, with Bell 47s, Vertol H-22s and HUP-2s, and Sikorsky S-55s entering service. Training used North American SNJ-5s.

By the end of the decade, seventy-five Breguet Br1050 Alize turboprop ASW aircraft entered service, mainly aboard the carriers, joined by 100 Dassault

Étendard strike aircraft. The early 1960s saw the US carriers returned, and the first two purpose-designed French ships, *Clemenceau* and *Foch*, join the fleet. With their longer angled decks, these ships were able to operate heavier fighter aircraft, and two squadrons of LTV F-8E Crusader interceptors were obtained. Throughout this period, *Arromanches* remained in service, operating as a helicopter and training carrier. A helicopter cruiser, *Jeanne d'Arc* also joined the fleet in 1963. Plans to replace the Étendards with Sepecat Jaguar strike aircraft during the mid-1970s had to be abandoned as the aircraft proved too heavy and underpowered for carrier operation, and an uprated Super Étendard had to be put in service. Nevertheless, the Neptunes were replaced by forty Breguet Br1150 Atlantique MR aircraft, in a joint venture with Germany. ASW helicopters entering service during the 1970s included twelve Sud Super Frelons, while the Sikorsky S-58 was built under licence. Lighter helicopters for liaison and training included the Alouette II and III. In common with most navies, frigates (often referred to in France as corvettes) and destroyers capable of carrying small helicopters entered service, and for these the Anglo-French Westland-Aerospatiale Lynx shipboard helicopter was developed.

Today, the *Aéronavale* accounts for 6,400 of the *Marine Nationale* or French Navy's personnel. After *Clemenceau* was withdrawn in 1997 defence cuts, her sister *Foch* remained in service until withdrawn in 2000 when *Foch* was sold to Brazil and replaced by the nuclear-powered *Charles de Gaulle*. The F-8s were withdrawn that same year. Many Super Étendards remain in service, being nuclear-capable and using R-550 Magic 2 AAM, while the first Rafale squadron became operational in mid-2001 and sixteen are operated in one squadron with up to thirty-eight on order. Both Rafale and Super Étendards operated from the *Charles de Gaulle* in early 2002 as part of a Coalition fleet following the New York and Washington terrorist attacks in September 2001. Most carrier versions of the Rafale have two seats for improved air-to-surface capability. Three Grumman E-2C Hawkeye AEW aircraft also operate from the carrier. Ashore, twenty-five upgraded Atlantique ATL2 aircraft operate in two squadrons. The *Marine Nationale* still has thirty-one Lynx HAS4 (FN) helicopters able to operate from twenty-eight frigates and four destroyers, which will soon be replaced by twenty-seven NH90 ASW and ASuV helicopters. Other helicopters include eight remaining Super Frelons and five SARs and four transport Dauphin IIs, with twenty-four Eurocopter AS565MA Panthers in course of delivery replacing thirty Alouette IIIs. There are five Nord N262 Fregate transports, and another eleven for training alongside six Dassault Falcon 10MERs, which also fill the liaison role. VIP operations and communications duties are covered by twelve Embraer EMB121AN Xingus. Basic training uses nine Rallye MS-880s, twelve Alouette Is and thirteen Alouette IIs. A second nuclear carrier is unlikely to be built, as plans for *Foch* to remain in maintained reserve as a standby were abandoned. There are also two Mistral-class helicopter carriers and there has been speculation that a third might be ordered. Exocet and ASMP ASM are used, with Laser, MICA and Magic 2 AAM.

FRENCH ARMY AVIATION/*AVIATION LÉGÈRE DE L'ARMÉE DE TERRE*

Founded: 1954

Originally founded to operate light aircraft and small helicopters in support of ground forces, the *ALAT* proved its worth during the Algerian emergency. After the withdrawal from Algeria, the force continued to increase and during its first twenty years managed to amass a total of 1,000 aircraft, of which 600 were helicopters, and an organisation based on a quota of forty aircraft to a division. Heavier helicopters such as the Vertol H21, SA321 Super Frelon and SA330 Puma were added to the Alouette II and III and the small number of Bell 47 Sioux. Fixed-wing aircraft during this period included Max Holste 1521M Broussard (Bushranger), Piper and Cessna O-1, while Nord 3200 and 3400 trainers were also used.

The French Army has been reducing in size over the past twenty years, and has fallen from 169,300 to 134,000 personnel since 2001 as the decision to abandon conscription takes effect. Few fixed-wing aircraft remain, and the number of helicopters has fallen from 500 to 418 since 2001. *ALAT* has forty collaborative programme Tiger attack helicopters in service with a requirement for up to 215 more, but also has 268 SA341/342s. Four AS532UL Cougar Horizon helicopters provide electronic surveillance. There are 105 SA330B/H Pumas and sixty-four AS532M Cougar transport and assault helicopters, with the first of thirty-four NH90 TTH helicopters entering service. Training takes place on eighteen AS555 Fennec helicopters. Fixed-wing aircraft include five PC-6B Turbo Porters and two Cessna-Reims Caravan IIs, as well as twelve TBM700s, all used on liaison duties. UAV in use include CL-289 and Sperwer.

GABON

- Population: 1.5 million
- Land Area: 103,000 square miles (266,770 sq.km.)
- GDP: $12.2bn (£7.8bn), per capita $8,026 (£5,130)
- Defence Exp: $134m (£85.7m)
- Service Personnel: 4,700 active

GABON AIR FORCE/*FORCE AÉRIENNE GABONAISE*

Formed: 1960

Originally part of French Equatorial Africa, Gabon became independent in 1960 and received the standard French military gift of three Max Holste 1521

Broussards, an Alouette II helicopter and a C-47. By the early 1970s, this force had reduced to two Broussards, but the force has been restructured since and has acquired combat aircraft, modelling itself on the French pattern. It now has 1,000 personnel. There are thirteen Mirage 5 aircraft for interception and ground attack duties. An EMB110P Bandeirante provides MR with a second for transport, with three C-130H Hercules and one each of CN235, Nord 262C Fregate, ATR 42F and a Bell 412SP helicopter, with a VIP flight consisting of a Douglas DC-8, Falcon 900EX and Gulfstream IVSP. An AS350 Ecureuil is used on communications duties.

The Presidential Guard has a COIN capability, with three Magister and four Beech T-34C-1 Turbo Mentor armed-trainers, as well as VIP Bandeirante, Falcon 900 and AS355 Ecureuil. The small army has five Gazelle armed helicopters and five Pumas.

GAMBIA

- Population: 1.8 million
- Land Area: 4,008 square miles (10,000 sq.km.)
- GDP: $735m (£470bn), per capita $425 (£271)
- Defence Exp: $7m (£4.5m)
- Service Personnel: 800 active

GAMBIA NATIONAL ARMY AIR WING

A small air wing within the army operates a single Su-25, a VIP Il-62 and two AT-802Ss.

GEORGIA

- Population: 4.6 million
- Land Area: 26,900 square miles (69,671 sq.km.)
- GDP: $12.8bn (£8.2bn), per capita $2,754 (£1,760)
- Defence Exp: $343m (£219.3m)
- Service Personnel: 21,150 active

GEORGIAN AIR FORCE

Formed: 1991

On the break-up of the USSR, the Georgian Air Force was established using aircraft stationed in the country from what had been the Soviet air forces.

Compared to many other former Soviet republics, Georgia has the advantage that the Su-25 was manufactured in the country, avoiding the problems that have afflicted many post-USSR air forces. The GAF has 1,330 personnel. Nevertheless, economic problems have affected serviceability and just eleven Su-25s are operational, although there are also eight Mi-24 attack helicopters. In addition to a VIP Tu-134A, transport is provided by six An-2s and two Yak-40s, plus sixteen Mi-8s and two Mi-14s as well as seven UH-1Hs. Training uses nine L-29 Delfins and an Su-25UB conversion trainer.

GERMANY

- Population: 82.3 million
- Land Area: 137,732 square miles (356,726 sq.km.)
- GDP: $3.4tr (£2.17tr), per capita $41,352 (£26,433)
- Defence Exp: $46.5bn (£29.7bn)
- Service Personnel: 250,613 active, plus 161,812 reserves

For many years Germany was on what could have so easily have become Europe's front line, but the country's own armed forces were not the strongest in Europe and had the Cold War become a hot war, defence would have relied heavily on American and British forces based in the country, augmented by those of Canada and France. Some economists lauded the country for putting economic growth before defence, but without the means of defending the country, the opposing argument would be that no matter how successful, the economy and the country as a whole remained vulnerable. Deterrent is not simply nuclear and the value of strong conventional forces is so often underrated by politicians and the wider public.

No doubt as a result of having been the instigator of two world wars, the two largest conflicts in history, there was also strong German reluctance to do anything outside the strict North Atlantic Treaty Organisation area. This was matched by those in the other Western democracies who opposed German reunification on the collapse of the Soviet Union and the Warsaw Pact, even though such objections were bound to fail and were impossible to justify on any rational grounds. The emotions that feared German resurgence were nothing compared to those of Germans wishing to be reunited.

Post-German reunification, the situation has changed considerably. Germany has Europe's strongest army, and while one might argue that her armed forces could be stronger, the same can be said about those of all the leading European powers. Germany is now participating in many UN operations, although there has been criticism of her troops in Afghanistan, who have been accused of sitting safely in Kabul and leaving American and British forces to face the full force of the Taliban.

The real danger is that while Germany is participating in all multinational corps active in Europe, the spread of these in itself undermines NATO, the most successful and enduring alliance of all time, and the one that ties the United States and Canada into Europe. These two countries can cope without Europe, but Europe cannot cope in the long term against threats still to come without US involvement at the very least, and the Second World War showed Canada to be another worthwhile ally. The attitude of Russia to her neighbours is not one to engender confidence and complacency would be dangerous for we should never forget that Russian expansion predated the Revolution. There is a danger that one day Germany may once again be on a potential front line between East and West, and if not actually on the front line, be just one step away from it behind Poland.

GERMAN AIR FORCE/*LUFTWAFFE*

Re-formed: 1955

German military air power dates from the purchase of a Zeppelin dirigible by the German Military Board in 1907. Further dirigibles followed, and in 1910, the German Army received its first aircraft: five Henri Farmans, five Wrights and an Antoinette. A Naval Air Service was formed in 1911 with two Curtiss seaplanes. In 1912, the army air arm became the Military Aviation Service. Both air arms grew quickly in the couple of years remaining before the outbreak of World War I, in which Germany was one of the Central Powers, allied with the Austro-Hungarian Empire and Turkey. The NAS obtained Rumpler-Etrich Taubes (Doves), Euler biplanes and had Farmans built under licence, having thirty-six aircraft by 1914. The MAS also favoured the Taube, and versions of this aircraft were built by Albatros, Aviatik, AEG, DFW, Euler, Gotha, LVG and Otto, so that by 1914, Taubes comprised half the MAS strength of 250 aircraft. Among these were the Stahltaube (Steel Dove) built by the Jeannin concern.

Although small bombs were dropped on Paris as early as August 1914, at first aircraft were used primarily for reconnaissance. The Zeppelins proved too vulnerable for reconnaissance duties, and were diverted to bomb targets in London and England's east coast ports, causing the British to divert aircraft from France to home defence. The MAS gained a technical advantage when it adopted the Dutchman Anthony Fokker's invention allowing a synchronised machine gun to fire through an aircraft's propeller disc, first introduced on the Fokker EI *Eindecker* monoplane in July 1915. It was to be early the following year before Allied aircraft incorporated such a feature. Tactical developments included the introduction of large formations – known as 'flying circuses' – of fighter aircraft in 1916, the most famous being that led by the 'Red Baron', Baron Von Richthofen. The NAS also remained active throughout the war, raiding the south-east of England. At sea, the commerce raider *Wolf* used a seaplane for reconnaissance.

German aircraft production soared under the demands of wartime, rising from 1,350 in 1914 and peaking at 19,750 in 1917, before the Allied blockade led to shortages of materials and eventual defeat. The MAS operated Fokker EI, EII, EIII and DIII, and Pfalz DIII and Roland DII fighters; Albatros CIII, Aviatik CII and LVG CIII reconnaissance-bombers; and AEG GIV, Friedrichshafen GIII and Gotha GIV twin-engined bombers. The NAS operated Gotha, Sublatnig, Brandenburg and Friedrichshafen seaplanes. In addition to their own needs, the German services provided aircraft and experienced personnel to other members of the Central Powers. Yet, during the closing stages of the war, despite constant technical improvement in the new aircraft introduced in the final year or so, Allied aerial supremacy grew. At the end of the war, the MAS had 4,000 aircraft in service, and another 15,000 under construction, and 80,000 personnel. Between September 1915 and September 1918, two months before the Armistice, the MAS suffered 11,000 casualties and lost 2,000 aircraft.

In June 1919, the Treaty of Versailles was signed, disbanding both the MAS and NAS in 1920, prohibiting the manufacture of military aircraft and even, until 1926, restricting the size of civil aircraft that Germany could build. The Treaty conditions were circumvented through training pilots in so-called civil gliding schools and at a centre established in the USSR in 1928, while aircraft factories were established in Russia, Sweden and Turkey. A nucleus of wartime pilots was also secretly retained.

After Adolf Hitler came to power in the early 1930s, the *Luftwaffe* was established in March 1935, as an autonomous air force taking absolute control over all German service aviation. A naval air arm existed at first, but it was merged into the *Luftwaffe* after World War II began. The new *Luftwaffe*'s initial aircrew and other key personnel were based on a core of wartime veterans, boosted by graduates from the flying schools. Aircraft design and development was achieved through bomber types being developed in the guise of airliners for *Deutsche Luft Hansa* (sic), the national airline. The *Luftwaffe* was soon equipped with Heinkel He51 fighters, He45 and 46 reconnaissance aircraft, Junkers Ju52/3M transports, Dornier Do11 and Do23 bombers and Focke-Wulf Fw44 Stieglitz and Arado Ar66 trainers. A research centre was established by the *Luftwaffe* to evaluate new aircraft, many of which were to play an important part in World War II. Manufacturers were given loans to expand production and as an incentive for engineering firms to move into aviation, of which the most notable example was Blohm und Voss. Within a short time, the young *Luftwaffe* had 2,000 aircraft.

New aircraft designs were soon feeding into the *Luftwaffe*'s operational units, including the Messerschmitt Bf109 and Me110 fighters, Junkers Ju87 Stuka dive-bombers, and Dornier Do17 and Heinkel He111 bombers. Invaluable operational experience came when the Spanish Civil War broke out, and Germany allied herself with the Nationalists. The forces of General Franco were provided with twenty Ju52/3Mg bomber/transports and six Heinkel He51 fighters, followed by He70 reconnaissance aircraft and He59 and He60 seaplanes, flown by the 'volunteers' of the German Condor Legion. The transports carried the troops of the Spanish Foreign Legion from North Africa to Spain in one of the major moves in the early part of the war. More up-to-date aircraft followed to counter new Soviet fighters also finding their way to Spain on the Republican side. In 1937, Messerschmitt Bf109B/C fighters replaced the He51s. The following year, He111B and Do17E bombers replaced the Ju52/3Mgs in the bomber role. The Condor Legion returned to Germany in 1939 when the war ended. The conflict had tested aircraft and tactics, and given personnel invaluable combat experience.

The annexation of Austria and the Czech Sudetenland in 1938 and the occupation of Bohemia and Moldavia in 1939 were all assisted by a massive show of *Luftwaffe* air power that made resistance pointless. The Austrian Air Force was merged into the *Luftwaffe*. This was followed by a reorganisation, replacing the original area groups, *Gruppenkommandos*, with air fleets, *Luftflotten*, one of which was commanded by an Austrian officer. *Luftflotten* were divided into *Luftgou*, air districts, and these were further divided into groups, *Gruppe*, and squadrons, *Staffelnen*. At the outbreak of World War II in Europe in September 1939, the *Luftwaffe* had a front-line strength of almost 4,000 aircraft, including 1,300 Bf109 series fighters, 350 Ju87 Stuka dive-bombers and 1,300 Do17 and He111 bombers.

Germany precipitated the war by invading Poland on 1 September 1939. Aerial supremacy was gained over the Polish Air Force at some cost, with the Poles fighting bravely, but outnumbered and their aircraft outclassed. The *Luftwaffe* then operated 'hit and run' bombing raids against British naval bases in the east of Scotland, but saw little further action until the invasion of Denmark and Norway the following spring. Denmark fell almost immediately, but the *Luftwaffe* met fiercer resistance in Norway, supported by British and French forces and by RAF and Fleet Air Arm aircraft. Even before the Norwegian campaign ended, the *Luftwaffe* supported the German Army as it raced through the Netherlands and Belgium and into France, which was largely occupied by the end of June, leaving just the Vichy zone unoccupied until late 1942. The *Luftwaffe* bombed the undefended Dutch city and port of Rotterdam, and was left to destroy the retreating British and French forces at Dunkirk, but failed to achieve this before they could be evacuated. These campaigns showed the *Luftwaffe's blitzkrieg* strategy at work, with dive-bombers and medium bombers operating in close co-ordination with ground forces – in complete contrast to the British and American concept of strategic bombing.

The *Luftwaffe* next attempted to destroy the RAF in the Battle of Britain, but failed to do so. This was due in part to the British radar network and to the short range of German aircraft even when operating from France, while a switch to bombing the cities also gave the RAF a breathing space. The bombing campaign, known as the *Blitz*, was hampered by the lack of a heavy bomber. It came to a premature end in 1941 as German attentions turned elsewhere, to Operation Barbarossa, the invasion of Russia. The *Luftwaffe* was by this time becoming heavily committed, operating in the Balkans and against Greece and then Crete, and against Malta, while also supporting German forces in North Africa. Even so, the *Luftwaffe* had almost 2,000 aircraft available for the invasion of the USSR, with a further 1,000 contributed by Germany's Romanian, Hungarian, Italian and Finnish allies, facing 9,000 Soviet aircraft. The operation opened on 22 June with raids against sixty-six Soviet airfields, crippling the Red Air Force. Here the *Luftwaffe* was to suffer the same fate as the ground forces, faced with long distances and unprepared for the worst of the Russian winter. It was after the invasion of the Soviet Union that the bases in Norway came into their own, enabling the *Luftwaffe* to shadow and attack the Arctic convoys mounted by the Allies to bolster the Soviet defences and then their counter-attack against German forces.

Unusually among air forces, the *Luftwaffe* was responsible for the German paratroop force, but squandered much of this in the ill-conceived invasion of Crete, after which Hitler forbade any further paratroop assaults.

New equipment continued to enter service, with the *Luftwaffe* benefiting from continued technical improvements. Earlier in the war, the *Blitz* on British cities had demonstrated superior navigational and bomb-aiming systems. New aircraft entering service included Focke-Wulf Fw190 fighters, Dornier Do217 bombers, Blohm und Voss reconnaissance flying boats, Focke-Wulf Fw200 Condor and Junkers Ju290 anti-shipping aircraft, and Heinkel He115 seaplanes. A heavy bomber, the Heinkel He177, arrived too late to be truly effective, and was also dangerously unreliable. Towards the end of the war, new weapons were introduced, including the FX radio-controlled missile and the Hs293 glider-bomb, both launched from Do217 bombers. The revenge weapons, the V1 and V2 missiles, were launched from occupied Europe towards British cities, including London, and could be regarded as the predecessors of the cruise missile and ballistic missile respectively. The Messerschmitt Me262 jet fighter entered service in 1944, too late to stem overwhelming Allied aerial superiority, especially once Hitler insisted that it be used as a bomber. There was also the Arado 234 jet reconnaissance aircraft and the Messerschmitt Me163 Komet rocket-powered fighter, and a few of the unusual twin-engined pusher-puller layout He219 fighter: all too late and in too small numbers.

April 1945 saw the *Luftwaffe* grounded by fuel shortages, and its personnel deployed as ground troops. It was disbanded immediately the war ended.

Germany was divided into four Allied Zones of Occupation, and the former capital, Berlin, also divided. In 1949, the three Western zones were amalgamated to form the Federal German Republic, West Germany, while the Soviet zone became the German Democratic Republic, East Germany.

The GDR was the first to re-establish an air force, with an air arm of the People's Police, the *Vokspolizei*, formed in 1950, using Yakovlev Yak-18, Polikarpov Po-2 and Fieseler Fi156C Storch light aircraft and former *Luftwaffe* personnel. This became the *Luftstreitskrafte* in 1955, and once again, 'flying clubs' produced members for the new air force. With the GDR a member of the Warsaw Pact, the USSR supplied MiG-15 fighters and MiG-15UTI and Yak-11 and Yak-18 trainers, and these were joined by MiG-17s, Ilyushin Il-14M and Tupolev Tu-104 transports by 1960, with Mi-1 and Mi-4 helicopters following later.

The *Luftwaffe* reformed in 1955. West Germany became a member of NATO and qualified for US military aid. Its first aircraft were thirty Piper Cub trainers, which arrived in 1956, and were followed by thirty North American T-6 Harvard trainers and some Lockheed T-33A jet trainers. US aid included 450 Republic F-84F Thunderstreak fighter-bombers and 100 RF-84F Thunderflash reconnaissance-fighters, as well as Canadair and Italian versions of the Sabre fighter. Hunting Pembrokes, de Havilland Herons, Douglas C-47s and licence-built Nord Noratlas provided a strong transport element. Potez Magister jet trainers and Piaggio P149D trainers, as well as Bell 47 Sioux, Vertol H-21Hs and H-25s, Sikorsky H-34s, Saro Skeeters and Bristol Sycamores, and Hiller UH-12C helicopters, and the indigenous Do27 liaison aircraft completed the support element.

This left a divided Germany with two air forces facing each other across a frontier throughout the Cold War. The *Luftwaffe* was not allowed into West Berlin, itself well inside East Germany.

The *Luftwaffe* continued to develop during the 1960s, with licence-built Lockheed F-104G Starfighters and Fiat G91 ground-attack aircraft replacing the F/RF-84s and the Sabres. The German version of the Starfighter was a multi-mission aircraft giving a heavy pilot workload, and after many accidents was nicknamed the 'Widow-Maker'. The next major aircraft into the inventory was the McDonnell Douglas RF-4 Phantom which replaced the RF-104 version of the Starfighter, and in 1973, F-4s were also introduced. Joint projects between Germany and France resulted in the Alphajet advanced trainer and the C-160 Transall transport. Dornier built large numbers of Bell UH-1H helicopters under licence. In the mid-1970s, the Anglo-German-Italian multi-role combat aircraft, the Panavia Tornado, entered *Luftwaffe* service, replacing the Starfighters, while the Fiat G91s had earlier been sold to Portugal.

Similar developments took place in East Germany, with the *Luftstreitkrafte* receiving successive marks of MiG aircraft, including the MiG-19, MiG-21, MiG-23 and continuing up to the MiG-29.

German reunification came in 1990. The first move was for the *Luftstreitkrafte* to be merged into the *Luftwaffe*, although the older Soviet-era aircraft were quickly retired, with many of their personnel, leaving the *Luftwaffe* with just a small number of MiG-29s to convert to NATO standards. Over the past thirty years, the two air forces have moved from a combined personnel strength of 125,000 to the present figure of 50,270 – down by a third since 2001, while average flying hours are 150 a year, probably the bare minimum to maintain standards.

Today, the *Luftwaffe* is the third largest NATO air force by personnel numbers, and the second largest in Western Europe. Its structure is divided into an Air Force Command and a Transport Command. Gone are the inherited *Luftstreitkrafte* aircraft. AFC is developing a Eurofighter 2000 force of 102 aircraft in four wings with a total of eight squadrons, replacing F-4F Phantom IIs, and another thirty-six aircraft may be added in due course. The strike capability relies on 156 Tornado IDS in three strike wings with a total of six squadrons, while a reconnaissance wing has two squadrons with thirty-three Tornado ECRs. Transport Command has four wings, including a special air mission wing with seven A310MRT/MRTTs, six CL01 Challengers and three VIP AS532U2 Cougar IIs. Three wings with a total of seven squadrons operate eighty-three C-160 Transall transports which are being replaced by sixty A400Ms, while the other four squadrons include three with seventy-six UH-1Ds and a squadron with four VIP versions of this helicopter, which is being replaced by an initial forty-two NH90s. Conversion training includes twenty-seven Tornados and a number of Typhoons. In addition there are thirty-five T-37B and forty T-38A Talons. Missiles include AIM-9 Sidewinder, AIM-120A/B and AMRAAM, AGM-65 Maverick and AGM-88A HARM, as well as GBU-24 Paveway III. There are Hawk, Roland and Patriot SAM.

GERMAN NAVAL AIR ARM/*MARINEFLIEGER*

Founded: 1957

While German naval aviation existed during World War I, it was disbanded after the war. The German Navy, or *Kriegsmarine*, maintained a small air arm during the 1930s, with shore-based He59s and He60s, Dornier Do18 flying boats and Heinkel He115 seaplanes, as well as aircraft aboard battleships and commerce raiders. This force was merged into the *Luftwaffe* after World War II started. Rivalry between the two services over control of naval aviation fatally delayed the completion of an aircraft carrier, *Graf Zeppelin*, launched in 1938, while work on a second ship, *Peter Strasser*, started but was abandoned. These ships would have operated Messerschmitt Bf109 fighters and Junkers Ju87 Stuka dive-bombers. *Graf Zeppelin*'s catapults were donated to the Italian Navy for their planned carrier. Post-war, *Graf Zeppelin* was seized by Soviet forces, and capsized and sunk while under tow to Russia, overloaded with captured German equipment.

The *Marineflieger* was formed in 1957 for shore-based air operations over the Baltic and the North Sea. Early equipment included Armstrong-Whitworth Sea Hawk strike aircraft, Fairey Gannet ASW aircraft, Hunting Pembroke transports and Bristol Sycamore helicopters. During the 1960s, the Sea Hawks were replaced by licence-built Lockheed F-104G Starfighters, with seventy-two aircraft in four squadrons, and the Gannets by the Franco-German Breguet Atlantique, while Sikorsky CH-34 helicopters were also introduced for SAR, before being replaced in this role by twenty-two Westland S-61N Sea King helicopters. Other aircraft operated included eight Grumman HU-16 amphibians, twenty Dornier Do28 Skyservant liaison aircraft, as well as some of the smaller Do27s and Potez Magister jet trainers.

During the 1970s, the Starfighters were replaced by more than 100 Anglo-German-Italian Panavia Tornados in the anti-shipping strike role, but in 1994, more than half of these aircraft were transferred to the *Luftwaffe* and the force has now lost its fast jets altogether.

The arrival of six Type 122 frigates, designed jointly with the Netherlands during the early 1980s, saw the *Marineflieger* acquire its first shipboard helicopters, ASW and ASuV Westland Lynxes.

Today, the *Marineflieger* accounts for 2,260 of the German Navy's personnel strength of 19,162. It did not gain any significant increase in strength through the reunification of Germany, since the East German Navy was primarily a coastal defence force. The two remaining Atlantiques are near the end of their lives and the main MR force consists of eight AP-3C Orions, while two Dornier Do228s handle pollution control. There are twenty-one Sea King Mk41s for SAR and twenty-two Sea Lynxes, upgraded to Mk88A standard, capable of carrying Sea Skua anti-shipping missiles. Both helicopter types will be replaced by thirty-eight European NH90 helicopters in 2015. Kormoran ASW missiles are also deployed.

GERMAN ARMY AVIATION/*HEERESFLIEGER*

Founded: 1957

The *Heeresflieger* was founded in 1957 to provide liaison, communications and AOP facilities for the new Federal German Army. Battlefield transport followed later using Bell UH-1D Iroquois, of which 200 were in service at one time, and Sikorsky CH-53 heavy-lift helicopters. North American OV-10Zs were operated at one time on target-towing duties. The move into operating attack helicopters came with 200-plus MBB PAH-1s, essentially Bo105P/Ms equipped with HOT anti-tank missiles, while another ninety-six Bo105s operated in the observation and liaison roles.

Today, the *Heeresflieger* has undergone substantial change. The European Tiger attack helicopter has entered service, with eighty on order. Combat helicopters use the HOT missile. The UH-1Hs are being replaced by up to eighty-two NH90 TTHs, which may leave a number of PAH-1s in service, as well as the Bo105 AOP and liaison machines. The ninety-one CH-53Gs include twenty that have been upgraded with long-range tanks, EW equipment and NVG compatible cockpits. Training units operate 111 Bo105s and fourteen Eurocopter EC635s (EC135). UAV include KZO and Luna X-2000 as well as micro UAV, the Aladin and MIKADO.

GHANA

- Population: 23.9 million
- Land Area: 92,100 square miles (238,539 sq.km.)
- GDP: $15bn (£9.6bn), per capita $628 (£401)
- Defence Exp: $264m (£168m)
- Service Personnel: 15,500 active

GHANA AIR FORCE

Founded: 1959

Ghana became an independent member of the British Commonwealth in 1957, but the Ghana Air Force was not founded until 1959, with Israeli and Indian assistance and a gift of Hindustan HT-2 trainers. A Piper Super Cub was also obtained. RAF personnel were seconded in 1960, when de Havilland Canada Chipmunk trainers were also provided, followed later that year by DHC-2 Beaver light transports. Early plans were to provide a transport and communications force, so DHC-3 Otters and DHC-4 Caribou were also introduced. Soviet attempts to influence Ghana involved the supply of Antonov An-12 and Ilyushin Il-18 transports in 1963, with Mil Mi-4 helicopters. These aircraft were returned to the USSR when the Nkrumah dictatorship was overthrown. Eventually, Whirlwind (S-55) and then Wessex (S-58) helicopters were obtained, while during the late 1960s, six Aermacchi MB326F armed-trainers and Short Skyvan light transports were introduced.

Ghana has in recent years been active in supporting UN and Organisation of African Unity operations. There are just 1,500 personnel in the GhAF and the emphasis is still on transport and communications, although there is a COIN element using seven MB339As and up to four KA-8 armed jet trainers. There are four Fokker F-27-400M Troopships with a VIP F-28-3000, a Britten-Norman BN2T Turbine Islander. Helicopters include four Mi-17s, a VIP AB412, as well as two SA319 Alouette IIIs and two A109s. Training is on Cessna 172s.

GREECE

- Population: 10.7 million
- Land Area: 50,534 square miles (132,561 sq.km.)
- GDP: $390bn (£249bn), per capita $36,280 (£23,190)
- Defence Exp: $6.45bn (£4.1bn)
- Service Personnel: 156,000 active, plus 237,500 reserves

HELLENIC AIR FORCE

Founded: 1931

Greece occupies an important strategic position offering control over the Balkans and eastern Mediterranean, and in which air power plays a vital part. The importance of air power was for many years heightened by the long-running feud with neighbouring Turkey, including territorial disputes and tension over the island of Cyprus and parts of the Aegean Sea. Greece itself was occupied by Turkey until 1829.

The Royal Hellenic Army established an air squadron in 1912 with four Farman biplanes, using these during the Balkan Wars of 1912–13. The Royal Navy provided assistance for the Royal Hellenic Navy to form an aviation service in 1914, using Farman and Sopwith seaplanes. The two air arms were merged into the Hellenic Air Service in 1916, operating against German and Turkish bases in Bulgaria before re-dividing into separate air arms in 1917, becoming the Royal Hellenic Naval Air Service and the Royal Hellenic Army Air Force. Post-war, both services were strengthened using surplus British and French aircraft. The 'Royal' prefixes were dropped when Greece became a republic in 1924. To counter the effects of the Depression, a Greek Air Force Fund was established and several towns raised funds to purchase aircraft, with Salonika buying twenty-five. New aircraft for the HNAS included Avro 504N trainers, Hawker Horsley torpedo-bombers, Greek-built Blackburn Velos (Dart) torpedo and reconnaissance seaplanes, and Armstrong-Whitworth Atlas general-purpose biplanes. French aircraft predominated, with Breguet Br19A and 19B reconnaissance-bombers and Morane-Saulnier trainers.

The two air arms were unified once again in 1931, this time into an autonomous Hellenic Air Force, becoming the Royal Hellenic Air Force with the restoration of the monarchy in 1935. Before the outbreak of World War II in Europe in 1939, new aircraft included PZL P-24 and Bloch MB151 fighters, Bristol Blenheim, Fairey Battle and Potez 63 bombers, Henschel Hs126 AOP aircraft, Avro Anson reconnaissance-bombers, and Dornier Do22 and Fairey IIIF seaplanes. Many of these types – including the Battle and Blenheim, and in the offensive role, the Anson – were to prove poor performers elsewhere. Even so, the RHAF managed to repel an Italian invasion in 1940, but Greece was overrun by German forces in 1941. Some personnel managed to escape to Egypt where fighter squadrons were formed with the help of the RAF using Hawker Hurricanes and a bomber squadron was equipped with Blenheims. Later, these were replaced with Supermarine Spitfire fighters and Martin Baltimore bombers.

Greece was liberated in 1944, but immediately Communists tried to seize power and the ensuing civil war lasted for five years. Units formed overseas returned to engage in COIN operations bringing SB2C-4 Helldiver dive-bombers, Anson, Oxford and C-47 transports, L-5 liaison aircraft and Tiger Moth and Harvard trainers. The civil war left Greece much weakened, but membership of NATO in

1952 opened the way for assistance under the US Mutual Defence Aid Programme. Personnel were trained in the USA and Germany. Substantial numbers of aircraft were delivered, including eighty F-86D Sabre Mk2/4 fighters, 250 F-84G Thunderjet fighter-bombers and RF-84F Thunderflash reconnaissance-fighters, and T-33A jet trainers, soon joined by additional C-47s, T-6G Texan trainers, and the RHAF's first helicopters, H-19D Chickasaws. Tension between Greece and Turkey, another NATO member, in the late 1960s saw a temporary cessation of arms deliveries, resumed in 1970 with F-102 Delta Dagger interceptors.

A military coup in 1967 eventually led to a republic. The 'royal' prefix was dropped.

While many Thunderjets and Thunderflashes remained in service with the Delta Dagger force, this was soon augmented with two squadrons each with eighteen F-104G Starfighter interceptors and four Northrop F-5A Freedom Fighter squadrons. Additional transports included Noratlas and C-119G Packets, plus an SAR squadron of HU-16 Albatross amphibians.

The tendency for the USA to interrupt arms supplies whenever tension rose between Greece and Turkey led to attempts to diversify supplies. While LTV A-7 Corsairs were added during the 1970s, followed by F-4E Phantoms, Mirage F1Gs and, later, Mirage 2000EGs, interceptors were bought as well: F-16C/D Fighting Falcon interceptors. C-130 Hercules were introduced, but other transports included Do28s and YS-11s, with CL215 and, later, CL415, amphibians being obtained for fire-fighting and MR, and many Agusta Bell helicopters.

Despite Greek sympathies with the Greek-Cypriot authorities in Cyprus, the HAF could do little to counter the Turkish invasion of Cyprus in 1973 as the only direct route lay over Turkish territory.

The HAF has 31,500 personnel, a third more than thirty years ago. It is divided into Air Defence, Tactical Air Force, Air Support and Air Training Commands. Fighter and ground-attack aircraft include four squadrons with up to seventy-one F-16CG/DG Block 30s, three with up to fifty-eight F-16CG/DG Block 52s, three with up to forty-five Mirage 2000EG/BGs or Mirage 2000-5s, two with thirty-five F-4E Phantom IIs and two with forty-three A/TA-7EH Corsair IIs. Reconnaissance is by a squadron of nineteen RF-4E Phantom IIs. AEW is provided by a squadron of six EMB145H Erieyes. Missiles used include AIM-7 Sparrow, AIM-9 Sidewinder, R-550 Magic 2, Super 530D and AIM-120 AMRAAM AAM, as well as AGM-65 Maverick and AGM-88 HARM. Greece had intended to buy up to ninety Eurofighter Typhoons, but deliveries have been delayed indefinitely due to the cost of hosting the Olympic Games in 2004. MR is provided by two squadrons with six P-3B Orions and two CL-415s with mixed HAF and Hellenic Navy crews.

Air Support Command has three transport squadrons operating fifteen C-130B/Hs and ten C-130Hs, one squadron with twelve Alenia/Lockheed-Martin C-27 Spartans; while there are also two C-47 Skytrains, six Do28s, two VIP RJ-135s, a Gulfstream V and a YS-11-200. Helicopters include two squadrons with AS332 Super Pumas for CSAR and SAR, while there are four VIP AB212 and twelve

AB205 utility helicopters. Training uses conversion trainer variants of the combat aircraft as well as seven Bell 47Gs. There are five fixed-wing training squadrons: two with forty-five T-6A/Bs which replaced T-41As and T-37B/Cs in 2002, another two operating forty T-2E Buckeyes, while the fifth has nineteen T-41Ds. There are Nike Hercules, Patriot PAC-3, Skyguard, Sparrow, Crotale and SA-15 SAM, as well as AMRAAM, Sparrow, Sidewinder, MICA and Magic AAM and Maverick and SCAPLP ASM, as well as Exocet anti-shipping missiles.

HELLENIC NAVY AIR ARM/*AEROPORIA ELLENIKI POLEMIKOU NAFTIKOU*

Founded: 1975

The Greek naval air arm was founded in 1975, initially with four Alouette IIIs for liaison duties, although these were soon supplemented by sixteen Agusta Bell AB212s for shipboard operation as the first of six Dutch-designed Kortenaer-class frigates entered service during the early 1980s. More recently, the Sikorsky S-70B Aegean Hawk has entered service, armed with the Penguin missile, and this also operates from frigates, with all fourteen currently in service able to operate helicopters. A number of personnel serve with the HAF's Orions.

Currently, 400 out of the Navy's 16,000 personnel are directly involved in naval aviation. There are eleven S-70B Aegean Hawks in the ASW and ASuV roles, and also able to provide SAR, while eight AB212s also undertake these roles with another two on electronic warfare duties. Two Alouette IIIs remain in service on SAR, liaison and training.

HELLENIC ARMY AIR ARM/*ELLENIKI AEROPORIA STRATOU*

Founded: 1975

Greek Army aviation was originally founded for liaison, communications and transport. It acquired six heavy-lift Meridionali-built Boeing CH-47 Chinooks during the late 1970s, by which time it also had eight Bell AH-1S HueyCobras for the anti-tank role. The main force at this time comprised fifty Agusta Bell 204 and 205 helicopters, with a small liaison and AOP force of Bell 47Gs.

Currently, the air arm is spearheaded by forty Boeing AH-64A/D Apaches, with fifteen CH-47Ds, 100 UH-1H Iroquois and fourteen AB206s for the transport and utility roles. Twenty NH90 are on order. Three VIP C-12A Hurons (King Air 200) and thirty-eight U-17As (Cessna 180) provide the fixed-wing element, with the latter mainly used for training with twenty-six Nardi-built Hughes 300Cs. UAV include between twelve and eighteen Sperwer.

GUATEMALA

- Population: 13.3 million
- Land Area: 42,042 square miles (108,889 sq.km.)
- GDP: $40bn (£25.6bn), per capita $3,011 (£1,924)
- Defence Exp: $156m (£99.7m)
- Service Personnel: 15,500 active, plus 63,863 reserves

GUATEMALAN AIR FORCE/*FUERZA AÉREA GUATEMALTECA*

A French aviation mission in 1919 with Avro 504 trainers marked the start of military aviation in Guatemala, forming an air unit within the Army. This became the *Cuerpo de Aeronautica Militar* in 1929 with a miscellany of French aircraft. Standardisation occurred in 1937, with Boeing P-26A fighters, Waco general-purpose aircraft and Ryan STM-2 trainers. During World War II, Guatemala allowed the Allies to use bases, and in return received Douglas C-47 transports, Boeing-Stearman PT-17 Kaydets, North American AT-6s and Vultee BT-15 trainers. An American aviation mission in 1945 provided a squadron of North American F-51D Mustang fighters. Guatemala joined the Organisation of American States in 1948, but did not receive new aircraft until ten Douglas B-26 Invader bombers and a number of Lockheed T-33A jet trainers were delivered in 1960. Later, the T-33As were joined by Cessna A-37B Dragonfly armed jet trainers, Pilatus PC-7 armed turboprop trainers, and Basler Turbo 67 turboprop conversions of the C-47 for COIN.

The *FAG* has 1,070 personnel, but receives basic logistic support services from the Army. Aircraft serviceability is believed to be no better than 50 per cent. Two combat squadrons have seven PC-7s and eight A-37Bs. Transport is provided by four Turbo 67s, two Fokker F-27-100/400s, four IA-201 Aravas, a Navajo, a Cessna 206 and two Beech 90/100s. Helicopters include eleven UH-1Hs, two 206B JetRangers and three Bell 212s. During 1999, five Pillan T-35B trainers were obtained from Chile, although one was lost, augmenting five Cessna R172Ks and six PC-7s.

GUINEA

- Population: 10.1 million
- Land Area: 94,927 square miles (245,861 sq.km.)
- GDP: $4.9bn (£3.1bn), per capita $495 (£316)
- Defence Exp: $55m (£35.2m)
- Service Personnel: 12,300 active

GUINEA AIR FORCE/*FORCE AÉRIENNE DE GUINEA*

Founded: 1958

The former French Guinea became independent in 1958 and almost immediately left the French Community. Close links with the USSR led to military aid being provided, including MiG-17 and MiG-21 fighters, Antonov transports and Mil helicopters. Today, the *FAG* has just 800 personnel. Its combat aircraft are no longer operational, other than four Mi-24 attack helicopters obtained in 1998 from the Ukraine. There are four An-14s, an An-24, an SA330 Puma and two Mi-8s, two Alouette IIIs, and an SA342K Gazelle for VIP use. Two MiG-15UTI trainers are also believed non-operational.

GUINEA-BISSAU

- Population: 1.5 million
- Land Area: 13,948 square miles (36,125 sq.km.)
- GDP: $479m (£306m), per capita $318 (£203)
- Defence Exp: $20m (£12.8m)
- Service Personnel: c.4,000 active, but being reduced

GUINEA-BISSAU AIR FORCE/*FORCE AÉRIENNE DE GUINEA-BISSAU*

Founded: 1974

Formerly Portuguese Guinea, Guinea-Bissau became independent in 1974, beset with economic problems and internal unrest, including an army mutiny in 1998. Internal conflict re-ignited in 1999. The small air force has 100 personnel with two MiG-17Fs and a MiG-15UTI, probably no longer serviceable, and has withdrawn MiG-21s from use. The fixed-wing transport aircraft have been withdrawn, leaving a Gazelle and two Alouette IIIs for liaison. The coastguard has an FTB337 (Cessna 337) for offshore patrol.

Manpower figures come from a UN-sponsored census of the armed forces in 2008.

GUYANA

- Population: 752,940
- Land Area: 83,000 square miles (214,970 sq.km.)
- GDP: $1.2bn (£0.77bn), per capita $1,534 (£980)
- Defence Exp: $67m (£42.8m)
- Service Personnel: 1,100 active, plus 670 reserves

GUYANA DEFENCE FORCE AIR CORPS

Founded: 1967

British Guiana became independent in 1967 and established a small Guyana Defence Force, of which 100 personnel are in the air corps. It is a purely transport force, necessary for jungle-covered terrain. At one time five Mi-8 helicopters were operated, but have been withdrawn leaving a Y-12, two Bell 412s and a Rotorway 162F.

HAITI

- Population: 9 million
- Land Area: 10,700 square miles (27,713 sq.km.)
- GDP: $7bn (£4.5bn), per capita $784 (£501)
- Security Exp: $49m (£31.3m)
- Service Personnel: none officially active

HAITIAN AIR CORPS/*CORPS D'AÉRIEN D'HAITI*

Formed: 1943

The Haitian Air Corps was founded in 1943 with US assistance within the Army to carry mail between the main towns. After the end of World War II, Haiti became a member of the Organisation of American States in 1948, and received North American F-51D Mustang fighters, and Beech, Boeing, Fairchild and Vultee trainers. Douglas C-47 and Beech C-45 transports arrived later, as did North

American T-6G Texan and T-28A Trojan trainers. In 1994, with US help a military dictatorship was replaced by a civilian administration after an invasion. The armed forces and police have disbanded since 2004 prior to a new National Police Force being formed. These plans will have been thrown into disarray following the severe earthquake of January 2010, and it is likely that the country will be dependent on the United Nations for internal security and much administration for some time to come.

HONDURAS

- Population: 7.8 million
- Land Area: 43,227 square miles (111,958 sq.km.)
- GDP: $14.6bn (£9.3bn), per capita $1,864 (£1,191)
- Defence Exp: $102m (£65.2m)
- Service Personnel: 12,000 active, plus 60,000 reserves

HONDURAS AIR FORCE/*FUERZA AÉREA HONDUREÑA*

Formed: 1948

The Honduras Army formed an Air Corps after the end of World War I using two surplus Bristol F2B fighters for AOP and liaison duties. Between the world wars, Boeing 95 mail planes were obtained for use as light bombers, but little progress was made. In return for support in safeguarding the Panama Canal during World War II, the USA provided Beech, Boeing, Fairchild, North American, Ryan, Vultee and Waco trainers as well as Beech C-17 and Noorduyn Norseman transports.

Further US military aid followed Honduran membership of the Organisation of American States in 1948, including Lockheed P-38 Lightning and Bell P-63 Kingcobra fighters, followed by North American F-51D Mustang and Republic F-47D Thunderbolt fighter-bombers, and Beech C-45 and Douglas C-47 transports. The fighter-bombers were replaced by Vought F4U Corsair fighter-bombers in 1958, when Douglas C-54 transports, Lockheed T-33A jet trainers and Cessna liaison aircraft entered service followed by the first helicopters, three Sikorsky H-19s. Later, Beech AT-11 and North American T-6G Texan trainers were added.

Jet combat aircraft followed during the 1970s, with COIN Cessna A-37B Dragonflies, and ex-Israeli Dassault Super Mystère fighter-bombers before Northrop F-5E Tiger II fighter-bombers and Lockheed C-130 Hercules were added.

The *FAH* has 2,300 personnel. Modernisation has been delayed by economic difficulties. Two FGA squadrons operate with eight A-37Bs and another with eight Shafrir AAM-equipped eleven F-5E Tiger IIs. There are nine Embraer EMB312

Tucano armed-trainers. A C-130A Hercules and two C-47 transports operate alongside Cessna and Piper light aircraft. Helicopters include five Bell 412EPs, five UH-1Hs, two MD500Ds and two UH-1Hs. Training uses two F-5Fs, nine EMB312 Tucanos and five T-41Ds (Cessna 172).

HUNGARY

- Population: 9.9 million
- Land Area: 35,912 square miles (93,012 sq.km.)
- GDP: $139bn (£88.9bn), per capita $14,082 (£9,002)
- Defence Exp: $1.86bn (£1.19bn)
- Service Personnel: 29,450 active, plus 44,000 reserves

HUNGARIAN ARMED FORCES AIR COMPONENT

Founded: 1949

Hungary was formerly part of the Austro-Hungarian Empire, one of the Central Powers during World War I, becoming an independent republic in 1918. As a former belligerent, Hungary was not allowed an air force under the Treaty of Versailles. Nevertheless, in 1936, a small military air arm came into existence using Fiat CR32 fighters, Meridionali Ro37 reconnaissance aircraft and Heinkel He46 AOP aircraft, with Hungarian-designed Weiss WM13 trainers. During 1938, German assistance was provided, including advisers, Junkers Ju86D and Heinkel He70 bombers, and Bucker Bu131 Jungmann trainers. The following year, Fiat CR42 fighters, Caproni Ca135 and C310 bombers and Nardi trainers joined what had become the Hungarian Army Air Force.

In 1940, Hungary entered World War II on the side of the Axis Powers. Hungarian units were sent to the Russian front, assisting in Operation Barbarossa. German aircraft entered service in substantial numbers, including Messerschmitt Bf109 fighters, Junkers Ju87D Stuka dive-bombers and Ju88A bombers, while there were also Italian Reggiane Re2000 fighters. Many aircraft were built under licence in Hungary, including aircraft for the *Luftwaffe*! Under the gruelling conditions of the Russian Front and the massive Soviet counter-attacks, Hungary and its armed forces suffered badly during the closing stages of the war, surrendering early in 1945.

Post-war, Hungary was forbidden more than a few fighter and transport aircraft until a Communist government took control in 1949, and Soviet military aid commenced, creating a new Hungarian Air Force. Yakovlev Yak-9 fighters, Lisunov Li-2 (C-47) transports and a variety of trainers were supplied. The first jets, MiG-15 fighters, arrived during the early 1950s with obsolete Ilyushin Il-10 ground-attack aircraft and Tupolev Tu-2 bombers. These were followed later by MiG-17 fighters and Ilyushin Il-28 jet bombers, Antonov An-2 transports, and additional Li-2s, Mil Mi-1 and Mi-4 helicopters, Yakovlev Yak-11 and -18 trainers, and MiG-15UTI trainers. Hungary became a member of the Warsaw Pact in 1955, but attempted to break away from Soviet influence in 1956, with Hungarian and Soviet forces opposing each other. The USSR quelled the revolution. In its aftermath the air force was stood down for some years, without new aircraft until the late 1960s, when MiG-21 interceptors, MiG-19 fighter-bombers and Sukhoi Su-7B ground-attack aircraft were introduced. These aircraft and the remaining MiG-17s equipped twelve squadrons, with one Ilyushin Il-28 bomber squadron. Ilyushin Il-14 transports joined the earlier An-2, Li-2, Mi-1 and Mi-4 helicopters. Later MiG-29 interceptors, Mi-24 attack helicopters, Mi-8 and Mi-17 transport helicopters were supplied. The main training aircraft became the Aero L-39, with a number of Yak-52s, while the MiG-21 force was upgraded.

The break-up of the USSR and the collapse of the Warsaw Pact caused a major re-orientation of Hungarian foreign and defence polices. Despite having the Soviet Bloc's strongest economy, the cost of new equipment has seriously hampered modernisation. Hungary became a member of NATO in 1999, but struggled to upgrade MiG-29s to become compatible with NATO. The MiG-21s were retired in 2000, as were Mi-2s and Yak-52s.

Today part of the unified armed forces, the air component has 5,664 personnel, having contracted considerably. One squadron has fourteen Gripen C/Ds and another five MiG-29A/UBs, with many more in store. An attack helicopter wing has seventeen Mi-24Ds. There are seventeen Mi-8/-17 transport helicopters; two EW Mi-17s are in store. Transport is provided by five An-26s. Training uses ten L-39ZOs obtained from the *Luftwaffe* and nine Yak-52s. Western missiles are replacing those of the Warsaw Pact era, with Sidewinder and AMRAAM taking over from *Alamo* and *Archer* AAM, and Maverick ASM are also in service.

ICELAND

- Population: 306,694
- Land Area: 39,758 square miles (102,846 sq.km.)
- GDP: $12.8bn (£8.2bn), per capita $41,704 (£26,658)
- Security Exp: $44.5m (£28.4m)
- Paramilitary Personnel: 130 active

ICELANDIC COAST GUARD

A member of NATO, Iceland does not maintain standing armed forces, due to the very small population and the presence of US forces. The Icelandic Coast Guard has three fishery protection vessels, augmented by a single MR F-27-200 Friendship, a Eurocopter AS332L Super Puma and an SA365N Dauphin II. Based at the international airport at Reykjavik, the helicopters can be deployed aboard the fishery protection vessels.

INDIA

- Population: 1,157 million
- Land Area: 1,262,275 square miles (3,268,580 sq.km.)
- GDP: $1,300bn (£831bn), per capita $1,124 (£718)
- Defence Exp: $359bn (£229bn)
- Service Personnel: 1,325,000 active, plus 1,155,000 reserves

Before the global economic crisis at the end of the first decade of the new century, India had one of the fastest-growing economies in the world, ending years of decline, and while it might be an overstatement to describe the country as prosperous, it has emerged in stronger shape than at any time since independence. While militant Islamism is a problem, it is also true that relations with neighbouring Pakistan have also improved over the past decade or so.

The country maintains strong armed forces backed by a steadily improving industrial base, so that the current construction of an aircraft carrier of some 40,000 tons can be seen as a logical progression from the use of redundant British warships. A strong military is essential given poor relations with the other neighbouring country, China, and the continuing development of Chinese military strength. This begs the question whether growing prosperity in India and the need to maintain strong armed forces will make the country the next regional power? There is a certainly a need to protect a substantial merchant fleet, especially with piracy on the increase as the result of the Western democracies turning a blind eye to this problem for far too long.

The world's most populous democracy now has nuclear weapons, and a Strategic Forces Command to oversee their use, with a Political Council and an Executive Council with the latter advising the former, which is chaired by the prime minister. Only the Political Council can authorise the use of nuclear weapons. One can but hope that as in the Cold War, the devastation that nuclear weapons can inflict will provide stability to South Asia, and beyond.

INDIAN AIR FORCE

Founded: 1933

Indian military aviation dates from 1933 when the Indian Air Force was formed under RAF control, initially operating just four Westland Wapiti general-purpose aircraft on army co-operation duties. It was not until 1940 that this force reached squadron strength, but considerable expansion occurred during World War II. Westland Lysanders replaced the Wapitis in 1941. The IAF expanded to seven

Hawker Hurricane fighter squadrons and two Vultee Vengeance dive-bomber squadrons. Pilots were trained mainly in Australia and Canada under the Empire Air Training Scheme, but a number were trained in India and the UK. Supermarine Spitfire fighters arrived in 1944. Throughout the war, the IAF operated mainly as a tactical air force in close support of ground forces. The 'Royal' prefix was granted in recognition of the IAF's wartime achievements in 1945. At the end of the war, a squadron was deployed to Japan as part of the Commonwealth Occupation Force.

India divided into India and Pakistan on independence in 1947. The RIAF was also divided. India's share was seven fighter squadrons with Supermarine Spitfires, Hawker Hurricanes and Tempests, and a Douglas C-47 transport squadron. Additional Spitfires and Tempests were delivered following independence, and Hindustan Aircraft Industries refurbished redundant Consolidated B-24 Liberator bombers to provide a small bomber force. These were followed by the first jets, de Havilland Vampire F3 jet fighters and FB9 fighter-bombers. De Havilland Devon (Dove) light transports were also added, while Percival Prentice trainers were built in India pending production of the Hindustan HT-2 trainer. The 'Royal' prefix was dropped in 1950 when India became a republic within the British Commonwealth.

Over 100 Dassault Ouragan jet fighter-bombers were bought in 1953, and the following year twenty-six Fairchild C-119G Packet transports were introduced. In 1955, the IAF acquired Vickers Viscount 700s and its first Soviet aircraft, Ilyushin Il-14 transports, as well as Auster AOP9 aircraft. More than 100 Dassault Mystère IVA fighters were ordered in 1956, while licence-production of the Folland Gnat light fighter started. The IAF continued rapid expansion throughout the late 1950s, introducing Hawker Hunter F56 fighters and T7 trainers, English Electric Canberra B8 jet bombers, de Havilland Vampire T55 jet trainers, de Havilland DHC-3 Otter transports, Bell 47G Sioux and Sikorsky S-55 helicopters, often in substantial numbers. North American T-6G Texan trainers were built under licence.

Throughout the 1960s and for the two decades that followed, Indian defence procurement relied heavily on Soviet armaments. Substantial defence forces were developed because of border clashes with Pakistan, often breaking into warfare, as well as several border clashes with Communist China. Licence-production of the MiG-21 interceptor started and the Hindustan HK-24 Marut fighter-bomber and Kirshak AOP aircraft entered production. The Gnat returned to production after proving to be particularly successful, earning itself the nickname of 'Sabre Slayer' during one war with Pakistan. The Indo-Pakistan War of 1971 saw East Pakistan break free from West Pakistan with Indian assistance to become the new state of Bangladesh. Antonov An-12 transports, Mil Mi-4 helicopters and de Havilland Canada DHC-4 Caribou transports were also introduced during the 1960s. Within the first quarter century after independence, the IAF grew to 90,000 personnel. Throughout this period, the IAF also provided AOP cover for the Indian Army.

The 1970s saw the IAF continue its preference for Soviet equipment. It bought the Anglo-French Jaguar strike aircraft, building this under licence, including a number for shore-based anti-shipping duties. The 1980s saw the IAF acquire

Dassault Mirage 2000 interceptors, joining MiG-23s, a small number of MiG-25s, and later, MiG-27s. The helicopter force also grew rapidly, with Mil Mi-8 and, more recently, Mi-17 transport helicopters, as well as Mi-24 and Mi-35 attack helicopters. The Alouette III was built in India as the Chetak, but the indigenous ALH is now in service. A substantial transport force includes the An-32 and Il-76. Recently, MiG-29A interceptors have been bought from Russia, and Sukhoi Su-30MKI strike aircraft and interceptors built under licence.

Despite occasional incidents, there have been fewer border clashes between India and Pakistan in recent years – the most recent being in 1999 when the IAF shot down two Pakistani MR aircraft that had strayed into Indian airspace. India's defence budget rose by 30 per cent between 2000 and 2001, and by a further 13 per cent between 2001 and 2002 and has continued to grow since.

Currently, the IAF has 127,200 personnel, with annual average flying hours having increased in recent years from 150 to 180. It is organised into five regional air commands: Central based on Allahabad; Western based on New Delhi; Eastern at Shillong; Southern at Trivandrum; and South-Western at Gandhinagar. There are thirty-five fighter and ground-attack squadrons: three operating forty-eight Su-30MKI *Flanker*; three with forty-eight MiG-29 *Fulcrum*; six with ninety-eight MiG-27ML *Flogger*; sixteen with 213 MiG-21bis/FL/M/MF *Fishbed*, more than half of which are being upgraded; three have thirty-six Mirage 2000H/THs, which have a secondary ECM role, and four have eighty Jaguar S(I), of which sixteen are used for anti-shipping strikes with Sea Eagle ASM. AEW and AWACs is provided by a squadron with three Il-76TDs, while reconnaissance is provided by a squadron of three Gulfstream IVSRAs. These are supported by a tanker squadron with seven Ilyushin Il-78 *Midas*. Tactical support of ground forces is by two squadrons with forty Mi-24/-35 *Hind*. UAVs are operated by five Searcher squadrons.

Air transport is provided by at least sixteen squadrons and a number of flights. No less than seven squadrons operate eighty-four An-32 *Sutlej*; another two have twenty-four Il-76 *Candid*; one has six Boeing 737s; two have Indian-built Do228s; there is a squadron of EMB145s and a flight of EMB135s; and there are twenty BAe748s still in service. Transport helicopters include fifteen squadrons with 174 Mi-8/-17 *Hip*; a squadron with eight heavy-lift Mi-26 *Halo*; and two squadrons with twenty ALH Druv, which should increase to eight with 150 aircraft, replacing the six squadrons with Cheetah (SA315B) and Chetak (SA316B).

This substantial force has extensive training support using conversion trainer versions of the combat aircraft, denoted by the 'U' suffix for MiG and Su types, and by 'TH' for the Mirage, although most of the two-seat MiG-21s are being withdrawn. Fast jet training is on the Hawk, with twenty-five operational being joined by up to eighty Indian-built examples. There are 162 HJT-16 Kiran and sixteen HJT-36s. Eight Super Dimona motor-gliders were introduced in 2000.

Missiles in use include the AS-7 *Kerry*; AS-11B anti-tank weapon; AS-12 and AS-30; Sea Eagle and AM 39 Exocet; and AS-17 *Krypton* in the ASM role. AAM include AA-7 *Apex*; AA-8 *Aphid*; AA-10 *Alamo*; AA-11 *Archer*; R-550 Magic; Super 530D. SAM missiles are used by thirty-one squadrons and ten flights equipped with a total of 280 launchers, including Divina (SA-2), Pechora (SA-3), SA-5 and SA-10. Searcher UAV are in service.

INDIAN NAVAL AVIATION

Founded: 1950

Indian Naval Aviation first emerged in 1950 with the creation of a Fleet Requirements Unit for communications and target-towing duties, equipped with Short Sealand amphibians, five Fairey Firefly target-tugs and three Hindustan HT-2 trainers. It was not until 1961 that the Majestic-class light carrier, HMS *Hercules*, was refitted and transferred to India as INS *Vikrant*. A total of thirty-five Sea Hawk fighter-bombers equipped two squadrons, with ten Breguet Br1050 Alize ASW aircraft, six Westland SH-3D Sea King and ten Alouette III helicopters. The carrier was able to carry sixteen Sea Hawks, four Alize and two Alouettes at any one time. Eventually, the number of Sea Kings rose to thirty-one, with later Mk42A/B versions able to carry both AS torpedoes and Sea Eagle anti-shipping missiles.

Aircraft from *Vikrant* played an important part in the Indo-Pakistan War of 1971, as a result of which East Pakistan broke away to become independent as Bangladesh. Sea Hawk fighter-bombers flew from the carrier to attack Chittagong and Cox's Bazaar in the then East Pakistan, helping to overwhelm Pakistani air defences in the final days of the conflict.

In the mid-1980s, *Vikrant* was joined and then replaced by the former HMS *Hermes*, renamed *Viraat*. *Viraat* came with a 'ski-jump' to operate the Sea Harrier FRS51 V/STOL fighters and T60/T4 conversion trainers acquired at the same time. A new aircraft carrier is being built in India, described as the Vikrant-class, of 40,000 tonnes standard displacement and is likely to be in service by 2015 when she will replace the *Viraat*, and later be joined by a second ship of the same class. The MiG-29 has been mentioned as a possible carrier-borne aircraft.

Today, aviation accounts for 7,000 of the Indian Navy's 55,000 personnel. There are thirteen Sea Harrier FRS51s in a squadron, while shore-based aircraft include eight long-range MR Tu-142M *Bear* with ten Dornier Do228s, seventeen BN2A/B Defenders, some of which are being upgraded with turboprop engines and ten BAe748s. Helicopters include thirty-five Sea King Mk42/A/B/Cs for ASW, ASUW, SAR and transport, with twelve Ka-28 *Helix*, seven Ka-25 *Hormone* and nine Ka-31 *Helix B* for ship-based ASW. The Kamov helicopters will be supplemented by up to forty ALH helicopters in the ASW and light transport roles. Searcher UAV are also deployed. Training uses three Harrier T60s and two T4s, as well as twelve Kiran I/IIs and eight Deepaks, while fifty Chetaks (SA-316B) operate in the training and communications roles. The Indian Navy controls the operations of the Indian Coast Guard, which has an additional 6,200 personnel. Coastal patrol is provided by two squadrons operating twenty-four Do228s, while another squadron has fifteen Chetak helicopters that are being replaced by up to forty ALHs.

INDIAN ARMY AVIATION CORPS

India has the second largest army with 1,100,000 personnel. The Aviation Corps is relatively young, having taken over AOP and liaison from the IAF in the 1980s, and has grown rapidly with an attack capability of 120 Lancers, an armed version of the Cheetah helicopter, for COIN operations in the Himalayas, as well as sixty Chetak (Alouette III) for observation and liaison duties, which will be replaced by the Indian ALH (Advanced Light Helicopter). Nishant and Searcher UAV are in service.

INDONESIA

- Population: 240.3 million
- Land Area: 736,512 square miles (1,907,566 sq.km.)
- GDP: $559bn (£357bn), per capita $2,328 (£1,488)
- Defence Exp: $3.5bn (£2.3bn)
- Service Personnel: 302,000 active, plus 400,000 reserves

INDONESIAN NATIONAL DEFENCE-AIR FORCE

Formed: 1949

Indonesian military aviation preceded independence in 1949 as an aviation division of the People's Peace Preservation Force was established in December 1945, following Japanese surrender and before Dutch forces could return in spring 1946. The new air arm had just five Indonesian pilots and used abandoned Japanese aircraft. These were soon put out of action by a Dutch air attack. India provided assistance before and immediately after independence, when Dutch assistance was also provided for what had become the Indonesian Air Force, AURI. Former Dutch aircraft included North American F-51D/K Mustang fighter-bombers and B-25 Mitchell bombers; Convair PBY-5A Catalina amphibians; Lockheed 12A transports; Piper L-4J Cubs, Auster IIIs and Aiglets for training and AOP duties. India provided Hindustan HT-2 trainers. Indonesia extends for more than 3,000 miles end-to-end with many islands, so the Catalina amphibians were important. Lockheed 12As also provided air services before the national airline, Garuda, started operations in 1950.

In 1955, Indian personnel were seconded to the AURI to help reorganise it. At this time, aircrew were being trained in the USA, UK and the Netherlands, with the first jets, de Havilland Vampire fighter-bombers, supplied by the UK that same year. Indonesia began to be drawn into the Soviet sphere of influence during the late 1950s, and in 1958, Czechoslovakia provided sixty MiG-17 fighters, a number of MiG-15 fighter-bombers and forty Ilyushin Il-28 light bombers, while pilots were trained in Egypt. Western aircraft continued to enter service, including DHC-

3 Otters, Cessna 180s, Grumman HU-16 amphibians, Fairey Gannet AS4 ASW aircraft, and T5 trainers, Lockheed C-130B Hercules transports and Sikorsky S-58 helicopters.

By 1960, Indonesian policy had become aggressive. An invasion of West Irian, formerly Dutch New Guinea, in 1962 brought Indonesian and Dutch forces into conflict, although the territory was ceded to Indonesia the following year. This period also marked a confrontation, which lasted until 1966, with the UK, Australia, New Zealand and Malaysia over the creation of the Federation of Malaysia, through which Malaya and Singapore were to become independent from the UK. Western supplies were cut off, so the USSR supplied thirty-five MiG-19 and fifteen MiG-21 fighters, with twenty-five Tu-16 bombers, Il-14 and An-2 transports. AURI's operations throughout the Malaysian confrontation were largely unsuccessful, and the end coincided with the downfall of the then dictator and president, Sukarno.

In 1975, Indonesia seized the Portuguese colony of East Timor. East Timorese attempts to gain independence and the use of the AURI against them caused controversy over arms sales to Indonesia. When, under UN auspices, East Timor did finally become independent in 1999, the AURI, renamed the Indonesian National Defence-Air Force, TNI-AU, had no role. Meanwhile, Indonesian foreign relations had become less focused on Russia. Western aircraft types were ordered once again, including F-5E/Fs, A-4 Skyhawks, Hawk trainers and attack aircraft, a few OV-10 Bronco COIN aircraft, and, later, F-16A interceptors. More modern variants of the C-130 Hercules were also introduced.

Economic problems have meant that the TNI-AU has contracted in recent years, and now has just 24,000 personnel, less than half the figure for the early 1970s, although reorganisation of Indonesia's armed forces has meant that the paratroops have now passed to the army. In 1998, orders for Mi-17 helicopters and Su-30Ks were cancelled. Upgrades to the F-16 and F-5 aircraft have been kept to the necessary minimum. Indonesia now builds aircraft, including a joint utility transport venture with CASA of Spain, the CN235. In future, the TNI-AU will be needed increasingly for COIN, as the widely disparate population of this sprawling nation seeks greater self-determination. It could also have a considerable role in disaster relief.

The TNI-AU is organised as east and west operational commands and a training command, although it is believed that less than half the aircraft are operational. Combat aircraft include a squadron with ten F-16A/Bs; one with twelve F-5E/Fs; two with thirty-five Hawk 109/MR209s; one with eleven A-4E/J/TA-4H Skyhawks, and a squadron with four Su-27/-30 *Flanker*. A flight of five OV-10F Bronco COIN aircraft is believed to be non-operational. MR is provided by just three 737-2X9 Surveillers. There are five transport and tanker squadrons operating a Boeing 707, two KC-130Bs and fifteen C-130B/H/L-300-30s, four F27-400M Troopships and three F28-1000/3000s, a VIP Boeing 707-320C, and ten NC212-100/200 Aviocars. A Skyvan 3M is used for aerial surveys. Communications duties are provided by two Cessna 172s and four 207s.

Helicopters include three squadrons with ten SAR S-58Ts, five NAS332 Super Pumas and eleven NAS330 Pumas, with two used for VIP duties. Training aircraft include twenty T-34 Turbo Mentors, six T-41s (Cessna 172), as well as thirty-nine AS-202s. Ten Bell 47G Sioux were replaced in 2003 by twelve EC120s.

Missiles include Maverick ASM and Sidewinder AAM.

INDONESIAN NATIONAL DEFENCE-NAVY

At one time, the Indonesian Navy operated MiG-19 and MiG-21 interceptors, but in recent years it has become more focused on transport and communications, SAR and ASW duties, while long-range MR is left to the air force. Its development has been affected both by the state of the economy and by sanctions. Before sanctions were applied, ex-Australian Army Nomads were introduced for a variety of roles, including transport and MR. The CN235 collaborative venture transport aircraft has also been introduced.

Currently, just 1,000 personnel out of a total of 45,000 are directly involved with the air arm. Patrol aircraft include two 235MPAs and twenty-five Searchmasters (Nomad). Transport is provided by a CN-235M, four C-212-200s, two DHC-5 Buffaloes and a number of light aircraft of Cessna and Piper manufacture. Helicopters include three NAS332L Super Pumas, eight Mi-2 'Hoplite', four Bell 412s, five Bo105s and three EC120Bs, while nine Wasp HAS1 are grounded. A variety of training types are in use, including six Piper Tomahawks and four Senecas, two Beech Bonanzas and a TB9 Tampico, as well as two Alouette II helicopters.

INDONESIAN NATIONAL DEFENCE-ARMY

Indonesian Army Aviation originated with transport and liaison duties, but in recent years has acquired combat helicopters despite economic difficulties.

The current helicopter force is spearheaded by six Mi-35 *Hind* combat helicopters supported by sixteen Mi-17 *Hip*. There are also eight Bell 205As, twelve NB-412s and seventeen Bo105s, as well as twelve Hughes 300C trainers.

IRAN

- Population: 66.4 million
- Land Area: 627,000 square miles (1,626,520 sq.km.)
- GDP: $359bn (£229bn), per capita $5,418 (£3,463)
- Defence Exp: $9.6bn (£6.1bn)
- Service Personnel: 523,000 active, plus 350,000 reserves

Since 1979, Iran has gone from being an ally of the West to a pariah state, with doubts over who really wields power in the country and the armed forces divided between those of the state and those of the Revolutionary Guard Corps. Western concerns about Iran developing nuclear weapons and the country's support for extreme Islamic groups, and especially for the Taliban in neighbouring Afghanistan, have led to an uneasy relationship, even though the serviceability of much of its equipment is in doubt and believed to be no better than 60 per cent for US-supplied equipment and 80 per cent for that from Russia. Disputes over the results of a presidential election in 2009 have also resulted in internal unrest, although for the present it seems that the security forces are able to keep this under control.

For the foreseeable future, Iran seems to be the state most likely to maintain a state of instability in the Middle East. Russian willingness to help with the nuclear power programme and with arms supplies could also see a reversion to the Cold War, while China has also supplied Iran with defence equipment. An ominous sign is that the Revolutionary Guard Corps Air Force controls the nation's 'strategic' missiles.

ISLAMIC REPUBLIC OF IRAN AIR FORCE

Formed: 1932

At one time possessing the most modern and most powerful air force in the Middle East, Iranian military aviation dates from 1922, when the Iranian Army created an Air Office. A Junkers F13 was purchased and a German pilot hired. The following year, an Aero A30 was bought for AOP duties, followed in 1924 by four DH4s and DH9s as well as additional F13s. The first Iranian pilots were trained in France during 1924, when the Air Office became the Iranian Air Force, but still part of the Army.

In 1932, the IAF became the Imperial Iranian Air Force, a separate service. Modernisation started, with eighteen de Havilland Tiger Moth trainers, followed by twelve Hawker Fury fighters in 1933, and later by Hawker Hart light bombers and thirty-two Audax AOP aircraft, while instructors were seconded from the Royal Swedish Air Force. Although thirty-eight Hawker Hurricane fighters and some Curtiss 75A Hawk fighter-bombers were ordered in 1938, just two Hurricanes reached Iran before deliveries were suspended following the outbreak of World War II in Europe. In 1941, Iran was invaded by British and Soviet forces to guarantee an overland supply route to Russia and prevent Iran joining the Axis powers. During the war, the IIAF stagnated, except for the delivery of a small number of Avro Ansons and some additional Tiger Moths after Iran declared war on Germany in 1943.

Post-war, the IIAF eventually received thirty-four Hurricane fighters, including the rare two-seat version. These were soon joined by the first of many American aircraft, including Republic F-47D Thunderbolt fighter-bombers, Douglas C-47 transports, Piper L-4 AOP aircraft, and North American T-6G Texan and Boeing-Stearman PT-13 Kaydet trainers. Until 1950, all aircrew were trained in the USA or West Germany. As a founder member of the Baghdad Pact (later renamed the

Central Treaty Organisation), Iran received American military aid throughout the 1950s, receiving Republic F-84F Thunderstreak and North American F-86F Sabre fighter-bombers, and Lockheed T-33A trainers. During the 1960s and early 1970s, these aircraft were supplemented and to some extent replaced by F-5A and F-4D Phantom fighter-bombers. Lockheed C-139E Hercules transports were also introduced, while helicopters included sixteen Meridonali-built CH-47 Chinooks, forty Bell UH-1Ds and forty Agusta Bell 205s, as well as 100 206As, sixteen Super Frelons and a number of HH-43B Husky. Short Tigercat SAM were also deployed. Many helicopters were operated on behalf of the Army and the Navy, although these were developing their own air arms. The paramilitary Imperial Iranian Gendarmerie also had five Agusta Bell 205 helicopters. By this time, there were 17,000 personnel.

The 1970s saw continued expansion while the IIAF kept pace with technological developments, using the rapid rise in the price of oil to buy sophisticated aircraft, with seventy-nine Grumman F-14 Tomcat interceptors and Northrop F-5E Tiger II fighter-bombers. Sea King ASW helicopters were acquired for naval use, with Sea Stallions for minesweeping and MR Lockheed P-3F Orions. The Hercules and Chinook fleets increased. In 1979, the Shah was deposed and an Islamic republic declared, with a change of name to the Islamic Republic of Iran Air Force. The revolutionary fervour of the new regime, including storming the US embassy and the taking of its staff hostage led to a complete breakdown in the relationship between Iran and the West, allowing the USSR to become Iran's main supplier. Naval and army air power finally moved from air force control at this time.

Iran's Islamic fundamentalism also raised tensions with its neighbours. This led to what some historians describe as the 'First Gulf War', when from 1979 and throughout the 1980s, Iran and Iraq were at war, and while the intensity of operations fluctuated, conflict was often heavy. The outcome of the conflict was inconclusive, with Iraq receiving some support from the West, and Iran moving closer to the USSR. Soviet aircraft, including Sukhoi Su-24Mk strike aircraft, Ilyushin Il-76 and Antonov An-74 transports, and in 1990, MiG-29A interceptors, were introduced. The conflict ended in 1989, with no clear outcome. A complete change in relations with Iraq appeared to come following that country's invasion of Kuwait in 1990. During the 'Second Gulf War' that followed, Iraq sent many aircraft to Iran for safety from Allied air attacks, with between 100 and 120 aircraft flown into Iran, which promptly retained the aircraft as reparations for earlier losses!

Over the last twenty years, relations with Iran have remained difficult for both the West and its neighbours, while the break-up of the USSR has also resulted in a less close relationship with Russia. Although tentative steps have been made towards an easier relationship with the rest of the world, US pressure on the CIS still resulted in a reluctance to provide up-to-date equipment for Iran, forcing the country to turn to China. Chinese aircraft include the Shenyang F-6 (MiG-19), F-7N (MiG-21), and a number of transport aircraft.

Despite a belligerent tone, the IRIAF has reduced in size over the past eight years or so, with personnel falling by a third to 30,000, with 12,000 of these engaged in air defence. Serviceability of aircraft is a growing problem, with many aircraft cannibalised for spares to keep others flying.

Mainstays of the combat force are two squadrons with twenty-five F-14A Tomcats; a squadron with seventeen A-7s and another with eighteen F-6s; two squadrons with twenty-five MiG-29As, some of which are ex-Iraqi aircraft, a squadron with thirty Su-24s and another with twenty-five Su-25s, while there are supposed to be four squadrons with sixty F-5Es and four with sixty-five F-4D/Es as well as six RF-4Es in a reconnaissance squadron. Serviceability of these last aircraft must be very poor. There are five Lockheed P-3F Orion MR aircraft. Transport includes five squadrons with a variety of aircraft, including seventeen Lockheed C-130E/H Hercules plus two reconnaissance RC-130H. There are four Boeing 747Fs, a 727 and three 707s, as well as thirteen F-27 Troopships, ten PC-6s and nine Y-12s. Some sources claim that a substantial force of as many as fifty-three Mi-17 helicopters is in service, while there are still some CH-47s operational, thirty AB-214s and a couple of Bell 206As. In addition to the conversion trainers, there are twenty Beech Bonanzas, fifteen Embraer EMB321 Tucanos, fifteen JJ-7s, and forty PC-7s. Hawk, Stinger, Rapier and Tigercat SAM are still believed to be deployed, as well as SA-7, while AAM missiles include a mixture of Western and Soviet types, including AIM-7 and AIM-9, AA-8, AA-10 and AA-11.

REVOLUTIONARY GUARD CORPS AIR FORCE
Founded: 1979

The Islamic revolution in Iran channelled the energies of its supporters into the Revolutionary Guard, initially while evaluating the loyalty and commitment of members of the former Imperial armed forces. This force has since acquired an air arm, in some cases using aircraft taken from the armed forces and, given the paucity of information from the country, there may be some double accounting of equipment. The Revolutionary Guard has more than 125,000 personnel, of which 100,000 are ground forces, 20,000 naval and 5,000 marines, but little is known about air arm numbers other than that they are commanded by an officer of one-star rank. Aircraft are believed to include twenty PC-7s used both on COIN and for training; twenty Mi-8AMTSH helicopters; and a fixed-wing transport force with six An-74s and twenty Shahed-5s (CASA-IPTN CN212).

ISLAMIC REPUBLIC OF IRAN NAVY

Although the IRIAF still operates fixed-wing MR aircraft, the Iranian Navy has control of its helicopters, without warships capable of operating these. Some 2,600 of the total of 18,000 naval personnel are engaged in naval aviation. There are three Sikorsky RH-53D Sea Stallion minesweeping helicopters and ten SH-3D Sea King ASW helicopters, with ten Agusta Bell AB212AS also for ASW, five AB205s for transport and two AB206 JetRangers for liaison. Fixed-wing aircraft include five Do228s, four F-27 Friendships and four Aero Commander 500/690 liaison aircraft.

ISLAMIC REPUBLIC OF IRAN ARMY

The Iranian Army has its own aircraft, although serviceability is likely to be low given the fact that most aircraft are of Western origin. There are twenty CH-47C Chinooks left, no more than fifty of the original 100 AH-1J Cobra attack helicopters plus twenty-five Mi-17 *Hip*, as well as fifty Agusta Bell AB214A/Cs, eighty-seven AB205A-1 utility helicopters, with forty AB206A/B JetRangers on observation and communications duties. There are two VIP Falcon 20Es, five communications Aero Commander 690s and ten liaison Cessna 185s.

IRAQ

- Population: 28.9 million
- Land Area: 169,240 square miles (435,120 sq.km.)
- GDP: $76bn (£48.6bn), per capita $2,627 (£1,679)
- Defence Exp: not known
- Service Personnel: 191,957 (mainly conscript) active, plus 386,312 Ministry of Interior forces

IRAQI AIR FORCE

Formed: 1931

Originally founded as the Royal Iraqi Air Force in 1931 to coincide with the return of the first Iraqi pilots from training in the UK, the RIAF initially had five de Havilland Gipsy Moth trainers, to which another four were soon added. In 1932, four Puss Moths fitted with bomb racks arrived to be deployed on 'air control', suppressing rebel uprisings. Further expansion followed in 1934, with additional Puss Moths as well as de Havilland Tiger Moths and Dragon Rapides and Hawker Nisrs. Breda Ba65 fighter-bombers were delivered in 1937 with Savoia-Marchetti SM79B bombers, while during 1938–40, Gloster Gladiator fighters, Avro Anson and Douglas DB-8A bombers were delivered, as well as additional Dragon Rapides and Dragonfly light transports. Initially, Iraq was unaffected by World War II, until a German-inspired uprising in 1941 saw the RIAF in conflict with the RAF, losing most of its aircraft.

Re-equipment started in 1946 with thirty Hawker Fury fighter-bombers. In 1948, four Bristol Freighters, some de Havilland Doves and Auster AOP6 and T7

119

aircraft were added. Iraq became a founder member of the Baghdad Pact and was able to acquire further British equipment throughout the early and mid-1950s. New aircraft included twenty DHC-1 Chipmunk T20 trainers in 1951; the first jets, twelve de Havilland Vampire FB52 fighter-bombers and six T55 trainers in 1953; and the first helicopters, two Westland Dragonflies (S-51), in 1955. Two de Havilland Heron light transports arrived in 1956, followed by a squadron of Hawker Hunter F6 fighters in 1957. In 1958, the royal family was assassinated, a republic declared and the RIAF dropped the 'royal' prefix. Iraq left the Baghdad Pact, which was re-named the Central Treaty Organisation, and the IAF turned to the USSR for help.

The USSR wasted little time. The first MiG-15 fighter-bombers arrived in October 1958, followed by Russian instructors and Ilyushin Il-28 jet bombers. Aid continued into the next decade, with MiG-17, MiG-19 and MiG-21 fighters and interceptors, Antonov An-12 transports and Mil Mi-1 and Mi-4 helicopters, but halted after defections by MiG-21 pilots, with their aircraft, during 1965 and 1966. Once again, Iraq turned to the UK for supplies, buying Hawker Hunter FGA9s, FR10s and T66/9s, Westland Wessex (S-58) helicopters and BAC Jet Provost T52 jet trainers. Soviet supplies resumed following the Arab-Israeli War of 1967, including Sukhoi Su-7B ground-attack aircraft. Iraq became a substantial air power in the region, with sixty MiG-21s, fifty MiG-19s and MiG-17s, fifty Su-7Bs, thirty-six Hunter FGA9s, ten Il-28s and eight Tu-16s, as well as a wide variety of transport aircraft, including An-2, An-12, An-24 and Il-14. Training aircraft included Yak-11 and Yak-18, and L-29 Delfins. Guided missiles were also obtained from the USSR, starting with the SA-2 SAM.

The IAF continued to expand, more than doubling in personnel over the next thirty years. Iraq used its considerable oil wealth to purchase more sophisticated Soviet equipment. MiG-23 and then MIG-25 aircraft were delivered, while a number of Chengdu F-7s (MiG-21) were also obtained. The start of the war with neighbouring Iran in 1979, which lasted throughout the 1980s, saw a slight improvement in relations with the West, and the IAF was able to obtain helicopters from France, the USA and Germany, and Mirage F1EQ attack aircraft.

Following the war with Iran, Iraq turned its attention to its much smaller neighbour Kuwait, long claimed as an Iraqi province. An Iraqi threat to Kuwait in 1961 had been countered by the deployment of British aircraft carriers and marines. In 1990, Iraq invaded. Under the auspices of the United Nations, a coalition of nations assembled forces in the area to first counter any threat to Saudi Arabia and then to liberate Kuwait from Iraqi occupation. The coalition ground assault was preceded by an air campaign, which the IAF attempted to counter for no more than a day or so, with many of its aircraft, including all the Tu-16s, either shot down or destroyed on the ground. Many surviving aircraft were sent to neighbouring Iran for safety, where the aircraft, between 100 and 120 in number, were seized as reparations.

To protect Kuwait, and also the Marsh Arabs in the south and the Kurdish population in the north, Iraq was banned from flying military aircraft north of the

36th parallel and south of the 33rd parallel, known as the 'no fly' zones. Strict economic sanctions were later relaxed to allow Iraq to sell oil to earn currency for food and medical supplies. Revenue from the oil sales was used mainly to upgrade Iraq's anti-aircraft defences, prompting repeated strikes by US and British aircraft against air defence radar and missile sites.

British and American concerns about Iraq's possession of 'weapons of mass destruction' resulted in an invasion on 20 March 2003, led by US forces supported by the United Kingdom, Australia and Poland with support from Kurdish irregulars in Iraqi Kurdistan. After the invasion was completed on 1 May, thirty-six other countries became involved. The IAF, which had 35,000 personnel at the time, was in a poor state to resist, with many pilots flying as little as twenty hours annually. Iraqi armed forces were disbanded once the invasion was completed but attempts are being made to rebuild them amidst continuing unrest and terrorism in the country.

Currently, the IAF has less than 2,000 personnel and its activities are confined to reconnaissance and transport. The reconnaissance function is fulfilled by a squadron with eight Cessna 208B Grand Caravans and another squadron using eight SB7L-360 Seeker UAVs. Another squadron operates three C-130E Hercules and a VIP King Air 350. The helicopter force includes twenty-eight Mi-17s, two PZL W-3Ws, and thirty-six UH-1Hs. Training is on eight Cessna 172s.

IRELAND

- Population: 4.2 million
- Land Area: 26,600 square miles (68,894 sq.km.)
- GDP: $266bn (£170bn), per capita $64,044 (£40,938)
- Defence Exp: $1.53m (£0.97m)
- Service Personnel: 10,460 active, plus 14,875 reserves

IRISH ARMY AIR CORPS

Founded: 1922

Most of Ireland became independent of the UK in 1922, and immediately the Irish Army Air Corps was formed to provide air cover for ground forces on internal security duties as a civil war raged between the Government and those opposed to the treaty with Britain. Early equipment included Avro 504K trainers, Bristol F2B and Martinsyde F4 Buzzard fighters, and DH9 bombers. The IAAC took over former RAF bases. De Havilland Tiger Moths were purchased for training, and by 1929, the Air Corps had 160 personnel. A variety of aircraft were introduced, all in small numbers, before World War II, including Avro 621s, 626s, 631s and Ansons, Gloster Gladiators, a Fairey IIIF, Miles Magisters and a de Havilland Dragon

Rapide. Although Ireland remained neutral throughout the war, the Air Corps grew to have three squadrons, one each with Gloster Gladiator fighters, Avro Anson patrol aircraft and Supermarine Walrus amphibians, as well as a small number of Vickers Vespa army co-operation aircraft. During the war, the Gladiators were replaced by Hawker Hurricanes and the Vespas by Westland Lysanders, supplied by the UK against the possibility of a German invasion.

Post-war, replacement and modernisation was slow, as Ireland remained neutral throughout the Cold War. Throughout the 1950s, the Air Corps operated Supermarine Spitfire fighters and twin-seat trainers, de Havilland Dove light transports and Vampire jet trainers, as well as DHC-1 Chipmunk basic trainers and a number of Percival Provost piston-engined trainers. The first helicopters, three Alouette IIIs, were not introduced until 1966. A small number of SIAI SF260WE armed-trainers were later introduced for training and anti-terrorist operations, while the growing importance of fisheries protection resulted in the acquisition of two CASA CN235MPs. Four SA365F Dauphin helicopters were acquired for naval liaison and SAR. Helicopters were chartered from commercial operators to enhance Ireland's SAR capability.

The Irish Air Corps has had the unusual distinction of being investigated by management consultants, who recommended greater standardisation of aircraft and its own SAR and medium-lift transport capability. Economic problems have meant that most of the recommendations have not been implemented.

The Air Corps has 850 personnel. MR is provided by two CN235MPAs, while transport includes a VIP Gulfstream GIV, Learjet 45, and a Beech 200. A Defender 4000 and an AS355N are used to support the police, while other helicopters include three EC135P2/T2s, six AW139s and three SA316B Alouette IIIs. Training now uses five Cessna FR-172Hs and eight PC-9Ms. One of the naval patrol craft has a helicopter landing platform.

ISRAEL

- Population: 7.2 million
- Land Area: 7,993 square miles (20,850 sq.km.)
- GDP: $198bn (£126.6bn), per capita $27,390 (£17,508)
- Defence Exp: $9.78bn (£6.25bn)
- Service Personnel: 176,500 active, plus 465,000 reserves

ISRAEL DEFENCE FORCE-AIR FORCE

Founded: 1951

The Air Force dates from 1951 and is part of the unified Israel Defence Force. Israel's military aviation history dates from the time when the country was still under a League of Nations mandate. The Zionist Haganah underground movement formed the Sherut Avir which used Auster and Taylorcraft light aircraft for AOP duties, obtained by cannibalising aircraft abandoned by the British Army during World War II. After independence in 1948, the Sherut Avir briefly became the Chel Ha'avir.

The need for more potent aircraft followed an attack by Egyptian Spitfires in 1949, and again the Chel Ha'avir was forced to cannibalise abandoned aircraft, Spitfires and de Havilland Mosquito fighter-bombers found on former RAF bases, followed by the purchase of some 300 war-surplus aircraft of both types, plus some Avia C210 fighters. These were followed by Boeing B-17G Fortress heavy bombers, Curtiss C-46, Douglas C-47 and C-54 transports, and Boeing-Stearman PT-17 Kaydet, North American T-6 Harvard and Avro Anson and Airspeed Oxford trainers, and in 1951, by some former Swedish North American F-51D Mustang fighter-bombers. Some of the transports were converted to bombers for raids on Cairo, the Egyptian capital, and some of the Harvards were adapted for ground-attack duties. Chel Ha'avir pilots came from a large number of air forces.

The IDF-AF was formed in 1951. The first jets, fourteen Gloster Meteor F8 fighters, were obtained in 1953, and were soon joined by six Meteor NF13 all-weather fighters and T7 trainers. In 1955, thirty new Dassault Ouragan fighter-bombers were delivered, joined by forty-five ex-*Armée de l'Air* Ouragans. Eight Nord 2501 Noratlas transports entered service, with forty-one Fokker S-11 Instructors. When Egypt received new equipment in 1955, Israel ordered twenty-four Canadair-built CL-13 Sabre 6 and twenty-four Dassault Mystère fighters, and increased the Mystère order to sixty aircraft when the Sabres were embargoed. During the Suez crisis of 1956, the IDF-AF gained aerial superiority for the loss of eleven aircraft against a numerically superior Egyptian force. The Mustangs had remained in service on ground-attack duties, but their water-cooled engines were vulnerable to ground fire, so they were replaced by licence-built Potez Magister trainers during the late 1950s. Twenty-four Sud Vatour II light bombers and a number of Dassault Super Mystère interceptors were also obtained at this time.

The 1960s saw sixty Dassault Mirage IIICJ fighter-bombers enter service, to prove highly successful in the June 1967 war with Israel's Arab neighbours, when the IDF-AF virtually wiped out the Egyptian and Jordanian air forces. A further fifty were ordered, but the French government embargoed the order. This marked a growing reliance on the United States for arms supplies, but it also led to Israeli development and production of a Mirage derivative, the Kfir. US aircraft introduced during the late 1960s and early 1970s included fifty McDonnell Douglas F-4E Phantom fighter-bombers and eighty-five A-4E/M Skyhawk strike aircraft. The IDF-AF had also acquired a substantial number of helicopters by this time, including twenty-five AB205s, twenty Alouette IIIs and twelve Super Frelons, as well as eight CH-53 Sea Stallions and fifteen H-34 Choctaws.

The easy victory in 1967 had lulled Israel into a false sense of security with Arab territory as a buffer zone. In October 1973, Arab forces once again attacked Israel, selecting 6 October, Yom Kippur, a Jewish day of fasting when Israeli forces would be at their most vulnerable. Learning the lesson of their earlier defeat, the

Arabs used heavy air strikes to neutralise the IDF-AF, with heavy SAM missile defences within an interlocking air defence system based on SA-2, SA-3, SA-6, SA-7 and SA-9 missiles. The IDF-AF was familiar only with the SA-2, but evading this brought aircraft into the range of the other missiles. Subsequently, the IDF-AF was to admit to losing 115 aircraft, although US sources have suggested that the true figure was nearer 200. To counter the improved Arab defences, the IDF-AF had to attack in greater strength, sending a squadron of aircraft at a time rather than a flight, and using stand-off or 'smart' weapons, such as the US-supplied Walleye. The war ended with a cease-fire on 22 October, although some Israeli operations continued until 24 October, before stopping under US pressure.

While US military aid has fluctuated, overall it has proved constant. The easing of tension between Israel and Egypt during the 1980s meant a less active existence for the IDF-AF, but tension has continued between Israel and Syria, with neighbouring Lebanon often being used as a battleground between the two nations or forces supported by them. IDF-AF operations have included attacks on terrorist camps inside Lebanon, but more usually have consisted of helicopter support for attacking ground forces. Equipment has kept pace with technical developments, with older aircraft upgraded using Israel's growing aerospace expertise. Airborne early warning support came with the delivery of four Grumman E-2C Hawkeye aircraft, superseded by Israel's own Boeing 707-mounted Phalcon radar. The IDF-AF became a major customer for the F-15 series of interceptors, following this with large numbers of Lockheed F-16s. An attack helicopter capability is based on the Peten, or AH-64 Apache.

The IDF-AF was not involved during the Gulf War of 1991, with Israel under strong US pressure to remain neutral for fear of losing the support of the Arab nations involved in the anti-Iraqi coalition. More recently, the IDF-AF has not been heavily involved with clashes between Israeli security forces and Palestinian militants operating from the Left Bank of the Jordan or from Gaza.

The IDF-AF has 34,000 personnel, rising to 89,000 on mobilisation. Large numbers of aircraft are in storage, either as reserves or awaiting disposal. Combat aircraft are often given extensive upgrades. Two fighter and strike squadrons have forty-four F-15A/C Eagles and a third has twenty-five F-15Is; eight squadrons have 142 F-16A/C Fighting Falcons and four have 102 F-16I Sufas; while some 200 A-4N, F-4E 2000 and Kfir are in stored reserve. AEW Boeing 707 Phalcons have been replaced by five Gulfstream G550 Eitrams, with another three for ELINT, also handled by ten IAI-201 Aravas, with another nine on transport duties, with eleven C-130E/Hs, twenty Beech 200s and fifteen Do28s. Tankers include five Boeing 707s and five KC-130Hs. Four squadrons operate forty-five AH-1E/F Cobras, thirty AH-64A Apaches and eighteen Sarats (AH-64D Apache). MR is handled by three IAI-1124 Seascan versions of the Westwind. There are seven AS565 Atalefs (Panther) on ASW, while transport and support is provided by forty-one CH-53D Sea Stallions and forty-eight UH-60A/Ls or S-70 Yanshufs (Black Hawk), fifty-five Bell 212s and thirty-four 206s in the utility role. Training uses forty Super Magisters, being replaced by A-4N, twenty-six TA-4H/Js, and seventeen Grob 120s. There are almost fifty Searcher UAV with many more in store. Hawk and Patriot SAM are deployed, with AAM including AMRAAM, Sparrow, Sidewinder and Python III and IV, with Hellfire, Shrike, Walleye, Maverick, Standard and Popeye SAM.

ITALY

- Population: 58.1 million
- Land Area: 116,280 square miles (301,049 sq.km.)
- GDP: $2.26tr (£1.44tr), per capita $39,000 (£24,929)
- Defence Exp: $23bn (£14.7bn)
- Service Personnel: 185,235 active, plus 41,867 reserves and 107,967 Carabinieri

ITALIAN AIR FORCE/*AERONAUTICA MILITARE ITALIANA*

Founded: 1945

The Italian Army formed an aeroplane company during the Italo-Turkish War in 1911, buying Bleriot XIs, Etrich Taubes, Maurice Farman S-11s and Nieuports for reconnaissance duties. These aircraft were credited with making the first bombing attack. In 1912, the *Battagliore Aviatori*, or Aviation Battalion, was formed, being renamed the Military Aviation Service by the end of the year. That same year, the *Servizio d'Aviazione Coloniale*, Colonial Aviation Service, was also formed. The MAS changed its name to the Military Aviation Corps, *Corpo Aeronautico Militare*, in 1914.

Italy was one of the Allies in World War I. The *CAM* rose from seventy aircraft at the outbreak of the war to 1,800 at the end. Wartime aircraft included Nieuport 17C-1 Bebe and 110, Spad SVII, Hanriot HD-1 and Macchi M14 fighters; Caproni Ca33, Ca40 and Ca46, and Macchi M7 and M8 bombers; and Ansaldo SVA4, SVA5, SVA9 and SVA10, Savoia-Pomilio SP3 and SP4, and Fiat R2 reconnaissance aircraft. Post-war, the *CAM*'s strength dropped sharply, with little new equipment.

Benito Mussolini's assumption of power in 1923 marked a revival for Italian military aviation. The *CAM* became the autonomous *Regia Aeronautica*. Steps were taken to increase its strength and boost morale through prestigious events, the most notable being a mass flight of twenty-four Savoia-Marchetti SM55X flying boats to New York and back in 1933. This was the same year that the *RA* reached a strength of 1,200 aircraft: thirty-seven fighter squadrons operated Fiat CR20s and CR30s; thirty-four bomber squadrons with Caproni Ca73s and Ca101s; thirty-seven reconnaissance squadrons operated Romeo Ro1s, Caproni Ca97s and Fiat R22s. Flying boat squadrons were equipped with Savoia-Marchetti SM55X, while transport squadrons used Caproni Ca101s, Ca111s and Ca133s, and Savoia-

Marchetti SM81s. The *RA* was involved in Italy's invasion of Abyssinia (Ethiopia) in 1935, and was accused of bombing as well as reconnaissance and transport against an opponent with little or no air defence. In 1936, the Spanish Civil War started, and the *RA* sent a strong contingent to fight alongside German and Nationalist forces.

Italy entered World War II in June 1940, shortly before the fall of France, as an ally of Germany. At the outset, the *RA* had some 3,000 aircraft, of which around 400 were obsolete or obsolescent types based in the African colonies. The *RA* was active mainly in the Mediterranean, bombing Malta, supporting ground forces in North Africa and the invasions of Yugoslavia and Greece. A token force of seventy-five Fiat BR20M Cicogna bombers and fifty CR42 and CR50 fighters was sent to Belgium for operations over England. The *RA* also provided aircraft for Operation Barbarossa, the German invasion of the Soviet Union, suffering heavy losses. A number of German aircraft entered *RA* service during the war, including Junkers Ju87 Stuka dive-bombers, and, later, Messerschmitt Bf109F and Me110G fighters and Dornier Do217 bombers. Daimler-Benz water-cooled engines replaced Italian air-cooled engines in a number of aircraft types, including Fiat CR52 and CR55 fighters, improving their performance by reducing drag.

Italy capitulated on 8 September 1943, splitting the *RA* as many units joined the Allies, becoming the Italian Co-Belligerent Air Force, while the remainder became the *Aviazone della Republica Sociale Italiano*.

Post-war, the *Aeronautica Militare Italiano* was formed as an autonomous air service. Initially, it used surviving wartime aircraft as well as Supermarine Spitfire and Bell P-39 Airacobra fighters, Martin Baltimore bombers and Douglas C-47 transports. As a former member of the Axis alliance, the 1947 Peace Treaty dictated that only 200 out of a permitted maximum of 350 aircraft could be combat types. These restrictions were removed when Italy became a member of the North Atlantic Treaty Organisation in 1949.

A major re-equipment programme started once Italy joined NATO. US military aid included eighty Lockheed P-38J Lightning fighters, 100 Beech C-45 transports and a number of Stinson L-5 liaison aircraft. In 1950, the first jets, de Havilland Vampire FB5 fighter-bombers were built under licence. US equipment predominated throughout the 1950s, with Republic F-47D Thunderbolt fighters being joined and then replaced by F-84F Thunderstreak fighter-bombers and RF-84F Thunderflash reconnaissance-fighters. Lockheed PV-2 Harpoon MR aircraft, Grumman S2F-1 Tracker ASWt, SA-16A Albatross amphibians, and North American T-6G Texan and Lockheed T-33A jet trainers entered service. De Havilland Vampire NF54 night-fighters and Canadair-built F-86 Sabres were introduced in 1955. Helicopters arrived during the late 1950s, including Sikorsky UH-19s and SH-34Js. Agusta built Bell helicopters under licence, starting with the 47G Sioux.

Aermacchi MB326 trainers replaced the Texan, while the 1960s also saw Lockheed F-104G Starfighter and Fiat G91 fighters enter service, with Fairchild C-119G Packet and Fiat G222 transports. Italy also entered collaborative projects, joining the UK and Germany to produce the Tornado, and Brazil to produce the AMX attack aircraft. Italy ordered the Tornado as a strike and reconnaissance aircraft, but leased F3 interceptor versions from the RAF while awaiting the Anglo-German-Spanish-Italian Eurofighter 2000. Many older aircraft have been extensively upgraded, including the Starfighter, Tornado, AMX and Atlantique.

The *AMI* has fallen from 73,000 personnel to 42,935 over the past thirty-six years following the ending of conscription. There are two fighter squadrons operating thirty Typhoon 2000s with another fifty on order which will replace three squadrons of leased F-16s, while to intercept slow-moving aircraft engaged in smuggling, there is a squadron of MB339CD armed-trainers. Three strike squadrons have seventy upgraded Tornado IDS/ECRs; three have sixty-eight AMXs. An MR Atlantic squadron of ten aircraft operates under naval control. ECR uses a squadron of sixteen Tornado ECRs, while AEW is by a squadron of G222VS. Two transport squadrons operate twenty-two C-130J Hercules and a third has twelve C-27 Spartans, while there are also four Boeing 767MRTT tanker-transports, three VIP A319CJs and a number of smaller business jets. Sixteen Airbus A400Ms are on order for 2014, and up to 131 F-35A/Bs will be assembled in Italy, while ten NH90 TTHs are on order for transport and CSAR. Three SAR detachments operate twenty-six HH-3F (SH-3) Pelicans, while there are thirty-two AB212s. In addition to conversion trainers of the combat aircraft, there are eighty-four MB339A/CDs and thirty-six SF260Ms, as well as fifty NH-500Ds. Predator UAV are in service. Missiles include Spada SAM, with AMRAAM, Sidewinder and Sparrow AAM, HARM ARM and Storm Shadow SAM.

ITALIAN NAVAL AVIATION/*MARINA MILITARE ITALIANA*

Until the post-war period, all aviation was in the hands of the air force, or *Regia Aeronautica*. During World War II, the liner *Roma* was converted to an aircraft carrier, re-named *Aquila*, and intended to operate up to fifty Reggiane Re2001 aircraft. Equipped with catapults intended for the *Graf Von Zeppelin*, *Aquila* was completed as Italy capitulated, fell into German hands, was bombed by the Allies, and attacked by human torpedoes before being scuttled.

Post-war, the *MMI*, or *Marinavia*, was established as a helicopter force, with Italian law dictating that the navy could not operate fixed-wing aircraft. After operations with Agusta Bell 47G Sioux, the force expanded to include Sikorsky SH-3D and SH-34J, Agusta A106 and Agusta Bell 204B helicopters. Rapid expansion resulted from the introduction of helicopter cruisers, including the *Andrea Doria*, *Caio Duilio*, and the larger *Vittorio Veneto*. Experiments with Harrier V/STOL jet fighters aboard the *Andrea Doria* in the early 1970s encouraged the Italian Navy to introduce its first aircraft carrier, the *Guiseppe Garibaldi*, during the late 1980s. The law was changed to allow the Navy to operate fixed-wing aircraft, although the *IMMI* uses the USMC's AV-8B rather than the Sea Harrier. A second aircraft carrier, the *Cavour*, entered service in 2007. For the future, Italy plans twenty F-35Bs.

Currently, some 2,200 of the navy's 34,000 personnel are involved in aviation. In addition to the carriers, there are two destroyers, twelve frigates and ten offshore patrol craft able to carry helicopters. There are fifteen AV-8B Harrier IIs and two TAV-8B conversion trainers. Helicopters include sixteen EH-101 Merlins and twelve SH/ASH-3D/H Sea Kings for ASW and transport, and a further four EH-101s provide AEW. There are thirty-three AB212 ASW helicopters aboard the escort vessels, which will be replaced by up to fifty-six NH90s.

ITALIAN ARMY AVIATION – 'AIR CAVALRY'/*CAVALLERIA DELL'ARIA*

Formed during the post-war period, Italian Army aviation was known as the *CAALE* until 2000, when the title Air Cavalry was adopted. Early aircraft included 150 Piper L-18/21 and Cessna O-1E light aircraft, replaced by Aerfer AM-3C and Savoia-Marchetti SM1019 liaison aircraft during the 1970s, as well as 125 Agusta Bell 47G/J Sioux and seventy 204B helicopters. The *CAALE* was one of the first export customers for the CH-47C Chinook, operating twenty-six licence-built versions during the early 1970s. In more recent years, a combat ability has been acquired using armed versions of the Agusta A109 helicopter, followed by the A129 Mangusta attack helicopter. The *CAALE* saw service in Somalia, deploying the Mangusta there on UN operations and, more recently, in Albania, Bosnia and Macedonia.

Currently, the Air Cavalry has fifty-nine upgraded A129ES Mangustas, and has cascaded nineteen A109As onto observation duties. There are twenty-two CH-47C Chinooks, forty-three AB206A2 JetRangers, sixty AB205A/Bs, eighteen AB212s and twenty-one AB412s, while up to fifty-nine NH90s are entering service which will replace the AB212s. Three VIP P180 Avantis and three ACTL-1s (Do228) light transports are operated.

JAMAICA

- Population: 2.8 million
- Land Area: 4,411 square miles (11,424 sq.km.)
- GDP: $127bn (£81.2bn), per capita $4,486 (£2,867)
- Defence Exp: $90m (£57.5m)
- Service Personnel: 2,830 active, plus 953 reserves

JAMAICA DEFENCE FORCE AIR WING

Formed: 1963

Part of an integrated defence force, the Jamaica Defence Force Air Wing came into existence in 1963, with a Cessna 185, later joined by two Bell 47G Sioux helicopters and a DHC-6 Twin Otter. It has 140 personnel, and operates a BN2A Defender, a Cessna 210 and two DA-40-280FP Diamond Star trainers, as well as three Bell 412EPs, three 407s and four Eurocopter AS335N Squirrels.

JAPAN

- Population: 127.1 million
- Land Area: 142,727 square miles (370,370 sq.km.)
- GDP: $5.3tr (£3.3tr), per capita $41,723 (£26,670)
- Defence Exp: $52.6bn (£33.6bn)
- Service Personnel: 230,300 active, plus 41,800 reserves

JAPAN AIR SELF-DEFENCE FORCE

Founded: 1954

Japanese military aviation dates from 1911 and the founding of both the Japanese Army Air Force and the Japanese Navy Air Force. The JAAF started with three Henri Farman and two Wright biplanes, an Antoinette and a Bleriot monoplane. The JNAF had two Maurice Farman and two Curtiss seaplanes. These aircraft were soon joined by Japanese-designed Tokogawa 1 and Sei Model 1 and 2 biplanes, licence-built Maurice Farmans, a Nieuport and a Rumpler Taube for the JAAF, with Otari and Ushioku biplanes and a Bleriot for the JNAF. Japan sided with the Allies during World War I, but saw little action other than the occupation of the German mandated port at Tsingtao on the Chinese mainland, although a few Japanese pilots flew with the French *Aviation Militaire*. Towards the end of the war, the JNAF received a number of Short reconnaissance-seaplanes, Sopwith seaplane-fighters, and Deperdussin seaplane-trainers, in addition to Yokosuka Model A seaplanes.

After the Armistice, the foundations were laid for a Japanese aircraft industry using the three main industrial giants: Mitsubishi, Nakajima and Kawasaki. A French aviation mission visited the country to advise on the structure and future of the JAAF, while the JNAF received similar aid from a British naval mission. Both missions had a considerable influence on Japanese aviation, and fostered interest in the aircraft carrier. The JAAF was completely re-equipped with Spad S13C and licence-built 20C, Nieuport 24C and 29C fighters, as well as fifty ex-RAF Sopwith 1½-Strutter and Pup fighters; Breguet Br14B, Farman F50 and licence-built F60

Goliath bombers; Salmson SA-2 reconnaissance aircraft; Nieuport 81E, 83E and 24C, and Hanriot and Caudron C6 trainers, as well as the Nakajima Type 5 advanced trainer. The JNAF received Gloster Sparrowhawk Mars II and III and Mitsubishi Type 10 shipboard fighters, with reconnaissance and training versions of both; Short F5 America, Schreck FBA17 and Tellier flying boats; Avro 504K and 504L trainers; plus Sopwith, Airco, Vickers, Blackburn and Supermarine Types. A Gloster Sparrowhawk Mars IV fighter was used in trials flying from a platform built over a forward gun turret of the battleship *Hamishiro* in 1922, leading to the conversion of an oil tanker into the first Japanese aircraft carrier, the *Hosho*, that same year. The ship entered service in 1923 with Mars IV and Mitsubishi Type 10 fighters.

The late 1920s saw two more carriers for the Imperial Japanese Navy: the *Akagi*, a converted battlecruiser, entered service in 1928, and the *Kaga*, a converted battleship, in 1929. Aircraft were developed or built under licence specifically for carrier operations, including Nakajima A1N1 Type 3 (Gloster Gambit) fighters, Mitsubishi B2M1 Type 89 (based on a Blackburn design) naval bombers and the Mitsubishi C1M2 reconnaissance aircraft. Seaplanes and flying boats continued to enter service, including Yokosuka E1Y1 and Aichi Type 2 (Heinkel HD25) seaplanes, and Hiro H1H1, H1H2 and H2H1 flying boats. During the early 1930s, these aircraft were followed by Nakajima A2N1 shipboard fighters and advanced trainers, E4N1 and Kawanishi E5K1 reconnaissance seaplanes, and Hiro H3H1 flying boats.

The JAAF meanwhile re-equipped with Mitsubishi Type 87 light bombers, Kawasaki Type 87 (Dornier F) heavy and Type 88 reconnaissance-bombers. These were followed during the early 1930s by Nakajima Type 91 and Kawasaki Type 95 fighters, Mitsubishi Type 92 (Junkers G38) and Type 93, and Kawasaki Type 93 bombers, and Nakajima Type 94 AOP aircraft.

In 1931, Japan invaded Manchuria, which became the protectorate of Manchouko. This was followed in 1932 by action against Shanghai, with the JNAF attacking from two carriers. This was the start both of Japanese territorial expansion that was to lead to World War II in the Pacific, and of major expansion of the armed forces. Expansion had to be achieved without British or French assistance once they realised the threat posed by Japanese ambitions.

Japanese aircraft designations were based on the year of the reign of the emperor, so that 7-Shi meant the seventh year of the Showa reign. In 1932, two aircraft of the 7-Shi range of prototypes, the Hiro G2H1 Type 95 naval bomber and Kawanishi E7K1 Type 94 reconnaissance aircraft were put into production. These were followed by the 9-Shi range of prototypes in 1934, including the Mitsubishi A5M1 Type 96 fighters and G3M1 Type 96 land-based long-range bomber and the Aichi D1A2 and Nakajima B4Y1 Type 96 carrier-borne bombers, the Watanabe E9W1 and Aichi E1A1 Type 96 reconnaissance seaplanes, and the Kawanishi H6K1 Type 97 four-engined flying boat, which all entered service around 1936. These aircraft were followed by Nakajima C5M1 Type 98 reconnaissance aircraft, and Hiro H5Y1 Type 99 reconnaissance flying boats.

During this period, the JAAF also selected and introduced new aircraft: Kawasaki Type 95 and Nakajima Ki27a and Ki27b Type 97 fighters; Mitsubishi Ki30 and Ki21, and the Kawasaki Ki48 Type 98 bombers; Nakajima Ki34 Type 97 transports; Tachikawa Ki36 AOP aircraft; Mitsubishi Ki15 and Ki51 Type 97 reconnaissance aircraft; and Ki51b ground-attack aircraft.

Operations in China resumed in 1937, with the JAAF and JNAF outnumbering Chinese forces by more than five to one. During the winter of 1937–38, Japanese forces advanced rapidly across China, although, outside the range of escorting fighters, Japanese bombers were subjected to devastating Chinese fighter attack. New aircraft continued to enter service, with the most notable being the Mitsubishi A6M2 Type O shipboard fighter, the famous Zero. New aircraft carriers had been entering service, giving Japan a strong naval air arm. By the end of 1941, China was virtually defeated.

On 8 December 1941, 353 JNAF aircraft operating from six aircraft carriers attacked the US naval base at Pearl Harbor, Hawaii, achieving complete surprise and inflicting major losses. The attack was a tactical success, but a strategic blunder: it brought the United States into the war, but failed to make Pearl Harbor unusable while the US Pacific Fleet's aircraft carriers were safely at sea. Japanese forces then swept into Hong Kong, and into Thailand and down the Malay Peninsula to take Singapore, followed by the Dutch East Indies before Japanese forces advanced into Burma, eventually being held in check both there and in New Guinea. JAAF and JNAF aircraft aided the fastest and best co-ordinated advances in military history. The Vichy French government allowed the Japanese to use bases in French Indo-China. On 10 December 1941, Japanese shore-based aircraft found and attacked the new British battleship HMS *Prince of Wales* and the elderly battlecruiser *Repulse* off the coast of Malaya, finding the ships without air cover in the first instance of warships at sea being sunk in an aerial attack. On 5 April 1942, aircraft from the carrier *Soryu* found and sunk the British heavy cruisers *Dorsetshire* and *Cornwall*. A few days later, on 9 April, they sunk the small British carrier *Hermes* and her escort, the Australian destroyer *Vampire*. Darwin in Australia and bases in Ceylon were bombed.

The US Pacific Fleet was quick to move to the offensive. Carrier-based bombers attacked Tokyo and other large cities, and while the damage was slight, the blow to Japanese confidence was immense, as was the damage to the reputation of the JAAF. The Battle of the Coral Sea was inconclusive, but in the Battle of Midway on 4 June 1942, the Imperial Japanese Navy lost all four carriers – *Akagi*, *Kaga*, *Hiryu* and *Soryu* – assigned to the operation. The IJN was never to recover from this crushing defeat.

Wartime aircraft operated by the JNAF included developments of the Zero fighter, including seaplane versions. Kawanishi H8K2 Type 2 flying boats, Nakajima J1N1 Type 2 fighter-reconnaissance aircraft, Aichi D4Y1 and D4Y2 and Mitsubishi G4M2 bombers were also used. Like the JNAF, the JAAF was also heavily dependent on pre-war designs and their developments, but new aircraft included the Nakajima Ki44 Type 2 and Ki84 Type 4, and Kawanishi Ki4 and Ki61

Type 3 fighters, Nakajima Ki49 Type 0 and Mitsubishi Ki67 Type 4 bombers, and Kokusai Ki49 Type 0 AOP aircraft.

In desperation towards the end of the war, *Kamikaze* (Divine Wind) suicide groups were formed by both services. Initially, standard combat aircraft were used, but *Oka* (cherry blossom) piloted bombs were also developed. These had to be carried close to the target by heavy bombers before being released: the concept was flawed because of the vulnerability of the mother aircraft. Suicide bombers proved effective against smaller warships, including escort carriers, but often failed to penetrate intense AA fire and were useless against the armoured decks of the new British carriers. Throughout the war, the relative lack of technical progress by the Japanese was a major weakness. As US forces advanced across the Pacific towards Japan, bases became available for USAAF heavy bombers, and the JAAF found that its aircraft were completely unable to reach these aircraft to shoot them down. Again the solution lay in suicide attack, with pilots using aircraft stripped of armament to 'ram' the bombers. The fire-bombing of Japanese cities, the Allied advance towards Japan, and finally the dropping of the first atomic bombs on Hiroshima and Nagasaki in August 1945 led to Japanese surrender and occupation by the Allies.

Japan's armed forces were disbanded after the surrender. In 1950, a National Police Reserve Force, a paramilitary organisation, was formed under US sponsorship to relieve the pressure on the occupation forces during the Korean War. This was followed in 1954 by the creation of land, air and sea 'self-defence' agencies, effectively re-establishing an army, air force and navy in all but name. These were formed with US assistance, largely because of Japan's proximity to Communist North Korea and China.

Initially, the Japanese Air Self-Defence Force operated North American T-6G Texan and Beech T-34 Mentor trainers, with a few Lockheed T-33A jet trainers and Curtiss C-46 Commando navigational trainers. The first students were mainly wartime veterans seeking refresher and conversion training. They started training in 1955, and by the end of the year the first North American F-86F Sabre fighter squadron was operational. Within three years, the JASDF had almost 300 Sabres, thirty-five C-46 Commando transports, and some 300 trainers, with many of the Sabres and T-33s built in Japan. In 1959, Japanese-designed Fuji T-1 jet trainers were in service, followed in 1960 by Japanese-built Lockheed F-104J Starfighters. McDonnell Douglas F-4EJ Phantom fighter-bombers were also built in Japan and entered service in 1970, by which time personnel numbers had reached 40,000 and the JASDF's first SAM missiles, three Nike-Ajax battalions, were operational. Japanese-designed aircraft entering service included military versions of the NAMC YS-11 transport. Most aircraft were licence-built US designs. Helicopters were also introduced, mainly for SAR, including Sikorsky H-19 and S-62, and Vertol 107. In 1972, NAMC C-1 tactical jet troop transports entered service, and these were followed a couple of years later by the Mitsubishi T-2 supersonic trainer. The Japanese-designed Mitsubishi F-1 strike aircraft entered service in the 1980s, followed by McDonnell Douglas F-15J Eagle interceptors, and later the Mitsubishi

F-2A/B, a development of the F-16 with a composite wing, was introduced as an F-1 replacement.

Japanese neutrality and sensitivity over Japanese military expansion has meant that the country avoided the post-war regional conflicts, and has had only minimal involvement in UN peace-keeping. The JASDF is a formidable force today, with 45,600 personnel. Mainstay of the combat aircraft today are seven squadrons with 150 F-15J Eagles, with three squadrons still operating seventy F-4EJ Phantom IIs and two with forty Mitsubishi F-2s. One reconnaissance squadron has ten RF-4Js. Electronic warfare is provided by two squadrons with a Kawasaki EC-1 and ten YS-11Es, with AEW by ten EC-2C Hawkeyes and four E-767s. SAR squadrons operate twenty U-125A (Hawker 800) 'Peace Kryptons', augmented by thirty UH-60J Black Hawks and ten KV-107s (BV-107). A tanker squadron operates four KC-767As. Transport is provided by a squadron of ten C-130H Hercules and two with twenty C-1s, while another squadron operates ten U-4s (Gulfstream IV), and one operates ten CH-47 Chinooks. Training uses 170 T-4s, twenty Mitsubishi F-2Bs, thirty T-7s and ten T-400s (Beech T-1A).

There are six SAM groups with twenty-four Patriot squadrons, while ASM missiles include the ASM-1 and ASM-2, with Sparrow and Sidewinder AAM.

JAPANESE MARITIME SELF-DEFENCE FORCE

Founded: 1954

The Japanese Maritime Self-Defence Force was founded in 1954 primarily as an anti-submarine force with US assistance. The initial equipment included four Bell TH-13 helicopters and some North American SNJ-6 trainers. The following year, twenty Grumman TBM-3W-2 and TBM-3S Avengers, seventeen Lockheed PV-2 Harpoons, ten Convair PBY-6A Catalinas and four Grumman Goose amphibians, and Sikorsky S-51 helicopters entered service. Rapid expansion was helped by the recall of many wartime JNAF veterans. In 1956, the first of forty-two Japanese-built Lockheed P2V-7 Neptunes started to replace the Harpoons, and in 1957, the Avengers were replaced by sixty Grumman S-2A Tracker ASW aircraft.

Licence-manufacture of US aircraft continued, but also allowed some development, with forty-six Kawasaki P-2J MR aircraft being produced as turboprop developments of the Neptunes which they replaced in JMSDF service during the early 1970s. Thirty-six Japanese-designed Shin Meiwa PS-1 turboprop flying boats at one time equipped three squadrons, and eleven of this aircraft's US-1 development provide SAR. Longer-range MR eventually passed to the Lockheed P-3C Orion. As with most navies, helicopters are operated from escort vessels, with a considerable expansion of the helicopter force in recent years as some destroyers carry three or four helicopters. The role of the JMSDF has developed beyond anti-submarine operations in recent years, and it now has the *Osumi*, officially classified as an LST, capable of carrying 330 troops, but with a through flight deck effectively making it a small helicopter carrier.

Currently, some 10,000 of the JMSDF's 44,100 personnel are involved with aviation. Six squadrons (one assigned to training) operate eighty Lockheed P-

3C/OP-3 Orion MR aircraft. Another seven EP-3C, NP-3C and UP-3D Orions are assigned to ELINT and radar calibration. SAR uses eleven Shin Meiwa US-1/1A flying boats and eighteen UH-60J Seahawk helicopters, while Gulfstream IVs are being introduced for anti-piracy patrols. Nine MH-53J Sea Stallion helicopters are used for minesweeping. Seven squadrons operate ninety-one shipboard SH-60J/K Seahawk helicopters. A transport squadron has four YS-11s and five LC-90s. Training uses nine OH-6D/DAs (MD500), thirty-eight T-5s and six YS-11s.

JAPANESE GROUND SELF-DEFENCE FORCE

Founded: 1954

The Japanese Ground Self-Defence Force was created in 1954 out of the National Police Reserve, which included a number of pilots trained from 1952 onwards by the US Army to fly Piper L-21 and Stinson L-5 Sentinel AOP aircraft and who formed the nucleus of the JGSDF's air element. Helicopters were soon introduced, including Japanese-built versions of the Vertol 107 (CH-46), Bell UH-1 Iroquois and forty-seven Sioux, Hughes OH-6A and Sikorsky S-62, although the earlier H-19 Chickasaw was supplied direct from the US. During the late 1960s and early 1970s, a massive expansion programme saw the number of helicopters rise from 290 to 400. Attack helicopters were introduced later, Fuji-Bell AH-1 Cobras.

The JGSDF has formed an air mobile brigade and in 2000 introduced the first of up to 100 Kawasaki OH-1 armed scout helicopters, replacing the large force of OH-6D/J Cayuse. In 2004, ten AH-64D Longbow Apaches augmented seventy AH-1F Cobra helicopters, while there are also twenty OH-60s (MD-500). Transport is provided by fifty CH-47J/JA Chinooks, thirty UH-60JAs and 140 UH-1H/J Iroquois. SAM is provided by Hawk and Chu-Sam missiles.

JORDAN

- Population: 6.3 million
- Land Area: 36,715 square miles (101,140 sq.km.)
- GDP: $21bn (£13.4bn), per capita $3,347 (£2,139)
- Defence Exp: $2.31bn (£1.48bn)
- Service Personnel: 100,500 active, plus 65,000 reserves

ROYAL JORDANIAN AIR FORCE

Founded: 1956

Originally established by the UK as the Arab League Air Force in 1949 following the first Arab-Israeli War, it was initially an army transport unit using two elderly de Havilland Rapide biplanes. Seconded RAF personnel filled key posts. During its first year, the ALAF received two de Havilland Tiger Moths and four Percival Proctor trainers. Training of Arab personnel began in 1950, using Auster light aircraft. Three Turkish MKEK4 Ugur trainers were operated briefly before being replaced by DHC-1 Chipmunks. During the early 1950s, four de Havilland Dove light transports, a Vickers Viking airliner and a Handley Page Marathon communications aircraft entered service, joined in 1956 by a Vickers Varsity. A gift from the British government in 1955, nine de Havilland Vampire FB9 fighter-bombers and two T55 trainers, were the first jets, joined by ex-Egyptian FB52s in 1956. Three ex-RAF North American T-6 Harvard trainers were delivered in 1955.

The ALAF became the autonomous Royal Jordanian Air Force in 1956. Jordan underwent a brief flirtation with Egypt, but with Saudi support the country returned to her former alliance with the UK. Twelve Hawker Hunter F6 fighters, a Westland Scout and four Alouette III helicopters, with additional Doves and a Beech Twin Bonanza were delivered during the late 1950s and early 1960s. Almost all of the aircraft were lost during the June 1967 Arab-Israeli War, when Jordan also lost the West Bank of the River Jordan. Re-equipment took place with British, American and Pakistani help, the latter providing four loaned North American F-86F Sabres. By the early 1970s, the RJAF had grown to 2,000 personnel. It was operating two squadrons with a total of thirty-six Lockheed F-104A Starfighters and a squadron with eighteen Hawker Hunter FGA9 fighter-bombers, as well as Westland Whirlwind (S-55) and four Alouette III helicopters, Douglas C-47 and de Havilland Dove and Heron transports. Short Tigercat SAM had been acquired for airfield defence. Pilots were trained in the USA.

Jordan occupies a difficult strategic position. The country did not take part in the October 1973 Yom Kippur War, although it did tie down considerable Israeli forces simply by mobilising. It did not take part in the Gulf War of 1991, but offended moderate Arab and Western opinion by siding with Iraq.

The Hunters were replaced by Northrop F-5E/F Tigers and the F-104s by Dassault Mirage F1B/C/E interceptors, while substantial tactical air mobility came with a large number of Bell UH-1H/L Iroquois helicopters. Anti-tank helicopters were also introduced, the Bell AH-1F Cobra. Over the past thirty years, the RJAF has grown considerably, and now has 12,000 personnel, but down from 15,000 in 2002. It introduced Lockheed F-16 interceptors during the winter of 1997–98, leasing these from the USAF.

There are fifteen F-16A/Bs in a fighter squadron, and twelve F-16AM/BNs in a ground-attack squadron while thirty Mirage F-1CJ/BJ/EJs are in two squadrons, and three strike squadrons operate thirty F-5E/F Tiger IIs. Two RU-38As are used for surveillance. A transport squadron operates four C-130H Hercules, two CN-235s and two C-212 Aviocars. A VIP Royal Flight operates an Airbus A340-211, a Lockheed TriStar 500, two Challenger 604, two Grumman Gulfstream IV and three S-70 helicopters. Attack helicopters are twenty upgraded AH-1F Cobras with TOW ASM, while transport helicopters include twelve AS332M-1 Super Pumas, thirteen EC635s and thirty-six Bell UH-1H/L Iroquois, while three Bo105s are operated for the police. Training uses sixteen T67M Fireflies which replaced Bulldog basic trainers, eleven C101CC Aviojet and six Hughes 500D helicopters. Six Seeker UAV are operated.

Sidewinder, Sparrow and Magic AAM are deployed, and Maverick ASM.

KAZAKHSTAN

- Population: 15.4 million
- Land Area: 1,048,070 square miles (2,778,544 sq.km.)
- GDP: $101bn (£64.6bn), per capita $6,550 (£4,186)
- Defence Exp: $1.33bn (£0.85bn)
- Service Personnel: 10,900 active

KAZAKHSTAN AIR FORCE

Founded: 1991

Kazakhstan became independent of the former USSR in December 1991. Equipment for the new Kazakhstan Air Force was acquired by taking Soviet equipment stationed at air bases within the country, not all of it suitable for the KAF. Former Soviet Tu-95MS Bears with a nuclear-strike capability abandoned at the airbase at Semipalatinsk were returned to Russia in exchange for new MiG-29s and Su-27s. The new state also found itself with a number of SS-18 ICBM in silos, although these were removed and the silos destroyed in 1996. Kazakhstan now has a significant air defence and ground-attack capability, although it is not known how many aircraft are in storage. Flying hours are claimed to be around 100 annually, far more than in most former USSR states, and the country is a member of the CIS joint air defence plan.

Currently, the KAF has 2,400 personnel, a massive reduction from the 19,000 personnel recorded in 2002. It has lost its fighter regiment with thirty-nine MiG-29 *Fulcrum*, and one with forty-two MiG-31 *Foxhound* and sixteen MiG-25 *Foxbat*, and is now dependent on a single regiment with L-39 Albatros, although most of the twenty-eight aircraft are in store, as are a number of MiG-21 *Fishbed*. There are now just nine of the forty-two Mi-24 *Hind* attack helicopters operational. Transport is provided by five An-24/26s and three An-12s. Helicopters include around twenty-three Mi-8/-17s, with some fifty in store as well as many UH-1Hs. A Boeing 757-200 and a Falcon 900 provide VIP transport. Training uses twelve L-39 Albatros and four Yak-18 aircraft. Missile defences include upwards of 100 launchers for SA-2, SA-3, SA-4, SA-5, SA-6 and S-300. ASM missiles include AS-7, AS-9, AS-10 and AS-11, while AAM are mainly AA-6, AA-7 and AA-8.

KENYA

- Population: 39 million
- Land Area: 224,960 square miles (582,646 sq.km.)
- GDP: $38.6bn (£24.7bn), per capita $990 (£632)
- Defence Exp: $696m (£445m)
- Service Personnel: 24,120 active

KENYA AIR FORCE

Re-formed: 1994

Kenya became independent from the UK in 1963, with British assistance in establishing national defence forces, including six DHC-1 Chipmunk trainers. The emphasis was on transport and communications using DHC-2 Beaver and DHC-4 Caribou, with an Aero Commander 500 for VIP duties. A limited combat capability came with the delivery of six BAC 167 Strikemaster armed trainers in 1971. BAe Bulldogs replaced the Chipmunks in 1972. During the late 1970s, the KAF received Northrop F-5E Tigers for interception and ground-attack, with DHC-5 Buffalo transports.

The KAF was involved with an attempted *coup d'état* in 1982, afterwards being disbanded and placed under army control. Reformed afterwards using loyal elements, it was renamed 'Kenya 1982 Air Force' to avoid any confusion with the disgraced former service. The present title was re-adopted in 1994. COIN remains important for the KAF, although there has been relatively little organised insurrection, and the Strikemasters have been replaced by BAe Hawk Mk52 and Embraer Tucano Mk51 trainers also capable of fulfilling the attack role. A new task has been anti-poaching patrols to preserve the country's wildlife using Hughes 500MD helicopters, which can also be armed.

Currently, the KAF has 2,500 personnel. It has twenty-two F-5E/F Tiger II interceptors; it can also use eight Hawk Mk52s, eleven Tucano Mk51s and many of the forty-one 500MD helicopters in the attack role. There are four DHC-5 Buffaloes as well as three DHC-8-100 Dash 8s and ten Y-12s with a VIP Fokker 70, while six Dornier Do28Ds are in store, as well as eleven SA330 Puma transport helicopters. TOW anti-tank and AGM-65 Maverick missiles are used.

KOREA, DEMOCRATIC PEOPLE'S REPUBLIC OF (NORTH)

- Population: 22.7 million
- Land Area: 46,814 square miles (121,248 sq.km.)
- GDP: Not known
- Defence Exp: Not known
- Service Personnel: 1,106,000 active, plus 4,700,000 reserves

KOREAN PEOPLE'S ARMY AIR FORCE

Founded: 1953

Korea was occupied by Japan from early in the twentieth century until the end of World War II, when the north was occupied by Soviet troops and the south by US troops. Almost immediately after Soviet occupation in 1945, a North Korean Aviation Society was formed, and in 1946 this became the Korean People's Army Aviation Division, becoming the Korean People's Armed Forces Air Corps in 1948. Considerable Soviet assistance was received, including Yakovlev Yak-9 fighters, Ilyushin Il-10 ground-attack aircraft and Yak-18 trainers, with most of these aircraft reaching North Korea between 1948 and June 1950, when North Korean forces invaded the south. Towards the end of 1950, MiG-15 jet fighters started to enter service, bearing the brunt of aerial action throughout the Korean War.

The end of the war in 1953 saw Korea divided into two nations, with North Korea becoming the Democratic People's Republic of Korea. The title of the Korean People's Army Air Force was also adopted at the time. Aircraft included MiG-15, Yak-9 and Lavochkin La-9 fighters, Tupolev Tu-2 bombers and Ilyushin Il-10 ground-attack aircraft, as well as Yak-11 and Yak-18 trainers. Ilyushin Il-28 jet bombers followed, and in 1957 MiG-17s were delivered, including a number of the Chinese-built version, the Shenyang F-4, with Antonov An-2 and Lisunov Li-2 (C-47) transports. The first helicopters, Mi-1s, arrived at this time and were soon joined by Mi-4s. A few MiG-19s were delivered before MiG-21 interceptors were supplied in 1966.

Despite attempts at easing the situation, the two Koreas continue an extreme version of the Cold War stand-off, aggravated by the very poor state of the North Korean economy, especially compared with South Korea. The situation has not been improved by North Korea conducting trials with ballistic missiles that threaten both her southern neighbour and Japan. North Korea spends massively on defence, with conscription lasting for up to ten years. Despite economic problems,

KPAAF personnel numbers have virtually quadrupled over the past thirty-six years from 30,000 to 110,000. Much of the equipment is obsolescent since the collapse of Communism, and the break-up of the Soviet Union means that it no longer enjoys Soviet aid. Aircraft have been acquired from air forces in the newly-independent former states of the USSR, including MiG-21s from Kazakhstan. The main emphasis is on attack, and this is reflected in the helicopter units with MiG-24 attack helicopters, with relatively few transport types. The most modern aircraft are MiG-29s and Su-25s, both acquired in 1988, although numbers of the former have been boosted in recent years.

The operational state of three regiments with eighty H-5 (Il-28 *Beagle*) bombers is unclear. For air defence, the KPAAF has thirty-five MiG-29A/Ss in one regiment, while there are thirty-four Su-25Ks in a ground-attack regiment. Three regiments operate 107 J-5s (MiG-17 *Fresco*); four have ninety-eight J-6s (MiG-19 *Farmer*); four have 130 J-7s (MiG-21 *Fishbed*); one has fifty-six MiG-23ML/P *Flogger* and one has eighteen Su-7 *Fitter*. There are twenty Mi-24 *Hind* attack helicopters. There are up to 300 Antonov An-2 or Harbin Y-5 light transports whose role is to infiltrate snipers in any future conflict. Apart from twelve An-24s, the remainder of the fixed-wing transport force consists of a varied selection of Antonov, Ilyushin and Tupolev types with each in small numbers. The most numerous helicopter is the Mi-2 *Hoplite* with 139, followed by eighty Hughes 500D/Es, with just fifteen Mi-8/-17 *Hip* and forty-eight Z-5s (Mi-4 *Hound*). Training uses 180 CJ-6s (Yak-18), ten CJ-5s, thirty MiG-15UTIs, known as the FT-2 in Korea, and twelve L-39 Albatros, as well as the conversion trainer variants of the combat aircraft. Average flying hours are believed to be around twenty, and doubtless much less for units other than those with the latest MiG-29A and Su-25K aircraft. Shmel UAVs are in service.

Missiles include SA-2 *Guideline* and SA-3 *Goa*, while AAM include AA-2, AA-7, AA-8, AA-10, AA-11, PL-5 and PL-7.

KOREA, REPUBLIC OF (SOUTH)

- Population: 48.5 million
- Land Area: 38,452 square miles (99,591 sq.km.)
- GDP: $882bn (£563.8bn), per capita $18,188 (£11,629)
- Defence Exp: $24.5bn (£15.6bn)
- Service Personnel: 687,000 active, plus 4,500,000 reserves

REPUBLIC OF KOREA AIR FORCE

Founded: 1949

The Republic of Korea was formed in 1948, with the Republic of Korea Air Force, RoKAF, formed the following year. The first pilots were Koreans who had flown with the Japanese forces during World War II. Plans for combat aircraft were overruled on the grounds that Korea was not of strategic significance, with the first aircraft mainly liaison and AOP types, including Piper L-4s, as well as North American T-6 Harvard trainers. North Korea invaded the south in June 1950, starting the Korean War in which the United Nations forces, led by the United States, attempted to protect South Korea from first North Korean and then Chinese attack. The RoKAF started to receive combat aircraft, with the first being North American F-51D Mustang fighters.

After the Korean War ended in an uneasy armed truce, US military aid continued at a high level. In 1956 the first jets, North American F-86F Sabre fighters, entered operational service, followed by Lockheed T-33A jet trainers in 1957. In 1965, Northrop F-5A/B fighter-bombers were introduced, and in 1970, McDonnell Douglas F-4D Phantoms. By the early 1970s, the RoKAF's personnel strength had reached 23,000. Modernisation since has taken the form of Northrop F-5E/F Tiger IIs, followed later by Lockheed F-16s, but many older types remain in service, including the now ageing F-5As and F-4 Phantoms. The ability to fund upgrades and replacements suffered during the Asian economic crisis of the late 1990s, which delayed development of the indigenous T/A-50 advanced trainer and light-fighter programme, a joint venture between Lockheed Martin and KAI, but modernisation is in hand with F-15Ks entering service.

The RoKAF has grown over the past thirty-six years to 64,000 personnel. Fighter and ground-attack formations are in seven tactical wings. One operates up to fifty-nine F-15K Eagles, two operate 165 Lockheed Martin KF-16C/D Fighting Falcons, one has seventy F-4E Phantom IIs, and three have 174 F-5E/F Tiger IIs. A reconnaissance group has four Hawker 800RAs, twenty KO-1s, eighteen RF-4Cs and five RF-5As. Close support is provided by A-50 Golden Eagles which are replacing twenty-seven A-37B Dragonflies, seven OV-10D Broncos, ten Cessna O-2As, and twenty O-1A Bird Dog AOP aircraft. Eight Hawker 800XPs provide reconnaissance and ELINT. Airlift capability includes ten Lockheed C-130H Hercules, twenty CN235-100/200s and six CH-47D Chinook helicopters, with a Boeing 737-300, two BAe748s, two AS332 Super Pumas, three S-992A Superhawks and three Bell 412s for VIP duties. SAR is provided by a helicopter squadron with five Bell UH-1D/Hs. Utility helicopters include twenty-eight UH-60 Black Hawks. There are seventeen BAe Hawk Mk67s, eighty-three KT-1s which have replaced the T-38 Talons, and twenty-five A-50/T-50s. UAV include Night Intruder. Missiles include Sparrow, Sidewinder and AMRAAM AAM, with Maverick and HARM ASM, and Nike Hercules, Hawk, Javelin and Mistral SAM.

REPUBLIC OF KOREA NAVY

The Republic of Korea Navy initially started using aircraft for liaison duties, before acquiring Grumman S-2 Trackers for shore-based MR. The introduction of warships with helicopter capability led to the arrival of the Westland Lynx. The Trackers were replaced by eight Lockheed P-3Cs in 1996 and the RoKN is believed to be looking for additional aircraft. The Westland Lynxes were upgraded to Super Lynx standard in the late 1990s after an order was placed for new Super Lynxes in 1997. Today the RoKN has eight Orions, twenty-four Super Lynx Mk99s for ASW and ASuW, as well as fifteen UH-60P/MH-60Ss. Communications duties are undertaken by a Bell 206B JetRanger and seven SA319s, as well as five Cessna Caravan IIs.

REPUBLIC OF KOREA ARMY

Initially, the Republic of Korea Army relied upon the RoKAF for its close support needs, but a significant transport and anti-tank capability has been developed with sixty Bell AH-1J/S/F Cobras and forty-five TOW-equipped MD500 helicopters, with another 130 AOP MD500s. There are eighteen CH-47D Chinook heavy-lift helicopters, as well as twenty Bell UH-1H Iroquois, with up to 130 Sikorsky UH-60P Black Hawks and twelve Bo105s. Three AS332L Super Pumas are used for communications duties.

KUWAIT

- Population: 2.7 million
- Land Area: 9,375 square miles (24,235 sq.km.)
- GDP: $123bn (£78.6bn), per capita $46,027 (£29,421)
- Defence Exp: $6.65bn (£4.25bn)
- Service Personnel: 15,500 active, 23,700 reserves

KUWAIT AIR FORCE

Re-formed: 1991

The Kuwait Air Force was formed with RAF assistance as an extension of the Security Department in 1960, initially for liaison and communications with five Auster AOP aircraft, some de Havilland Doves and a Heron. Threatened by its neighbour Iraq, which claimed Kuwait as an Iraqi province, it quickly acquired sophisticated aircraft. Six Hawker Hunters and six Agusta Bell 204s were followed by six BAC 167 Strikemaster armed-trainers, and then by twelve BAC Lightning interceptors, with two conversion trainers. These aircraft were replaced in due course and expansion continued during the late 1970s and 1980s including

Dassault Mirage F1CK interceptors and McDonnell Douglas A-4KU Skyhawk ground-attack aircraft.

Plans were laid for the acquisition of F/A-18C Hornets and Short Tucano trainers when, on 2 August 1990, Iraqi forces invaded. The small country was quickly overrun, but the KAF managed to resist for up to forty-eight hours before the remaining aircraft were flown to bases in neighbouring Saudi Arabia. About forty aircraft, including helicopters, were caught in the Iraqi invasion. The KAF became the Free Kuwait Air Force, and was able to join the Coalition Forces in the 1991 Gulf War, Operation Desert Storm, in the air campaign that preceded the ground assault to liberate Kuwait, losing at least one aircraft in the campaign. As Kuwait was liberated, retreating Iraqi forces destroyed captured Kuwaiti aircraft.

The Kuwait Air Force was reformed, and aircraft ordered before the invasion were soon delivered, with the first of sixteen Tucanos arriving in Kuwait in October 1991, and the first Hornets delivered early the following year. Altogether, forty Hornets were delivered, including eight F/A-18D twin-seat aircraft, and surviving Mirage F1s were refurbished. The Skyhawks were sold to Brazil. Other new aircraft included BAe Hawk 64 armed jet trainers.

The KAF has 2,500 personnel. There are two squadrons with forty upgraded F/A-18C/D Hornets. There are eleven Hawk 64s and eight Tucanos, for training and close support. Helicopters include sixteen AH-64D Longbow Apaches, equipped with Hellfire anti-tank missiles, displacing a similar number of Eurocopter SA342K Gazelles onto AOP and police duties. Four AS332 Super Pumas operate in the anti-shipping role. Transport includes three L-300-30 Hercules and eight SA330Hs. For VIP use, there is a 737-300 and a DC-9.

KYRGYZSTAN

- Population: 5.4 million
- Land Area: 77,181 square miles (199,900 sq.km.)
- GDP: $5.1bn (£3.26bn), per capita $946 (£604)
- Defence Exp: $43m (£27.5m)
- Service Personnel: 10,900 (many conscripts) active

REPUBLIC OF KYRGIZIA AIR ARM

Founded: 1991

Formerly a member of the USSR, Kyrgyzstan became independent in 1991, creating its own air force from aircraft and helicopters belonging to the former Central Soviet Air Force Training School. The potential to continue training students for other former USSR states seems to have been missed, possibly due to their funding problems. The country is a member of the CIS joint air defence treaty. The original equipment included almost seventy L-39 jet trainers, Mi-8 and Mi-24 helicopters, and up to fifty MiG-21 aircraft in various states of repair.

Today, there are 2,400 personnel. Operational aircraft include a regiment with twenty-eight L-39 Albatros, although many are in store, two An-12 and two An-26 transports, with nine Mi-24 attack and twenty-three Mi-8 transport helicopters. The fifty MiG-21s are all believed to be unserviceable. In 1999 Russia offered Su-24 and Su-25 ground-attack aircraft to deal with unrest in Central Asia, but these could not be afforded, although air strikes were launched against rebels in September 2000. SA-2, SA-3 and SA-4 SAM launchers are available in small numbers.

LAOS

- Population: 6.8 million
- Land Area: 88,780 square miles (231,399 sq.km.)
- GDP: $5.3bn (£3.4bn), per capita $792 (£506)
- Defence Exp: $17m (£10.8m)
- Service Personnel: 29,100 active

LAOS PEOPLE'S LIBERATION ARMY AIR FORCE

Founded: 1975

Laos was formerly part of French Indo-China and became an independent member of the French Union in 1949. In 1953, the country was invaded by communist Viet Minh forces from North Vietnam, and although these withdrew, the country was then subjected to repeated attempts to overthrow the government by communist Pathet-Lao forces. American assistance brought the Laotian Army Aviation Service into existence in 1954, initially for counter-insurgency and AOP duties. The first aircraft included twenty Cessna O-1 Bird Dog AOP aircraft, ten Douglas C-47s, three DHC-2 Beavers and four Aero Commander 520 transports. Sikorsky S-55 helicopters and North American T-6G Texan trainers followed later, with T-28D armed-trainers for COIN operations. In 1960, the Royal Lao Air Force title was adopted.

Unrest in neighbouring Cambodia and Vietnam, coupled with US withdrawal from Vietnam destabilised Laos, and in 1975, the government was overthrown and a republic declared, with the RLAF becoming the Laos People's Liberation Army Air Force. Soviet military aid was received, including MiG-21PF interceptors and Mi-8 helicopters. Despite a defence pact between Laos and Russia in 1997, it seems that the MiG-21 force is no longer serviceable. An attempt to return the aircraft to service with an overhaul failed as the structural life of the aircraft had expired. Russia supplied Mi-17s in 1998 and Kamov Ka-32T helicopters in 2000.

The LPLAAF has 3,500 personnel. The twenty-two MiG-21PF/Us are unserviceable, but nominally equip two squadrons. There is a Yak-40 VIP transport, four Antonov An-2s, three An-24s and an An-74, a Y-12 and five Y-7s (An-24 helicopters include a single Mi-6 heavy-lift machine), as well as twenty-one Mi-8/-17s and six Kamov Ka-32T *Helix*, with three SA360 Dauphins for SAR.

LATVIA

- Population: 2.2 million
- Land Area: 25,590 square miles (66,278 sq.km.)
- GDP: $29bn (£18.5bn), per capita $12,977 (£8,295)
- Defence Exp: $358m (£228m)
- Service Personnel: 5,745 active, plus 10,866 reserves

REPUBLIC OF LATVIA AIR FORCE

Founded: 1994

During World War II, Latvia was annexed by the USSR. The Latvian Air Force was formed in 1994, after the last Soviet forces departed, as a communications and liaison force as plans to acquire combat aircraft have been thwarted by a lack of funds. The RLAF has just 480 personnel and operates three An-2s and an L410 Turbolet, as well as four Mi-17 helicopters and two Mi-2s. Five PZL Wilgas are operated on behalf of the National Guard.

LEBANON

- Population: 4 million
- Land Area: 3,400 square miles (8,806 sq.km.)
- GDP: $30.7bn (£19.6bn), per capita $7,642 (£4,884)
- Defence Exp: $911m (£582m)
- Service Personnel: 59,100 active

LEBANESE AIR FORCE/*FORCE AÉRIENNE LIBANAISE*

Founded: 1949

Lebanon was under Turkish rule until the end of World War I, then administered by France under a League of Nations mandate until independence in 1943. The *Force*

Aérienne Libanaise was formed with British assistance for internal security duties. The first aircraft were two Percival Prentice trainers, but these were soon joined by three Savoia-Marchetti SM79 tri-motor transports, a de Havilland Dove and some DHC-1 Chipmunk trainers in 1950. Former French air bases were used, and by 1955, de Havilland Vampire FB52 fighter-bombers and T55 jet trainers, North American T-6 Harvard and additional Chipmunk trainers were in service, with an Aermacchi MB308 communications aircraft. AOP duties were filled by some ex-Iraqi de Havilland Tiger Moths. The *FAL*'s combat capability grew throughout the 1950s, with five Hawker Hunter F6 fighters and five FB9 fighter-bombers. During the 1960s, twelve Dassault Mirage IIIC fighter-bombers were introduced, while Potez Super Magisters took over the advanced jet training role, and BAe Bulldogs replaced the Chipmunks as primary trainers.

Civil war erupted in the Lebanon during the mid-1970s, with Israeli and Syrian intervention, during which some aircraft were destroyed, often on the ground. Some equipment continued to arrive throughout this period, mainly helicopters. The return of peace during the late 1990s saw Bell UH-1H helicopters donated by the USA, but the remaining Mirage IIIs were sold to Pakistan.

The *FAL* has 1,000 personnel, and the combat squadron of five Hunter F9s is inactive with the aircraft in store, as are the Magisters. The sole combat capability rests with a squadron of eight SA342L Gazelle anti-tank helicopters with five more in storage. Only twenty out of forty-five other helicopters are operational, including sixteen UH-1H utility machines and four R-44 Raven trainers.

LESOTHO

- Population: 2.1 million
- Land Area: 11,716 square miles (30,344 sq.km.)
- GDP: $2.4bn (£1.53bn), per capita $1,129 (£721)
- Defence Exp: $36m (£23m)
- Service Personnel: 2,000 active

LESOTHO DEFENCE FORCE – AIR WING

Formed: 1978

The former Basutoland in southern Africa, land-locked Lesotho became independent within the British Commonwealth in 1966, but an air wing for the defence force was not formed until 1978. Early aircraft included a Cessna 182 and a Westland-built Bell 47G Sioux helicopter, later joined by Bo105 helicopters. Serious unrest in 1998 was quelled by the arrival of troops from South Africa and Mozambique, who were withdrawn in 1999. The status of the air wing is uncertain, but it is believed to have 110 personnel with three CASA C-212 Aviocar transports and a GA-1 Airvan, as well as three Bell 412SP/EPs and a Bo105S.

LIBERIA

- Population: 3.4 million
- Land Area: 43,000 square miles (99,068 sq.km.)
- GDP: $1.6bn (£1.02bn), per capita $465 (£297)
- Defence Exp: not known
- Service Personnel: 2,400 active

LIBERIAN ARMY AIR RECONNAISSANCE UNIT

In 1847, Liberia became the first independent African state. Its long history was relatively stable until a presidential assassination in 1990 sparked off internal strife. At the time, the small army had an air arm for reconnaissance and communications duties, but all the aircraft were destroyed. There are plans to rebuild the armed forces, retaining an integrated defence force, but meanwhile the air arm is believed to have some 300 personnel, while an IAI-101B Arava is believed to have survived the upheaval.

LIBYA

- Population: 6.3 million
- Land Area: 679,358 square miles (1,759,537 sq.km.)
- GDP: $37.2bn (£23.8bn), per capita $5,880 (£3,758)
- Defence Exp: $800m (£511m)
- Service Personnel: 76,000 active, plus 40,000 reserves

LIBYAN ARAB REPUBLIC AIR FORCE

Founded: 1959

The Libyan Air Force was formed in 1959 with two training aircraft supplied by Egypt, accompanied by two Auster AOP6s provided by the UK. In 1963, the USA provided a Douglas C-47 and two Lockheed T-33A jet trainers as part payment for the use of Libyan bases. Later, seven Northrop F-5A fighter-bombers were supplied. A major arms deal with the UK, including SAM, was scrapped after the monarchy was overthrown and a revolutionary government installed in 1970, breaking the links with the UK and USA. This led to arms procurement from France and the USSR. Substantial numbers of Mirage V and F1 aircraft were matched by MiG-21s, MiG-23s and MiG-25s, as well as Su-20s, Su-22s and Su-24s, Tu-22A bombers and Mi-24 attack helicopters, transport and training aircraft. The new regime engaged in border disputes with many of its African neighbours.

The last Soviet aircraft were fifteen Su-24s supplied in early 1989. The UN imposed sanctions in April 1992 cutting Libya off from new aircraft supplies until these were lifted in 1999 following the arrest of the suspects in the Lockerbie air disaster in which a bomb planted aboard a Pan Am Boeing 747 caused it to explode in mid-air. Libya has since discussed having its MiG-23s, MiG-25s and Su-24s upgraded, and purchasing MiG-29 and MiG-31 interceptors, as well as long-range SAMs. Meanwhile, serviceability is believed to be very low, with many aircraft in storage.

Throughout its post-revolutionary history, the LARAF has maintained a very high ratio of aircraft to personnel. It uses 'advisors', effectively mercenary pilots, from North Korea, Pakistan and Syria.

The LARAF has 18,000 personnel, a drop of almost a quarter since 2002. Flying hours average eighty-five annually, limited by a shortage of spares. A bomber squadron operates the seven surviving Tupolev Tu-22A bombers, plus two Tu-22U conversion trainers. There are nine fighter squadrons, which between them operate forty-five MiG-21s, seventy-five MiG-23s, ninety-four MiG-25s, with three MiG-25U trainers and fifteen Mirage F1EDs. Another thirteen squadrons are in the FGA role, with a total of forty MiG-23BNs, fourteen Mirage F1ADs, six surviving Su-24MKs and eighty Su-22s. There are twenty-three Mi-24/-35 attack helicopters. There are seven transport squadrons with twenty Lockheed C-130H/L-100-20/30 Hercules, two An-124s, twenty-three Antonov An-26s, twenty-five Il-75s, sixteen G222s and fifteen L-410OUVPs, while helicopters include four CH-47 Chinooks, thirty-five Mi-8/-17s, forty-six Mi-2s, two AB212s and four AB205s. Training aircraft include ninety G-2 Galebs, a number of L-39 Albatros and twenty SF260WL Warriors as well as conversion trainers for the combat aircraft. Missiles include *Atoll*, *Acrid*, *Apex*, *Aphid* and Magic AAM, with *Kerry*, *Kilter* and Swatter ASM.

Army aircraft include forty SA342L Gazelles for anti-tank and liaison duties, with the latter role also falling to five AB205s and five AB206s, and ten Cessna O-1E Bird Dogs.

Naval aircraft include seven SA321 Super Frelon ASWs and SAR helicopters, and twelve Mi-14PL ASW helicopters.

LITHUANIA

- Population: 3.6 million
- Land Area: 25,170 square miles (65,201 sq.km.)
- GDP: $42.5bn (£27.2bn), per capita $11,951 (£7,639)
- Defence Exp: $501m (£320m)
- Service Personnel: 8,850 active, plus 6,700 reserves

LITHUANIAN AIR FORCE

Re-formed: 1992

Lithuania's history of military aviation dates from 1919, as the country fought for its independence in the wake of the Russian Revolution and the civil war that followed. A school of military aviation was established that year, while in 1920 the fledgling Lithuanian Air Force found itself fighting briefly against Polish forces. A number of Lithuanian aircraft were built during the 1920s and 1930s, initially with three trainer designs, the ANBO-1, II and III. Next came the ANBO-IV reconnaissance aircraft of 1932, also used as a light bomber and which made a round-Europe tour in 1934. It was the most numerous Lithuanian aircraft, with fourteen built, plus twenty of its ANBO-41 development. The Lithuanian Air Force was too small to counter the massive might of the Soviet Union when it invaded in 1940, annexing the country and making it part of the USSR. Lithuanian personnel were absorbed into the Soviet armed forces.

After the USSR broke up, Lithuania seized her independence. In early 1992, the Aviation Service, *Aviacijos Tarnyba*, was formed, but adopted its current title in 1993. Most Soviet aircraft were removed before independence, but four L-39 Albatros were obtained from the training school force in Kyrgyzstan, and in 1999 these were joined by new aircraft bought direct from the manufacturer. The LAF has a requirement for up to twelve fighters, but this has been affected by funding.

The LAF has 860 personnel. It now has just two L-39ZA Albatros for training and light attack duties, while nine Mi-8s undertake transport and SAR. Other transport aircraft include three C-27J Spartans and an An-26.

MACEDONIA, FORMER YUGOSLAV REPUBLIC

- Population: 2.1 million
- Land Area: 10,229 square miles (27,436 sq.km.)
- GDP: $9.1bn (£5.8bn), per capita $4,387 (£2,804)
- Defence Exp: $167m (£106.5m)
- Service Personnel: 8,000 active, plus 4,850 reserves

MACEDONIAN ARMY AIR WING

Formed: 1998

A small air corps was formed within the Macedonian Army as the country broke away from the former Yugoslavia during the late 1990s, initially operating four Zlin 242 trainers and an Antonov An-2, as well as four Mi-17 helicopters. This force is now being augmented by a handful of Su-25K/UBs after up to twenty ex-Turkish Northrop F-5As were rejected in favour of the more familiar Russian equipment. In 2000 Germany donated two MBB Bo105s and two Bell UH-1H helicopters, and Ukraine has also provided four surplus Mi-8/Mi-17 transport and four Mi-24 attack helicopters. An Mi-8 was shot down by Albanian rebels in 2001.

The personnel strength is believed to be around 1,130, more than half as many again as in 2002. Current aircraft include twelve Mi-24V/K *Hind* attack helicopters, seven Mi-8/-17 transport helicopters and two UH-1H utility helicopters, while the four Su-25K/UBs are in storage.

MADAGASCAR

- Population: 20.7 million
- Land Area: 228,600 square miles (590,002 sq.km.)
- GDP: $9.4bn (£6.1bn), per capita $471 (£301)
- Defence Exp: $103m (£65.8m)
- Service Personnel: 13,500 active

MALAGASY AIR FORCE/*ARMÉE DE L'AIR MALGACHE*

Founded: 1960

The island of Madagascar became independent from France in 1960, and received the standard French military aid package of three Broussards (Bushrangers), a C-47 and an Alouette III helicopter. This was augmented with additional aircraft so that it grew to include three Douglas C-47s, six Broussards, two Dassault MD315 Flamant light transports, with two helicopters, a Bell 47G Sioux and the Alouette III.

A combat capability was created later using Soviet military aid, with the arrival of first MiG-17s and then MiG-21s, but these aircraft have been abandoned and it is now a transport and liaison force. Transport is provided by an An-26, two C-212s and a BN2 Islander, with four VIP Yakovlev Yak-40s and two Cessna 337 Skymasters. A helicopter squadron has five Mi-8s. A Piper Aztec and four Cessna 172s are employed for training. New equipment is needed, but is unlikely to be afforded in the near future.

MALAWI

- Population: 15 million
- Land Area: 36,686 square miles (93,182 sq.km.)
- GDP: $3.3bn (£2.1bn), per capita $220 (£140)
- Defence Exp: $42m (£26.8m)
- Service Personnel: 5,300 active

MALAWI ARMY AIR WING

Formed: c.1992

Malawi, the former British colony of Nyasaland, became independent in 1963, with a small army which eventually acquired an air wing. A Police Air Wing also operated aircraft at first, but this was disbanded. Currently, 200 personnel are involved in aviation, which is confined to transport and communications using two Basler Turbo 67s (turboprop C-47), four Do228s and a VIP Hawker 800, with an SA330F Puma, a VIP SA332 Super Puma and an AS350B Ecureuil.

MALAYSIA

- Population: 25.7 million
- Land Area: 128,693 square miles (326,880 sq.km.)
- GDP: $222bn (£141.9bn), per capita $8,792 (£5,620)
- Defence Exp: $4.0bn (£2.57bn)
- Service Personnel: 109,000 active, plus 51,600 reserves

ROYAL MALAYSIAN AIR FORCE

Founded: 1958

Malaysia came into existence in 1963 as a federation of the former British colonies of Malaysia, Sabah (North Borneo), Sarawak and Singapore, although the latter withdrew in 1965. Malaysian military aviation had started much earlier with the creation of the Straits Settlements Volunteer Air Force in 1936, equipped with Hawker Audax aircraft. In 1940, it became the Malayan Volunteer Air Force, using civil aircraft, mainly de Havilland Tiger and Leopard Moths and Rapides that had been requisitioned on the outbreak of World War II in Europe in 1939. This force escaped from the Japanese invasion of Malaya, but it ceased to exist during the invasion of Sumatra.

In the immediate aftermath of the war, military aviation in Malaya was left to the RAF until 1950, when a Malayan Auxiliary Air Force was formed. At first, this was a training organisation using Tiger Moths, which were joined by North American T-6 Harvard trainers. The first combat aircraft were Supermarine Spitfire F21 fighters. The Tiger Moths were replaced by DHC-1 Chipmunks in 1956.

Malaya (but not the other territories) became independent in 1958, and the MAuxAF became the Royal Malayan Air Force, flying Scottish Aviation Twin Pioneer transports and the Chipmunk trainers. With the formation of the Federation of Malaysia in 1963, the present title was adopted. New aircraft included Handley Page Herald, DHC-4 Caribou, de Havilland Devon (military Dove) and Heron transports, and Canadair CL-41 Tutor armed jet trainers, known locally as the *Tebaun* (Wasp). The new Federation was born at a time of crisis, threatened by the territorial ambitions of neighbouring Indonesia, and the RMAF fought alongside the RAF and Fleet Air Arm against Indonesian infiltration. Afterwards, although Singapore left the Federation, the RMAF also had the support of the RAF, RAAF and RNZAF, all of whom maintained forces in Malaysia for some years afterwards. Co-operation with Singapore continued on defence.

Australian-built Commonwealth CA-27 Avon-Sabres were delivered in 1969, and these were joined by ten Dassault Mirage Vs during the early 1970s, by which time the RMAF had grown to 4,500 personnel and was operating twenty Alouette

IIIs and ten Sikorsky S-61A helicopters. Training by this time was using eighteen BAC Provost T-51s and fifteen Scottish Aviation Bulldog trainers.

Malaysia has accorded a high priority to defence since the formation of the federation. Lockheed C-130 Hercules were obtained to enhance the transport capability, followed by Northrop F-5E/F Tiger II interceptors and reconnaissance aircraft, and later by F/A-18D Hornets. Trainers included the PC-7, with MB339A armed-trainers and Hawk light fighters in the ground-attack role. Growing variety in procurement sources was underlined in 1995 when the RMAF acquired MiG-29N air superiority fighters. It has supplemented the surviving PC-7s with the MkII version, and the C-130 force has also been upgraded, including stretching some of the aircraft.

Defence expenditure was checked during the late 1990s by the Asian economic crisis. CASA CN-235 transports were delivered three years late in 1999, while a decision on new helicopters was postponed.

The RMAF has almost doubled its personnel strength since 2002 to 15,000. Its combat aircraft shows a complete lack of standardisation which must add to training and support costs, but includes a squadron of sixteen MiG-29N *Fulcrum*, which are to be withdrawn; a squadron with eighteen Su-30MKMs; one with eight F/A-18D Hornets, and a fourth squadron with thirteen F-5E Tiger IIs, used as a fighter lead-in force. The strike role is handled by two squadrons with fourteen Hawk 208s, which can be augmented by fifteen Hawk 108s from the advanced training school and by sixteen MB339AB/Cs. Four King Air 200Ts operate limited MR. There are four transport squadrons including two with a total of twelve C-130H/H-30 Hercules and two KC-130H tankers; a VIP squadron with a Boeing 737-700BBJ, an Airbus A319CT, an F-28 Fellowship, a Falcon 900 and a BD700 Global Express, as well as a squadron with six CN-235s, plus nine Cessna 402Bs, of which two are used for aerial survey work. Helicopters include twenty S-61s for ASW and another four on transport duties with four S-70 Black Hawks, while utility helicopters include eight SA316 Alouette IIIs and an A109. Trainers include forty-eight Pilatus PC-7/PC-7 MkIIs, twenty MD3-160s, as well as use of the MB339A, Hawk and Cessna 402B aircraft. UAV include Eagle and Aludra.

Missiles include *Alamo* and *Archer* as well as the Western Sparrow and Sidewinder AAM, with ASM Maverick and Harpoon.

ROYAL MALAYSIAN NAVY

Founded: 1988

The naval air arm was formed in 1988 with twelve Westland Wasp helicopters obtained from the Royal Navy for shipboard ASW duties. This small force grew to seventeen Wasps, with 160 personnel. The Wasps have been replaced by six AS555N Fennecs and six Super Lynxes. Two frigates can support helicopters, while two corvettes also have helicopter landing platforms.

ROYAL MALAYSIAN ARMY

Formed: 1977

A small air arm was formed by the Royal Malaysian Army in 1977 with ten SA316B Alouette III helicopters for observation, liaison and communications duties. Ambitious plans exist for the force to develop into a fully capable strike and transport force with attack and transport helicopters. The CSH-2 Roivalk was selected during the mid-1990s, but not ordered due to the economic crisis, as was a planned purchase of CH-47 heavy-lift helicopters. Today the force operates eleven A109s and nine SA316 Alouette III helicopters.

MALDIVES

• Population: 250,000
• Land Area: 115 square miles (298 sq.km.)

MALDIVES DEFENCE FORCE

The Maldives Defence Force charters aircraft from Air Maldives, the national airline, as required, while an Indian Air Force Mi-8 helicopter is also on loan.

MALI

• Population: 13.4 million
• Land Area: 464,875 square miles (1,204,350 sq.km.)
• GDP: $7.3bn (£4.7bn), per capita $559 (£357)
• Defence Exp: $180m (£115.1m)
• Service Personnel: 7,350 (includes conscripts) active

MALIAN REPUBLIC AIR FORCE/*FORCE AÉRIENNE DE LA REPUBLIC DU MALI*

Founded: 1960

The former French colony of Mali became independent in 1960, and received the standard French post-colonial package of a Douglas C-47 and two Max Holste 1521 Broussards (Bushranger), to which the USA added two C-47s and the USSR six obsolete MiG-15 fighter-bombers. Initially known as the Mali Air Force, the present title was adopted later. Further aid came from the USSR during the late

1960s, with the delivery of MiG-17s, followed by MiG-21s in 1986 following a brief six-day war the previous year with neighbouring Burkina Faso. Transport aircraft were also supplied. This intermittent flow of obsolete aircraft ceased with the collapse of the USSR, since when the only development has been the conversion of a C-47 as a Basler Turbo 67.

Today, the *FARM* has 400 personnel. It is believed that the eleven MiG-21s are non-operational, as are six L-29 Delfins, two Yakovlev Yak-18s and four Yak-11 training aircraft. It has become a transport and communications force, with two Antonov An-2 *Colt*, an An-24 and an An-26, with four helicopters, two Z-9s (AS365 Dauphin), an Mi-8 and an A350B Ecureuil.

MALTA

- Population: 405,165
- Land Area: 122 square miles (316 sq.km.)
- GDP: $8.2bn (£5.24bn), per capita $20,408 (£13,045)
- Defence Exp: $54m (£34.5m)
- Service Personnel: 1,954 active

ARMED FORCES OF MALTA AIR WING

Malta became independent of the UK in 1964, basing its armed forces on former Maltese units within the British Army and locally raised personnel of the Royal Navy. An aviation element was not formed immediately within the integrated defence force, but the aviation element later became part of No2 Composite Regiment. Currently, almost 100 personnel are allocated to aviation. There are no combat or transport aircraft, although the MR element has been strengthened with the addition of a C-212 to the two BN2B Islanders that have provided patrol and SAR duties and a King Air 200 is being introduced. Cessna O-1 Bird Dogs were replaced in the training role in 2000 by four ex-RAF Bulldog T1s, with a fifth added in 2001. Helicopters include a new NH500 and two EC145s, which may replace some of the five SA316B/3160 Alouette helicopters on communications and SAR duties, with two Agusta Bell 47G Sioux for training and communications. Two SAR Italian AB212 helicopters are manned jointly by Maltese and Italian AF personnel.

The future for many air forces and at least one navy lies in the Lockheed Martin F-35, also known as the Joint Strike Fighter, JSF, or Lightning II. The nationality of the main customers can be seen from the flags displayed on the side of the fuselage. (*Lockheed Martin*)

The present for many air forces is the Lockheed Martin F-16 Fighting Falcon. This is a USAF aircraft, but it is also in service with many air forces in Europe and elsewhere, including Norway, Denmark, the Netherlands and Belgium. (*Lockheed Martin*)

A head-on shot of one of the RAAF's Boeing F/A-18 Hornets, the backbone of Australia's air defence and strike capability. (*RAAF*)

Advanced jet training for the RAAF is on the BAe Hawk, such as this one. The Hawk is one of the most widely used advanced jet trainers. (*RAAF*)

The NH90 in Army service. The Australian Army will operate forty of these machines. (*Australian Army*)

Embraer, the Brazilian aircraft manufacturer, has a strong niche in building turboprop trainers and light attack aircraft, with one of the main customers for the new Super Tucano being the Brazilian Air Force. (*Embraer*)

A joint venture between Brazil and Italy for a strike aircraft, this is the AMX, seen here in formation. (*Embraer*)

A Brazilian Air Force airborne early warning and control aircraft based on the MB145 regional jet airliner. (*Embraer*)

A Douglas A-4 Skyhawk of the Brazilian Navy approaches the aircraft carrier *Sao Paulo* (the former French *Clemenceau*) tailhook down ready to catch the arrester wires. (*Brazilian Navy*)

A pair of Boeing CF-18 Hornets of the Canadian Armed Forces in formation. Today, the Air Command more usually calls itself the Canadian Air Force. (*CAF*)

Canada was amongst the first export customers for the Agusta-Westland AW101 Merlin, which it h renamed the Cormorant, but further helicopter deliveries will be of the Sikorsky S-92.
(*Agusta-Westland*)

A Canadian Boeing C-17 Globemaster II on approach to land at Kandahar in Afghanistan. (*CAF*)

The Sea Eagle is a specially adapted version of the Gulfstream 550 used for SIGINT duties by the Israeli Defence Force Air Force. (*ISDAF*)

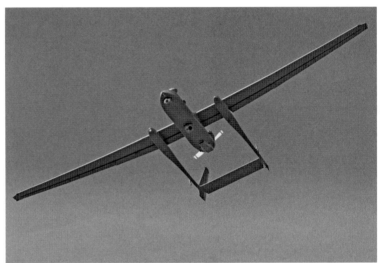

Air forces, armies and even navies are becoming great users of unmanned air vehicles, or UAVs; this is an Israeli Defence Force Air Force Shoval. (*ISDAF*)

Ground-attack capability in the Italian Air Force relies on the Italian-Brazilian joint venture AMX, shown here. (*Italian Air Force*)

A close-up of an Italian Air Force Lockheed-Martin F-16 Fighting Falcon fighter. (*Hellenic Air Force*)

A Harrier takes off from the Italian aircraft carrier *Guiseppe Garibaldi*. (*Italian Navy*)

An Agusta-Westland AW101 Merlin helicopter of the Italian Navy, with the carrier *Guiseppe Garibaldi* in the background. (*Italian Navy*)

Another joint venture helicopter is the NH90, and this is one of the first of around sixty for the Italian Army. (*Italian Army*)

An Agusta AW109 combat helicopter of the Italian Army in flight. (*Italian Army*)

A flight of three C-1 tactical transports in formation. (*Japan Air Self-Defence Force*)

The Japanese Mitsubishi F-2A fighter is a development of the Lockheed-Martin F-16. (*Japan Air Self-Defence Force*)

The Republic of Korea Air Force, or RoKAF, is one of the few outside the United States to be equipped with the Boeing F-15 Eagle, using the variant known as the F-15K. (*RoKAF*)

A Westland Lynx of the Republic of Korea Navy takes off from a shipboard platform. (*Agusta-Westland*)

Troops unload supplies from a Mexican Air Force Mi-17 helicopter. Mexico is one of the few Western countries to operate Russian aircraft. (*Fuerza Aérea Mexicana*)

This Mexican Navy CASA C-212 is one of eight aircraft used on maritime-reconnaissance. (*Aviación De La Armada De México*)

A Eurocopter Panther is one of the aircraft that can operate from Mexican destroyers and frigates. (*Aviación De La Armada De México*)

A Royal Netherlands Air Force Lockheed Martin F-16 Fighting Falcon making a guest appearance at an air show in Greece.
(*Hellenic Air Force*)

The Netherlands has followed the current fashion of placing all of its helicopters into a helicopter command, with one of the main helicopters being the NH90, as shown here.
(*Koninklijke Luchtmacht*)

The Royal Norwegian Air Force operates six Lockheed P-3C Orion aircraft on maritime-reconnaissance along the country's heavily indented coastline.
(*Lockheed Martin*)

The Royal Norwegian Air Force operates NH90 helicopters on behalf of the Royal Norwegian Navy from its frigates. (*Kongelige Norske Luftvorsvaret*)

The Royal Air Force of Oman operates fifteen Agusta-Westland Super Lynx helicopters on a variety of duties. (*Agusta-Westland*)

The *Força Aérea Portuguesa*, or Portuguese Air Force, operates twelve Agusta-Westland AW101 transport helicopters. (*Agusta-Westland*)

A close-up of a Romanian Air Force Antonov An-26. (*Hellenic Air Force*)

Two Mil Mi-28 combat helicopters fly in formation. (*Mil*)

The A-50 is the Russian Air Force's AEW development of the Ilyushin Il-76 transport. (*Ilyushin*)

An A-50 AEW aircraft refuels in mid-air from a tanker Il-76. (*Ilyushin*)

A close-up of a Royal Saudi Air Force Eurofighter 2000 Typhoon. Saudi Arabia has been a major user of both UK and US equipment. (*BAe Systems*)

The Spanish Navy's *Juan Carlos*, a helicopter carrier but with a ski-jump so that AV-8, or Harrier, aircraft can be operated, seen on trials. Two ships of this class have been ordered for the Royal Australian Navy and will give the service much enhanced power projection. (*Navantia-Ferrol*)

A Sikorsky MH-60 lands aboard a frigate of the Singapore Navy. (*Government of Singapore*)

One of no less than three fighter aircraft in service with the Spanish Air Force is the Dassault Mirage F1M some of which may be replaced by Typhoons. (*Ejército Del Aire*)

Amongst the transport types in service with the Spanish Air Force is this Airbus A310. (*Ejército Del Aire*)

Sweden's Saab Gripen lightweight fighter is a multi-role combat aircraft and an example of the way in which the small country manages to be highly self-sufficient in defence equipment. (*Flygvapnet*)

Defence self-sufficiency is extended further by the ability to produce a low-cost airborne early warning system using the Saab 340 regional airliner, with an example seen here escorted by two Gripen. (*Flygvapnet*)

Sweden was first to start the concept of an integrated helicopter command for all three services, although it now comes under the *Flygvapnet*, or air force. This is an Agusta-Westland AW109 attack helicopter. (*Helicopterflottij*)

Switzerland does not have heavy-lift helicopters, with the largest in service being the Eurocopter Puma, seen here. (*SAF*)

A Thai Saab Gripen takes to the air. (*Saab*)

Lockheed Martin F-16s refuel from a Turkish Air Force Boeing KC-135 tanker. (*Turk Hava Kuvvetleri*)

A CN-235 tactical transport precedes the larger C-160 Transall as aircraft queue to take off. (*Turk Hava Kuvvetleri*)

A Royal Air Force Eurofighter 2000 Typhoon shows its paces in a vertical climb. (*BAe Systems*)

The UK's sole remaining tactical strike aircraft are its mixed force of Harrier GR7 and GR9, shared with the Royal Navy in Joint Force Harrier. Here a Harrier hovers. (*BAe Systems*)

A Royal Air Force Boeing CH-47 Chinook. The RAF's heavy-lift helicopter force is being increased in response to concern over a shortage of such helicopters in Afghanistan. (*Boeing*)

The British Army Air Corps operates almost seventy Boeing Longbow Apache helicopters, built under licence in the UK by Agusta-Westland. (*Agusta-Westland*)

The United States Air Force has devoted much effort to the introduction of 'stealth' aircraft, such as this Lockheed F-117A Nighthawk. (*Lockheed Martin*)

Part of the United States Air Force's global reach relies on the Boeing C-17 Globemaster IIs of Military Airlift Command, seen here. (*Boeing*)

In a crisis, the USAF is strengthened by its reserve formations and the Air National Guard Units, which are organised on a state by state basis. This is a California Air National Guard Lockheed Martin C-130J Hercules. (*Lockheed Martin*)

Helicopters such as the Sikorsky MH-60R Sea Hawk vastly improve the anti-submarine capabilities of smaller warships such as frigates. Here a USN MH-60R deploys its dunking sonar. (*Sikorsky*)

The solution to the slower speed of the helicopter lies in aircraft such as the Bell Boeing MV-22 Osprey, seen here hovering above the deck of an assault ship. *(Boeing)*

While the Osprey fleet builds up, the mainstay of the US Marine Corps tactical transport helicopte is the Sikorsky CH-53 D Se Stallion, a heavy-lift machine. *(USMC)*

Longer-range search and rescue by the United States Coast Guard uses specially adapted versions of the Lockheed C-130H Hercules. *(USCG)*

MAURITANIA

- Population: 3.1 million
- Land Area: 398,000 square miles (1,085,210 sq.km.)
- GDP: $3bn (£1.9bn), per capita $967 (£618)
- Defence Exp: $20m (£12.8m)
- Service Personnel: 15,870 active

MAURITANIAN ISLAMIC AIR FORCE/*FORCE AÉRIENNE ISLAMIQUE DE MAURITANIE*

Founded: 1960

The former French colony of Mauritania became independent in 1960, and received the standard parting gift of a Douglas C-47 and two Max Holste 1521M Broussards. Uncertainty over sovereignty of the region bordering Morocco, disputed with Polisario guerrillas, led to acquisition of combat aircraft, with five Britten-Norman BN2A Defender aircraft for reconnaissance and COIN in the mid-1970s, with four Cessna FTB337Fs for the same roles. Plans to acquire six IA-58 Pucaras from Argentina were abandoned after a coup in 1978. Despite withdrawal from the region bordering Morocco, patrols have continued to avoid Mauritania being drawn into the conflict between the Polisario and Morocco.

The *FAIM* has 250 personnel. It has five BN2A Defenders and four Cessna FTB337Fs for COIN and patrol duties, a Basler Turbo 67, Navajo and Cheyenne II for patrol and two Harbin Y-12 transports, which were added in 1995. Five SIAI SF260E trainers were acquired in 2000.

MAURITIUS

- Population: 1.3 million
- Land Area: 720 square miles (2,038 sq.km.)
- GDP: $9.0bn (£5.75bn), per capita $6,970 (£4,455)
- Defence Exp: $41m (£26.2m)
- Service Personnel: 2,000 paramilitary active

MAURITIUS COAST GUARD

Mauritius maintains two paramilitary organisations, the so-called Special Mobile Force and the Coast Guard, with 500 personnel and an aviation unit. Currently, it operates two Do228s and a BN2T Defender on EEZ patrols. A police air wing operates four SA316B Alouette III helicopters.

MEXICO

- Population: 111.2 million
- Land Area: 760,373 square miles (1,972,360 sq.km.)
- GDP: $888bn (£567bn), per capita $7,985 (£5,104)
- Defence Exp: $4.41bn (£2.82bn)
- Service Personnel: 267,506 active, plus 39,899 reserves

MEXICAN AIR FORCE/*FUERZA AÉREA MEXICANA*

Founded: 1924

Mexico was among the first Latin American countries to become involved with military aviation, as revolutionary and counter-revolutionary groups used aircraft flown by mercenary pilots in 1910. During World War I, an aircraft industry was founded with government backing to overcome the shortage of aircraft. In 1924, the *Fuerza Aérea Mexicana* was formed with Bristol general-purpose biplanes, American-built DH4B bombers and Douglas O-2 AOP aircraft, for policing and COIN. In 1930, Avro 504K trainers were obtained, followed by the Mexican-designed Azcarate-E. Vought O2U Corsair biplanes were obtained during the early 1930s, but little further development occurred until the end of the decade when Grumman G-23 fighters, Waco D-6 general-purpose aircraft, Consolidated Fleet 21 and Ryan S-T trainers entered service.

Although neutral initially, Mexico joined the Allies in 1942, placing bases at their disposal. The USA supplied Douglas A-24 Dauntless ASW aircraft, Vought-Sikorsky OS2U AOP and North American NA-16 training aircraft. Mexico's aircraft industry provided the Tezuitlan trainer. In 1945, a *FAM* Republic F-47D Thunderbolt fighter-bomber squadron was due to join the Allied forces in the Pacific, but Japanese surrender came before its arrival. Post-war, Douglas C-47, Lockheed Lodestar and Beech C-45 transports arrived, with Fleet, Fairchild and Vultee trainers. Further US military aid followed Mexico's membership of the Organisation of American States in 1948.

Jet aircraft did not arrive until the late 1950s with de Havilland Vampire F3 fighters and T55 trainers, joined in 1961 by Lockheed T-33A armed-trainers. By this time, Bell 47G Sioux helicopters were in service, followed later by Alouette II helicopters. Beech T-11 Kansan and T-34 Mentor aircraft were also acquired. During the 1970s, Northrop F-5E/F Tiger II fighters were obtained, while the transport fleet was upgraded with Lockheed C-130A Hercules. Substantial numbers of Pilatus PC-7 armed-trainers were stationed at strategic locations.

Mexico has no obvious regional military threat, with a powerful and friendly northern neighbour and the nearest threat, Cuba, with economic problems seriously undermining military capability. COIN remains important due to the existence of five significant rebel groups. A new role is that of operations against drug traffickers while EEZ operations have also grown in importance, with two maritime patrol Embraer EMB145s and a third for AEW in 2004. Transport and liaison have always been important.

The *FAM* has almost doubled its personnel to 11,545 over the past thirty years. It was unusual until recently in maintaining its own paratroops, a battalion of 2,000 men, but these have now been transferred to the Army. A fighter squadron has eight F-5E Tiger IIs and two F-5Fs. Four close-support COIN squadrons are positioned in strategic locations with a total of sixty-six PC-7s and two PC-9s. Reconnaissance is provided by a squadron with four C-26Bs, two EMB145RSs and two Schweitzer SA237s, with another EMB145 for AEW. A squadron for spraying drug plantations has eight Cessna T206Hs and fifty Bell 206 JetRangers. There are seven liaison squadrons operating sixty-two Cessna 182s, almost entirely used in support of the Army. Substantial numbers of helicopters of many different types are operated in the COIN, transport and utility roles, including a squadron of armed MD530MF. Transport helicopters include twenty-nine Mi-8/-17s and an Mi-26, four S-65Cs, six S-70 Black Hawks, fourteen Bell 206s and seven 206Ls, four Bell 412s, two SA330S Pumas and two AS332L Super Pumas, as well as a VIP fleet of four AS332L Super Pumas and two EC225s. The fixed-wing transport role is filled by seven C-130E/K/L-100-20 Hercules, two An-32Bs, eleven IAI-201/202 Aravas, four Boeing 727s; a Beech 200, four PC-6Bs, a Turbo Commander 680, a Cessna Citation and a Learjet, as well as a 'Presidential Group' with a 737 and a 757, and a Gulfstream III. Training uses five squadrons operating twenty-eight PC-7s, twenty-six SF260s and twenty-eight Beech F-33C Bonanzas. UAV include Hermes and Skylark.

Missiles are mainly AAM Sidewinder.

MEXICAN NAVAL AVIATION/*AVIACIÓN DE LA ARMADA DE MÉXICO*

Mexican naval aviation was established after World War II as a land-based MR and SAR force. For many years it operated a number of Consolidated Catalina PBY-5A amphibians, later adding Alouette II and Bell 47J Sioux helicopters. Today, it has 1,250 out of the Mexican Navy's 55,961 personnel and its role has expanded to include COIN aircraft, while helicopters are operated from a destroyer and six frigates, and twenty patrol craft have helicopter platforms. There is little standardisation, with many types available in ones and twos. A squadron operates three E-2Cs and two Sabreliner 60s on AEW, while for MR there are seven squadrons with one operating eight CASA C-212PMs, another with seven L-90 Redigo armed-trainers, while five squadrons operate a mixture of two CN235MPAs, four Beech F-33C Bonanzas and four Barons, a Cessna 404, twelve MX-2s and six Lancair IV-Ps. Helicopters include five squadrons with Mi-8/-17s, of which eight are armed; two squadrons with two Mi-2s, two AS555s and four AS565MBs and six MD902s; and two squadrons with eleven Bo105s. Trainers include eight Z-242Ls, four MD500s, four Schweizer 300Cs and an R-44.

MOLDOVA

- Population: 4.3 million
- Land Area: 13,000 square miles (34,188 sq.km.)
- GDP: $5.1bn (£3.26bn), per capita $3,188 (£2,037)
- Defence Exp: $22m (£14.1m)
- Service Personnel: 5,998 active, plus 66,000 reserves

MOLDOVAN AIR FORCE

Founded: 1991

In common with many other former 'republics' within the USSR, Moldova acquired the Soviet aircraft based within its territory, in this case the most significant being thirty-four MiG-29s of a naval fighter wing. These aircraft were sold to Eritrea and the Yemen, while the USA bought twenty-one to prevent them being sold to Iran, at a price reported to be around US $40 million in 1997. This has left the Moldovan Air Force as a transport and communications unit, with 850 personnel and equipment that includes two Antonov An-2s, an An-26 and two An-72s plus six Mi-8 helicopters. Combat helicopters may be purchased in the future.

MONGOLIA

- Population: 3 million
- Land Area: 604,095 square miles (1,564,360 sq.km.)
- GDP: $4.7bn (£3bn), per capita $1,556 (£995)
- Defence Exp: $513m (£327m)
- Service Personnel: 10,000 active, plus 137,000 reserves

MONGOLIAN AIR FORCE

Founded: 1926

Part of the Mongolian Army, the Mongolian Air Force dates from 1926, when four Russian-built aircraft were supplied and a flying school was established in 1927. There is little factual history about the early years, although by 1933, around 100 aircraft of Soviet origin were in service, and Soviet aid continued to counter the growing Japanese threat throughout the rest of the 1930s. The peak strength of 450 aircraft was reached by 1938, and these included Polikarpov I-15 and I-16 fighters, Tupolev TB-3 bombers and R-5 AOP aircraft. There was an encounter with Japanese forces in 1939. During World War II, a Lavochkin La-5 fighter unit saw action against the Germans while fighting alongside Soviet forces within the USSR.

Post-war, aircraft supplied by the USSR included Polikarpov Po-2s and Antonov An-2s, followed during the 1950s by Ilyushin Il-14 and Lisunov Li-2 (C-47) transports, as well as Yakovlev Yak-11 and Yak-18 trainers, while MiG-15 jet fighters entered service towards the end of the decade, by which time they were obsolete. Mil Mi-4 helicopters were also introduced.

The collapse of the USSR cut the Mongolian Air Force, for many years known as the Air Force of the Mongolian People's Republic, off from its main source of arms. Despite its geographical position, Communist China has not been an alternative source other than for a small number of transports. Its current strength is 800 personnel. Ten MiG-21s were in a fighter squadron, but have been abandoned in the open and are not operational leaving the service as mainly a transport and communication force, although there are a number of combat helicopters. There are eleven Mi-24 attack helicopters, believed to be stored, while the remaining aircraft are transports, including an Airbus A310-300 and a Boeing 737, six An-2s and an An-26, plus thirteen Mi-8/-17 helicopters.

MONTENEGRO

- Population: 672,180
- Land Area: 5,019 square miles (13,812 sq.km.)
- GDP: $3.1bn (£1.98bn), per capita $4,554 (£2,911)
- Defence Exp: $61m (£39m)
- Service Personnel: 4,500 active

ARMY OF MONTENEGRO

Founded: 2006

Created on the separation of Serbia and Montenegro in June 2006, the country has been offered membership of NATO, although it is not yet a member, and is developing its own armed forces and defence infrastructure. The air element is under army command, at least for the foreseeable future, and consists of a squadron of nine G-4 Super Galebs, with six more in store, and a helicopter squadron with flights for transport, utility and army support which operates fifteen SA341/342 Gazelles, although only half are believed to be airworthy, plus three Mi-8Ts which are in store.

MOROCCO

- Population: 31.3 million
- Land Area: 171,388 square miles (466,200 sq.km.)
- GDP: $95.2bn (£60.9bn), per capita $3,043 (£1,945)
- Defence Exp: $3.19bn (£2.04bn)
- Service Personnel: 195,800 active, plus 150,000 reserves

ROYAL MOROCCAN AIR FORCE/*FORCE AÉRIENNE ROYAL MAROCAINE*

Founded: 1956

Morocco became independent of both France and Spain in 1956. An air force was formed, *Aviation Royale Chérifienne*, using this title for the next twenty years. Initially, personnel were trained in France and Spain, with *Armée de l'Air* personnel seconded to Morocco. The primary role was army support, using six Morane-Saulnier MS500 Criquet AOP aircraft. In 1957, a de Havilland Heron, two Beech Twin Bonanzas, three Max Holste 1521M Broussards and a Bell 47 Sioux helicopter were delivered for communications work.

Morocco's important strategic position with Mediterranean and Atlantic coastlines soon led to military aid from both the USSR and the USA. The USSR provided MiG-17 fighter-bombers, Ilyushin Il-28 jet bombers and MiG-15UTI trainers in 1961. In 1966, the USA provided Northrop F-5A fighter-bombers, Fairchild C-119G Packet and Douglas C-47 transports, Sikorsky H-34 and Kaman HH-43 Huskie helicopters, and North American T-6 Texan trainers. Aircraft were acquired from other sources, including Agusta Bell 205 helicopters and Potez Magister jet trainers. Relations with the USSR cooled during the late 1960s and early 1970s, with the USSR favouring Algeria instead. By the early 1970s, a shortage of spares saw the MiG-17s withdrawn. Relations with Spain also cooled because of Moroccan claims for the return of the two Spanish enclaves of Melilla and Ceuta.

Algeria supported an uprising by Polisario guerrillas in the Western Sahara, with the *FARM* supporting ground forces from 1975 until a UN-brokered cease-fire was agreed in 1988. Despite peace negotiations in 1989, the cease-fire has been breached on many occasions. Morocco has built a barrier several hundred kilometres long to keep the Polisario out, and the *FARM* patrols this barrier.

During the 1980s, F-5As were joined by F-5Es and Dassault Mirage F1 interceptors, and by specialised COIN aircraft including the Rockwell OV-10A Bronco and Franco-German Alphajet armed-trainers. A substantial force of Lockheed C-130 Hercules transports entered service, including tankers. Helicopters acquired at this time included Boeing CH-47C Chinook and Aerospatiale Puma, with Gazelle helicopters having an anti-tank and a COIN capability. An offer of surplus F-16 aircraft by the USA in 1991 was allowed to lapse. The Mirages received an extensive upgrade in 1996–97, but an attempt to acquire additional Alphajets surplus to *Luftwaffe* requirements to cover attrition fell foul of strict export rules.

Over the past quarter-century, the *FARM* has more than trebled in size to 13,500 personnel. Flying hours are relatively high for a developing country, at 100 plus. Six fighter and ground-attack squadrons operate nineteen FC-1CH interceptors in one squadron and another has fourteen Mirage F-1EH strike aircraft, while two more squadrons have twenty-three F-5E/F Tiger IIs and one has ten F-5A/Bs, which are being replaced by eighteen F-16Cs. Reconnaissance is by a squadron with four OV-10 Broncos and two C-130H Hercules have been fitted with side-looking radar. ELINT is undertaken by two C-130Hs and two Falcon 20s. Given the long coastline, MR is surprisingly limited, with just two Do28D aircraft, although eleven Defenders are operated by the fisheries ministry on EEZ patrols. In addition to two KC-130H Hercules tankers, transport is provided by fifteen C-130Hs, six CN235s and nine Beech 100/200s, while there is a VIP Boeing 707, Falcon 50 and two Gulfstream IIs, and four C-27J Spartans are entering service. Helicopters include a squadron of twenty SA342 Gazelles, twelve armed with HOT anti-tank missiles and eight with cannon. Transport helicopters include eight CH-47D Chinooks and twenty-four SA330 Pumas, while utility helicopters include twenty-five AB205s, three AB212s, eleven AB206s and two UH-60 Black Hawks, while interest has been shown in up to twelve EC725s. Training uses nine T-34Cs, twenty K-8s which are replacing T-37Bs, seven AS202 Bravos, twenty-four Alphajets and conversion trainer versions of the F-5 and F-16. An aerobatic team has nine CAP10B/231s. Missiles are mainly Sidewinder, R-530 and R-550 Magic AAM, Walleye ASM, and HOT anti-tank missiles.

The paramilitary *Gendarmerie Royale* provides a wide range of functions, including a coastguard service, and has an *Escadron Aérien*. This operates two Sikorsky S-70A Black Hawk helicopters, as well as two Eurocopter SA315B Lamas, six SA330C Pumas, six SA342K Gazelles and two AS365N Dauphins.

MOZAMBIQUE

- Population: 21.7 million
- Land Area: 302,250 square miles (771,820 sq.km.)
- GDP: $9bn (£5.75bn), per capita $417 (£266)
- Defence Exp: $70m (£44.7m)
- Service Personnel: c.11,200 active

MOZAMBIQUE AIR FORCE/*FORÇA AÉREA DE MOÇAMBIQUE*

Founded: 1975

Mozambique became independent from Portugal in 1975 and a civil war started that lasted until 1995. An air force was established almost immediately, the *Força Populare Aérea de Libertação de Moçambique*, with equipment provided by the USSR and operated by Cuban 'advisers'. The *FPALM* received more than thirty MiG-17F fighter-bombers, a number of MiG-15UTIs and seven Zlin 326 trainers, as well as nine Antonov An-26 transports, several Mi-8 and several Mi-24 helicopters. These were in addition to Cessna and Piper training and liaison aircraft, and four Alouette III helicopters. At one time, some fifty MiG-21s were stationed in the country, flown by Cuban and East German pilots, and these were incorporated into the *FPALM* in 1992 after an attempt to sell the aircraft failed.

The collapse of the USSR brought major problems for the *FPALM*, still engaged in fighting a civil war. Spares were abruptly cut off, serviceability levels dropped, and the Cubans and East Germans left. The civil war continued, and the attrition rate of aircraft remained high, despite support from Zimbabwe which deployed forces in support of the government. The *FPALM* quickly deteriorated into a transport and communications force.

The present title was adopted in 1998. The *FAM* has 1,000 personnel. It has a nominal combat force of MiG-21s which are non-operational, as are four Mi-24 helicopters, while the remaining two Mi-8 transport helicopters were conspicuous by their absence during the severe floods of 1999. Transport aircraft are two An-26s, down from the original eight and again availability is in doubt. Liaison duties are covered by a Cessna 182 and four Piper Cherokee Six light aircraft. There are still seven Zlin 326s for training.

MYANMAR

- Population: 48.1 million
- Land Area: 261,789 square miles (676,580 sq.km.)
- GDP: $26.7bn (£17.1bn), per capita $555 (£354)
- Defence Exp: believed to be around $7bn (£4.5bn)
- Service Personnel: 406,000 active

MYANMAR AIR FORCE

Founded: 1955

Although Burma became independent from the UK in 1948, it was not until 1955 that an air force, the Union of Burma Air Force, was established with British assistance. The UK provided a gift of ex-RAF Supermarine Spitfire Mk18 fighters, de Havilland Mosquito Mk6 fighter-bombers and Airspeed Oxford and de Havilland Tiger Moth trainers. This initial force was soon expanded with Spitfire Mk9s, and later by Hawker Sea Fury FB11 fighter-bombers, Bristol 170 Mk31M Freighters, Douglas C-47s and Beech D18s. Cessna 180s were introduced for liaison duties. All of these aircraft were provided in small numbers. During the late 1950s, thirty Hunting Provost T53 trainers were introduced, with some de Havilland Vampire T55 jet trainers. In 1960, the UBAF received DHC-3 Otters, and Japan supplied Kawasaki-Bell 47 Sioux helicopters. North American F-86F Sabre fighters and Lockheed T-33A armed trainers, Alouette III and Kaman Huskie helicopters were introduced in the late 1960s, with DHC-1 Chipmunk basic trainers.

In 1989, the country changed its name to Myanmar, and the present title was adopted. Three Mi-4 helicopters supplied by the USSR were followed in 1990 by Chinese aircraft, including F-6s (MiG-19) and F-7Ms (MiG-21), as well as the Polish WZL W-3 Sokol light helicopter.

Burma is a military dictatorship spurned by the West. Despite defence being given a high priority to quell internal dissent, the country's poor economic condition has meant few modern aircraft until Russia supplied a force of MiG-29s. Aircraft are bought by barter, as with the trading of teak to Yugoslavia in payment for twenty Super Galeb armed-trainers delivered in 1991. In 1998–99, twelve Sino-Pakistani K-8 trainers were introduced.

As part of the integrated defence force, the air arm has 15,000 personnel, having trebled since 1972. The most modern component is the squadron of eight MiG-29B *Fulcrum*, while another two squadrons operate fifty F-7s (MiG-21) and there are two ground-attack squadrons with twenty-two A-5Ms. The dual roles of COIN and training are covered by ten remaining Super Galeb G4s and sixteen PC-7s and ten PC-9s. Transport is provided by two An-12 *Cub*, three F-27s and four FH-227 Friendships, five PC-6B Turbo Porters and four Shaanxi Y-8Ds (An-12). Helicopters include eleven Mi-17s, with ten PZL W-3s and eighteen Mi-2s, augmented by twelve Bell 205As and six 206s, with nine SA316 Alouette IIIs. Training uses armed trainers and twelve K-8 Karakorums.

NAMIBIA

- Population: 2.1 million
- Land Area: 318,261 square miles (824,296 sq.km.)
- GDP: $11.8bn (£7.5bn), per capita $5,619 (£3,591)
- Defence Exp: $305m (£195m)
- Service Personnel: 9,200 active

NAMIBIA DEFENCE FORCE AIR WING

Formed: 1994

Formerly South-West Africa, Namibia was mandated to South Africa following World War I by the League of Nations and then by the United Nations. It became independent in 1991, creating a small defence force with an air wing forming in 1994. The first aircraft were two Hindustan SA315 Cheetahs (Alouette II), lost in a mid-air collision in 1999, and two SA316B Chetaks (Alouette III). During the intervening period, six ex-USAF Cessna O-2A Super Skymasters and a Cessna F406 Caravan II were acquired for surveillance. Two HAMC Y-12 transports were purchased later. VIP transport is provided by a Falcon 900 and a Learjet 31. Four K-8 trainers entered service in 2001, with two Mi-8s leased from Moldova. There have been reports of small numbers of MiG-23s being supplied, possibly on loan and doubtless with mercenary pilots, as well as F-7s (MiG-21). There are two Mi-25 armed helicopters, two Mi-17s and two SA319 Alouette IIIs, while transport aircraft include two An-26s, two Y-12s, as well as a VIP Falcon 900 and a Learjet 36, and there are also five Cessna 337 Skymasters.

The Fisheries Ministry is responsible for the Coast Guard service, which has a Cessna F406 Caravan II and an SAR Sikorsky S-61L helicopter.

NATO

NATO AIRBORNE EARLY WARNING AND TRANSPORT FORCE

Founded: 1980

The high cost of sophisticated airborne early warning and control aircraft led NATO to create a force of AEW aircraft which started to enter service in 1980. The force is theoretically based in Luxembourg, and unusually the aircraft all have that country's civil registrations, although operationally the aircraft are based in Germany. There are seventeen Boeing E-3A Sentries in the force, operating in three squadrons and a training unit, with crews drawn from Belgium, Canada, Denmark, Germany, Greece, Italy, Luxembourg, the Netherlands, Norway, Portugal, Turkey and the USA. NATOAEWF can also call upon the RAF's AEW force when necessary, although the latter also has a national as well as a NATO role. The USAF and the *Armée de l'Air* also have their own AEW, or AWACS, forces. The former has a NATO and a national role, while the latter is strictly national.

While Luxembourg also has an Airbus A400M on order that will be operated within a Belgian transport squadron, NATO has also bought three C-17 transports which are used by a twelve-nation NATO/Partnership for Peace group.

NEPAL

- Population: 28.6 million
- Land Area: 54,606 square miles (141,414 sq.km.)
- GDP: $5bn (£3.2bn), per capita $174 (£111)
- Defence Exp: $200m (£128m)
- Service Personnel: 95,753 active army, plus 62,000 paramilitary
  (The figures for personnel are provisional as attempts are being made to integrate the 23,500 members of the Maoist People's Liberation Army into the existing army, but the process is taking far longer than expected.)

NEPALESE ARMY AIR WING

Founded: 1971

Nepalese military aviation started in 1971, when the Royal Nepalese Army received its first aircraft, a Short Skyvan 3M, although the royal family had an Alouette III helicopter, which was joined in 1970 by a VIP version of the Short Skyvan. The new Royal Nepalese Army Air Service took over the royal family's aircraft and evolved as a communications and transport operation, but despite growing pressure from Chinese-backed Communist insurgents found it difficult to develop a true combat capability. The mountainous terrain makes surface communications difficult and the construction of major airfields costly. In 2008, a new Maoist government swept into power and is attempting to integrate its own 23,500-strong People's Liberation Army into the established armed forces.

Currently, the Air Wing has 320 personnel, up by almost half since 2002. The only fixed-wing aircraft are a BAe748 and a Skyvan. Helicopters include three Mi-17s, three AS332 Super Pumas and an AS350 Ecureuil, as well as two Bell 206Ls, two SA316B Alouette IIIs and an HAS315B Lama.

NETHERLANDS

- Population: 16.7 million
- Land Area: 13,959 square miles (36,175 sq.km.)
- GDP: $860bn (£549.7bn), per capita $51,430 (£32,875)
- Defence Exp: $13bn (£8.3bn)
- Service Personnel: 46,882 active, plus 3,339 reserves

ROYAL NETHERLANDS AIR FORCE/*KONINKLIJKE LUCHTMACHT*

(including Netherlands Defence Helicopter Command)

Founded: 1953

Dutch military interest in air matters started with an artillery observation balloon operated for a brief period in 1886, but sustained involvement did not occur until 1911, when aircraft and balloons were acquired for manoeuvres. It was another two years before the Royal Netherlands Army formed an Aviation Division with two van Meel and three Farman F-22 aircraft. In 1914, the Royal Netherlands Navy obtained two Martin TT seaplanes, buying another twelve in 1917. Meanwhile, the Royal Netherlands Army had obtained additional Farman F-22s, bringing its total of this type to twenty by 1915. In 1917, twenty Nieuport 17C-1 Bebe and ten Fokker DIII fighters were obtained, followed by additional Nieuports and some Caudron GIII reconnaissance aircraft in 1918.

The Netherlands remained neutral throughout World War I, and unlike the air forces of the combatant nations which contracted once peace returned, continued to grow, albeit slowly, in the post-war years, acquiring former German aircraft, forty Rumpler CVs and thirty-six Trumpenburg Spijke trainers. Ambitious post-war re-equipment plans were pruned as the economic situation deteriorated. New equipment included sixteen Thalin K and twenty Fokker DVII fighters, and fifty-six Fokker CI reconnaissance aircraft. Throughout the 1920s, new aircraft consisted of fifteen Fokker DXVI and ten DXVII fighters, thirty-two CIV, 100 CVI, four CVIII and five CIX reconnaissance aircraft, three FVIIA Trimotor transports, and thirty SIV trainers – with many of these lasting well into the following decade. The Royal Netherlands Indies Army received more modern equipment. Its initial post-war equipment included twenty-five DH9 bombers and twenty-five Avro 504K trainers. During the 1920s, ten Vickers Viking amphibians, six Fokker DVII and four DCI fighters, and twenty CIV reconnaissance aircraft entered service. In 1930, nine Curtiss P-6E Hawk fighters were delivered, and were followed during the late 1930s by more than 100 Martin 139-W bombers.

Neglect of Dutch air defences was founded on a policy of neutrality in the event of any future war in Europe. No plans were made for co-operation with Belgium, the UK or France, the most likely allies.

A last-minute attempt to re-arm came in 1938. Reorganisation turned the Aviation Division into the Army Air Service. Aircraft were ordered from Fokker and US manufacturers, but there was insufficient time. On 10 May 1940, German forces rushed into the Netherlands using paratroops and air-landed troops, but the AAS had too few aircraft. It had thirty Fokker DXXIs, some DXVIIs and twenty G1 fighters, sixteen Fokker TV bombers, eleven Douglas DB-8A attack aircraft, forty Fokker CV and CX reconnaissance aircraft, and a few Koolhaven trainers on

AOP duties. Overwhelmed by the *Luftwaffe*, this small force fought valiantly for five days, after which the survivors and the flying school escaped to England.

The Royal Netherlands Indies Army received many of the aircraft ordered for the AAS. In the eighteen months before Japan entered the war, the RNIA received twenty-four Curtiss Hawk 75-As, twenty-four Curtiss-Wright CW-22 Falcons, and seventy-two Brewster 339 and 439 fighters; a few Douglas DB-7 bombers, thirty-six Consolidated PBY-5 Catalina flying boats and forty-eight Sikorsky S-43 amphibians, and twenty Lockheed Lodestar transports. Once the Netherlands government-in-exile declared war on Japan in December 1941, this force fought the advancing Japanese, but by March 1942, the remnants were in Australia, with Japan in control of the Dutch East Indies.

In Europe, Dutch squadrons operated alongside the RAF and the Royal Navy's Fleet Air Arm, and formed the nucleus of post-war Dutch military and naval aviation following the liberation of the Netherlands. Aircraft included Supermarine Spitfire fighters, North American B-25 Mitchell bombers, Douglas C-47 transports and Auster AOP3s. Post-war re-equipment started in 1947, with priority given to training aircraft, including North American T-6 Harvard, de Havilland Dominie (Rapide) and Tiger Moth, Percival Proctor, Avro Anson and Airspeed Oxford trainers; a total of 350 aircraft. Lockheed 12 and 14 transports were also supplied. In the Far East, Japanese surrender saw the RNIA dealing with Indonesian rebels as it tried to regain control over the East Indies, using North American F-51D Mustang and Curtiss P-40N Warhawk fighters, B-25 Mitchell bombers and C-47 transports.

The first jets were delivered in 1948: 200 Gloster Meteor F4 and F8 fighters, many built under licence by Fokker, which also supplied forty of its S-11 trainers in the same year.

The Netherlands became a member of the North Atlantic Treaty Organisation in 1952.

In 1953, the AAS became the Royal Netherlands Air Force, *Koninklijke Luchtmacht*, an autonomous air arm. That year, it received the first of 200 Republic F-84E Thunderjet fighter-bombers, later replaced by F-84F Thunderstreaks, Hawker Hunters and North American F-86F Sabres. These were accompanied by Piper L-18 Super Cubs for liaison duties, DHC-2 Beaver light transports, Hiller H-23 helicopters and twenty Fokker S14 jet trainers, while Lockheed T-33A trainers were also delivered. The first of 120 licence-built Lockheed F-104G Starfighter fighter-bombers entered service in 1963, followed later by 105 Canadian-built Northrop F-5A/Bs as Thunderstreak replacements. Surface-to-air missiles were deployed for the first time in the 1960s, with Nike-Ajax and Nike-Hercules. The Starfighters were augmented by RF-104G reconnaissance aircraft for a single squadron. A squadron of twelve Fokker F-27M Troopships provided most of the transport and replaced earlier types, while the Beavers operated under army control.

The Netherlands was among the European NATO members participating in the Lockheed F-16 fighter-bomber programme, replacing the Starfighters and F-

5A/Bs, with the first of more than 200 entering service in early 1981. An unusual role for the *KLu* during the 1970s through to the mid-1990s was the operation of two Fokker F-27 Maritimes in the Netherlands Antilles, a role normally handled by the Dutch Navy. The F-27M Troopships were eventually replaced by a mixture of aircraft, including second-hand converted KDC-10A Extender tanker/transports, Lockheed Hercules and Fokker 60s. The provision of aircraft for the Army has continued, with Apache anti-tank helicopters and Chinooks for heavy lift, as well as smaller transport and liaison machines.

In common with the air forces of most of the Western democracies, the *KLu* has suffered from repeated defence cuts, its personnel strength reducing by around 60 per cent over the past forty years from 23,000 to 9,586 today. There were major cuts in 1991, 1993 and 1999, with the last of these cutting the F-16 force from seven squadrons to six and then more recently to just five. The aircraft have benefited from the *KLu*'s participation in the F-16 European Mid-Life Update programme. Flying hours remain reasonably good, at an average of 180. Today, the *KLu* has five multi-role squadrons operating eighty-seven F-16MLU AMs on fighter, ground-attack and reconnaissance duties. There are thirty AGM-114K Hellfire-equipped Boeing NAH-64D Apache helicopters in one squadron. Other helicopters include a squadron with eleven Boeing CH-47D Chinook heavy-lift helicopters, one with seventeen AS352U2 Cougars and nine SA3160 Alouettes and a squadron with twenty NH90s, all operated for the Army and as with all Netherlands helicopter units are officially part of the new Netherlands Defence Helicopter Command. Three Bell 412SPs provide SAR. Transport is provided by a single composite squadron with a DC-10, four Lockheed C-130H/H-30 Hercules, two Fokker 50s, and a VIP Gulfstream IV. The main trainer is the PC-7 Turbo Trainer, with thirteen owned by the *KLu* but operated by a contractor. Fast jet training takes place in the USA, as does training for the Apache, but other training is in the Netherlands. SAM includes Hawk, Patriot and Stinger, with AMRAAM and Sidewinder AAM and Hellfire and Maverick ASM.

ROYAL NETHERLANDS NAVAL AIR SERVICE/*MARINE LUCHTVAARTDIENST*

Re-formed: 1944

The Royal Netherlands Navy's air arm dates from 1917, when it was formed with six Martin seaplanes and three Farman F-22 aircraft. The country was neutral during World War I, but the many islands and vast territorial spread of the Netherlands East Indies gave naval aviation the chance to prove its worth. During the 1920s, the *MLD* operated some forty Hansa-Brandenburg W-12 seaplanes in the Netherlands East Indies, later augmenting these with Dornier Wal flying boats which remained in service up to the Japanese invasion of 1942. In the immediate pre-war period, Dornier Do24K flying boats were also acquired for operations in the East Indies, but the rapid advance of Japanese forces after the invasion saw the *MLD*'s personnel escaping to Ceylon (Sri Lanka) and Australia. In the Netherlands, the *MLD* operated Fokker TVIIIW seaplanes, many of which were flown to France as the Netherlands fell, and from there to England before the French surrender in June 1940.

For the rest of the war, Dutch naval air personnel in the UK flew with the Royal Navy's Fleet Air Arm. They flew Fairey Swordfish on anti-submarine duties from MAC-ships, merchant aircraft carriers, which were grain ships and tankers modified with a wooden flight deck and a side island for convoy escort duties in the North Atlantic, and able to carry three or four Swordfish. Once the MAC-ships were withdrawn, the Swordfish were replaced by Barracudas, in turn replaced by thirty Fairey Firefly fighter-bombers. In 1946, the Royal Navy escort carrier HMS *Nairana* was loaned to the *MLD*, re-named *Karel Doorman*, until the Royal Netherlands Navy's own light fleet carrier, also named *Karel Doorman*, formerly HMS *Venerable*, could enter service in 1948. Both ships were named in memory of the Dutch admiral lost during the Battle of the Java Sea in 1942. At first, the new carrier operated later versions of the Firefly, replaced by Hawker Sea Furies; themselves replaced in due course by Armstrong-Whitworth Sea Hawk jet fighter-bombers. The Fireflies and Sea Furies were accompanied aboard by ASW Grumman TBM-3W and TBM-3S Avengers, until these were replaced by Grumman S-2 Trackers, of which there were eventually three squadrons.

Shore-based MR remained with the *MLD* in the post-war years. A squadron of Convair PBY-5A Catalina amphibians was replaced in 1951 by Lockheed PV-2 Harpoon landplanes, themselves replaced in turn by fifteen P2V-5 Neptunes in 1953. Nine Breguet Br1150 Atlantique MR aircraft replaced the Neptunes in 1970. By this time, *Karel Doorman* had been withdrawn and sold to the Argentine Navy in 1969, and while the Sea Hawks were withdrawn, the three squadrons of thirty-six Trackers became shore-based. Naval aviation continued at sea, however, for by this time eight guided-missile frigates of the Van Speijk-class (based on the British Leander-class) had entered service with Westland Wasp helicopters, of which the *MLD* operated twelve. Eight Sikorsky SH-34J and six Bell UH-1 Iroquois helicopters were also in service by this time.

The Wasps were eventually replaced by Westland Lynx helicopters, initially aboard the eight Kortenaer-class frigates designed jointly with Germany. The Lynxes were modernised by 1989, but have been replaced by twenty European NH90 helicopters from 2007 onwards. The Atlantiques and Trackers were replaced by Lockheed P-3C-II Orion aircraft in the late 1980s, but MR has since been abandoned with the disbandment of the two squadrons and the closure of their base in 2006. The helicopter force itself is officially part of the Netherlands Defence Helicopter Command.

Today, the *MLD* operates from four Zeven Provincien and two remaining Karel Doorman frigates as well as from the LPD *Rotterdam*, which can take four NH90 helicopters.

NEW ZEALAND

- Population: 4.2 million
- Land Area: 103,736 square miles (268,676 sq.km.)
- GDP: $132bn (£84.4bn), per capita $31,412 (£20,079)
- Defence Exp: $2.07bn (£1.32bn)
- Service Personnel: 9,702 active

ROYAL NEW ZEALAND AIR FORCE

Formed: 1937

New Zealanders played an active part in World War I, flying with the RFC and RNAS, and ultimately, the RAF. New Zealand did not have any military aircraft of her own until 1923, when the New Zealand Permanent Air Force was formed within the New Zealand Army. This move was prompted by the 1920 British offer of surplus RAF aircraft to the dominions, with New Zealand's share of the 'Imperial Gift' meant to be 100 aircraft. Indecision meant that there was little choice left by the time plans were finalised, and less than half the aircraft offered were accepted. Some aircraft were leased to commercial operators, leaving the NZPAF with ten Bristol F2B fighters, ten DH4 and nine DH9 bombers, and four Avro 504K trainers. The aircraft were shared with the New Zealand Air Force, the official title given to the reserves!

Economic difficulties limited new aircraft purchase during the 1920s to five Gloster Glebes and a small number of de Havilland Puss Moth and Moth trainers, with a few additional 504K trainers. Limited expansion started during the 1930s, when a Saunders-Roe Cutty Sark flying boat and ten Fairy IIIFs were followed in 1935 by twelve Hawker Vildebeest torpedo-bombers and some Avro 626 trainers. In 1936, the NZPAF became the Royal New Zealand Air Force, but remained part of the army until 1937, when it became a separate service on the recommendation of a seconded RAF officer. A major expansion programme was implemented in the short time left before the outbreak of World War II in Europe. Reserve squadrons gained ex-RAF Blackburn Baffins, with new aircraft for the regular squadrons. The priority was for trainers, Airspeed Oxfords, Fairey Gordons, Hawker Harts and Vickers Vincents. An order for thirty Vickers Wellington bombers was due for delivery on the eve of war, but these were transferred to the RAF with their crews to form a New Zealand element. The RNZAF's main wartime role was the training of pilots and other aircrew for the Empire Air Training Scheme, but it did field twenty-seven combat squadrons, including twelve fighter squadrons, initially with Curtiss Kittyhawks and then with Chance-Vought FG1 Corsairs. Six bomber squadrons operated Lockheed Hudsons, followed by Venturas. Two flying boat squadrons operated Short Singapores, and then replaced these with Consolidated PBY-5A Catalina amphibians and Short Sunderlands. There were two Vickers

Vincent general reconnaissance squadrons, a Douglas Dauntless dive-bomber squadron and a Grumman Avenger squadron. Other squadrons operated transport aircraft, including Douglas C-47s and Lockheed Lodestars. The RNZAF mostly fought in the Pacific, and, post-war, formed part of the British Commonwealth Occupation Force in Japan.

The inevitable peacetime reduction in strength occurred, to five regular and four reserve squadrons, the latter equipped with North American F-51D Mustang fighters. During the 1950s, one regular squadron had de Havilland Vampire FB9 jet fighter-bombers, two had de Havilland Mosquito bombers until these were replaced with a single squadron of English Electric Canberra B1 bombers, while a fourth operated MR Short Sunderlands until replaced by Lockheed P-3B Orions in 1967. The fifth squadron operated transport aircraft, a mixture of Douglas C-47s, Bristol 170M freighters and Handley Page Hastings C3 transports. The reserve squadrons were reduced to Harvard armed-trainers before being disbanded in 1957. A squadron of de Havilland Venom fighter-bombers was leased from the RAF for a few years. The Venoms were replaced by ten McDonnell Douglas A-4K Skyhawk fighter-bombers, with additional aircraft delivered later. Lockheed C-130H Hercules updated the transport squadron, while Bell 47G Sioux and UH-1H Iroquois helicopters were introduced. BAC 167 Strikemaster and Victa Airtourer trainers were used for basic and advanced training, with TA-4Ks for Skyhawk conversion.

New Zealand was for many years a member of the South East Asia Treaty Organisation, with the UK, USA, France, Pakistan, the Philippines and Australia. A squadron was based in Singapore during the 1960s and 1970s. SEATO, intended as a Far Eastern NATO, never fully realised its potential and lacked a formal command structure. The alliance was replaced by a series of bilateral arrangements with Australia and the United States.

The Skyhawks were updated during the late 1980s, when they provided two strike squadrons. The close support Strikemasters were replaced during the early 1990s by Aermacchi MB339C armed-trainers. Ex-RAF Andover transports replaced the Bristol Freighters. The RNZAF was pulled out of Singapore, but stationed part of a squadron of Skyhawks at Nowra, near Sydney, to provide maritime air defence training for the Royal Australian Navy.

In 1999, it was decided to replace the Skyhawks with twenty-eight leased ex-USAF F-16 Hornets, but a change of government led to this plan being abandoned, before deciding to scrap the fast jet combat squadrons altogether in 2001, along with the seventeen MB339CB armed-trainers. Plans to upgrade the Orion force were rejected, and while the aircraft have been upgraded, their role is now one of surveillance.

Today, the RNZAF provides SAR, EEZ patrols and transport, as well as operating five SH-2G Seasprite helicopters operated on behalf of the RNZN from two frigates and a new support ship which also provides an amphibious capability for the Army. It has 2,504 personnel. There are six upgraded P-3K Orions in one squadron, and a transport squadron operates two Boeing 757-200s and five C-130H Hercules, which have been upgraded rather than replaced by C-130Js in a joint RAAF/RNZAF order. A helicopter squadron has had its fourteen UH-1H Iroquois helicopters, operated on SAR and to support the Army, replaced by eight NH90s. Training uses Airtrainer CT-4Es and Beech 200s as well as five AW109s.

ROYAL NEW ZEALAND NAVY

During World War II, many New Zealanders served with the Royal Navy's Fleet Air Arm, but naval aviation was not introduced in New Zealand until the arrival of two Westland Wasp helicopters in the 1960s to operate from a new frigate. Although the frigate force peaked at four, with seven Wasp helicopters, it was decided that this was too small for an independent air arm. Five Kaman SH-2G helicopters are now operated by the RNZAF from two frigates.

NICARAGUA

- Population: 5.9 million
- Land Area: 57,143 square miles (148,006 sq.km.)
- GDP: $6.3bn (£4.03bn), per capita $1,063 (£679)
- Defence Exp: $40m (£25.6m)
- Service Personnel: 12,000 active

NICARAGUAN AIR FORCE/*FUERZA AÉREA NICARAGUA*

Re-formed: 1996

Nicaragua became involved with military aviation shortly after the end of World War I, when the National Guard received four Curtiss JN-4 'Jenny' trainers and some war-surplus DH4 bombers from the USA in return for assistance in the Panama Canal Zone. Although the US Army provided some pilots, the small force dwindled so that by 1927, it had fewer than twenty personnel and just three Swallow biplanes. A second attempt came in 1938, with US assistance in establishing the National Guard Air Force, or *Fuerza Aérea de la Guardia Nacional*. Aircraft provided included Grumman G-23 fighters and Waco D biplanes, used as light bombers. A flying school was supplied with Boeing-Stearman PT-13A Kaydet, Fairchild PT-19 and North American AT-6 trainers, with USAAF instructors.

During World War II personnel fell to less than seventy. Post-war, the US provided surplus aircraft, including a number of Lockheed P-38 Lightning twin-engined fighters and some trainers. Nicaragua became a member of the

Organisation of American States in 1948, and the title of the Nicaraguan Air Force, or *Fuerza Aérea de Nicaragua*, was adopted. More aircraft were supplied, including one squadron each of North American F-51 Mustang and Republic F-47D Thunderbolt fighter-bombers, and Douglas C-47 transports. The *FAN* grew steadily to some 1,500 personnel over the next twenty-five years. Obsolete aircraft continued to predominate, including Douglas B-26 Invader bombers acquired during the 1950s, and COIN Lockheed T-33A armed-trainers. Beech C-45 transports and AT-11 trainers and North American T-28 Trojan trainers followed, as well as a Hughes 269 and four OH-6A helicopters for army co-operation. A civil war started during the 1960s, ending in 1970, before a more prolonged civil war from 1982 to 1998. Left-wing Sandinista guerrillas formed a government, and the *FAN* became the Sandinista Revolutionary Air Force, *Fuerza Aérea Sandinista*, operating against US-supported guerrillas. The new regime built and improved airfields, while pilots were trained in Bulgaria to fly the MiG-21, most of which deployed in Cuba and never arrived in Nicaragua, although the *FAS* did receive many Mil helicopters and Antonov transports.

Elections led to a change of government in 1990. Fighting continued, but the new regime immediately reduced the *FAS*, selling Mi-24 attack helicopters to Peru. The title *Fuerza Aérea Nicaragua* was re-adopted in 1996.

The *FAN* has 1,200 personnel. Its sole combat capability is sixteen to twenty Mi-17 armed transport helicopters, while there are also supposed to be a few MD500s. The main role is transport, with an Antonov An-2 and four An-26s, as well as a VIP Cessna 404 Titan. Other aircraft undertake training and communications work, including a number of T-41D Mescaleros (Cessna 172).

NIGER

- Population: 15.3 million
- Land Area: 458,596 square miles (1,253,560 sq.km.)
- GDP: $5.4bn (£3.45bn), per capita $364 (£232)
- Defence Exp: $67m (£42.8m)
- Service Personnel: 5,300 active

NIGER NATIONAL AIR SQUADRON/*ESCADRILLE NATIONALE DU NIGER*

Founded: 1960

A former French colony, Niger became independent in 1960, with the standard French package of aircraft for communications and transport duties – a Douglas C-47 Dakota transport and three single-engined Max Holste 1521M Broussards. From the outset, it has been part of the Army. Three ex-*Luftwaffe* Noratlas transports eventually replaced the C-47, while the Broussards were replaced by two Cessna F337s and two Dornier Do28s. A Douglas C-54 Skymaster was also operated.

The *ENN* has not developed a combat capability. In 1991, two Lockheed C-130H Hercules replaced the Noratlases, but just one remains, badly damaged after taxiing off a runway in 2000. Two Dornier Do228s were also acquired for light transport duties, although again just one of these survives; with a single Do28 and a VIP Boeing 737-200. There are 100 personnel. In addition to the aircraft mentioned, there is one An-26, acquired from Libya.

NIGERIA

- Population: 149.2 million
- Land Area: 356,669 square miles (923,773 sq.km.)
- GDP: $135bn (£86.3bn), per capita $909 (£581)
- Defence Exp: $1.49bn (£0.95bn)
- Service Personnel: 80,000 active

NIGERIAN AIR FORCE

Founded: 1964

The Federation of Nigeria dates from independence from the UK in 1960. The Federal Nigerian Air Force was formed in 1964 with assistance from India and Germany. Initial equipment included ten Nord Noratlas transports, thirty Dornier Do27 liaison aircraft and twenty-six Piaggio P149D trainers. A civil war erupted during the late 1960s as Biafra attempted to break away from the Federation. Although air power played a relatively small role during the conflict, Soviet military aid started, including MiG-15 and MiG-17 fighter-bombers, and a handful of Ilyushin Il-28 bombers, all using Egyptian pilots.

Successive regimes in Nigeria have given defence a high priority, especially after a military coup in late 1983. Nevertheless, competing demands for finance and oil price fluctuations have meant that contracts for new equipment have frequently had to be renegotiated and options have lapsed. Doubts persist over whether some aircraft have been paid for, notably fifteen Jaguar strike aircraft delivered during the late 1970s. There have also been problems with serviceability due to the climate and the scarcity of skilled labour. Attempts to sell aircraft to raise funds for new equipment have come to nothing. At one time, the Jaguar force was due to return to the UK for refurbishment and transfer to the RAF to cover attrition losses. In 1999 it was announced that the Jaguars, G222s and MiG-21s would be sold, but in 2000 plans were announced to keep the MiG-21s and upgrade these in a programme similar to that conducted in India. In 2000, US aid enabled eight C-130H Hercules and twelve Alphajets to return to operational use. Equipment purchases have continued, and have included Mi-24 *Hind* attack helicopters and Mi-34 training helicopters delivered by Russia during 2000, and in 2001 twenty CN235s were ordered on a barter deal for crude oil. AIEP Air Beetle trainers have been assembled from kits using local labour.

Nigeria has considerable regional ambitions, and since the late 1990s has deployed substantial forces in Sierra Leone to assist internal security. Defence continues to receive a high priority, but often funding lags behind planned purchases.

Now known simply as the Nigerian Air Force, it has 10,000 personnel. Equipment serviceability is estimated to be around 50 per cent, with a high attrition rate. Two fighter squadrons include one with fourteen Alphajets and one with seventeen MiG-21MFs, but the Jaguar Ns that should form a third squadron are non-operational. Many of the seventeen Aero L-39MS Albatros trainers are armed. There are ten Bo105D armed helicopters in a single squadron, as well as nine Mi-24/-35 *Hind.* Two transport squadrons have eight C-130H/H-30 Hercules, a G222, seven Dornier 228s (including VIP), with two Eurocopter SA330 Pumas and seven AS332 Super Pumas. There are a few Do28Ds and eight Do128s in the utility role. A presidential flight has a Boeing 727, a BAe 125-1000, two Falcon 900s and a Gulfstream V. Training uses three FT-7s (JJ-7), six Alphajets, twelve MB339ANs, fifty-eight AIEP Air Beetle T18s (Vans RV-6A), of which as many as twenty are awaiting repair, thirteen Hughes 300Cs and three Mi-34 helicopters.

The only guided missile listed is the *Atoll,* which would be used by the MiG-21s.

NIGERIAN NAVY

Formed: 1988

The Nigerian Navy has formed a small air arm, initially with three Westland Lynx ASW helicopters to operate from a German-built frigate. One of the aircraft crashed in 1989 and the two remaining are believed to be unserviceable. Two A109s were acquired in 2001.

NORWAY

- Population: 4.7 million
- Land Area: 125,379 square miles (322,600 sq.km.)
- GDP: $435bn (£278bn), per capita $93,335 (£59,662)
- Defence Exp: $5.94bn (£3.79bn)
- Service Personnel: 24,025 active, plus 45,250 reserves

ROYAL NORWEGIAN AIR FORCE/*KONGELIGE NORSKE LUFTFORSVARET*

Founded: 1944

Norwegian military aviation dates from 1912, when the Royal Norwegian Navy was given a Taube and the Royal Norwegian Army was given a Maurice Farman. Official support followed in 1915, when the Naval Air Service, *Marinens Flyvevæsen*, and Army Air Force, *Hærens Flyvapen*, were formed. Both services had their own aircraft factories so that neutral Norway was not affected by the wartime aircraft famine for non-combatant nations. The factories built aircraft under licence, including Maurice Farman types, Bristol F2B fighters and Hansa Brandenburg W33 seaplane fighters.

Post-war, the peacetime strength of the two services was established at thirty-six fighters and thirty-six bombers for the *HF*, with twenty fighters, twenty torpedo-bombers and twenty-four reconnaissance aircraft for the *MF*. Norwegian aircraft were provided for both services, including MF9 fighters, MF11 reconnaissance seaplanes and the MF8 and MF10 seaplane trainers. The *MF* also received Douglas DT-2B torpedo-biplanes and Heinkel He115 seaplanes. The *HF* received thirty Curtiss Hawk 75As and Gloster Gladiator fighters, Caproni Ca310 and Ca312 bombers, Douglas DB-8A attack aircraft, and Fokker CV and CVD reconnaissance-bombers. The small size of the two services, and their elderly aircraft, meant that the spirited resistance mounted against the German invasion of April 1940 was to no avail, despite the support of the RAF and the Fleet Air Arm. Aircraft and personnel escaped to the UK. *HF* personnel in the UK were formed into two fighter squadrons, initially flying Hawker Hurricanes, but these were later replaced by Supermarine Spitfires, with one of the squadrons becoming the highest scoring Allied unit during the war, and also having the lowest accident rate. *MF* personnel went to Canada to train on Northrop N-3PB seaplanes, ordered before the invasion, and afterwards operated these from British bases under the control of RAF Coastal Command.

The two air arms were merged in 1944 to form the Royal Norwegian Air Force, *Kongelige Norske Luftforsvaret*. On its creation, the new *KNL* had three fighter squadrons, two bomber squadrons, a reconnaissance squadron and a transport squadron. Aircraft included Spitfire IXs, de Havilland Mosquito VIs, Consolidated PBY-5 Catalinas, Airspeed Oxfords, Avro Ansons, ex-BOAC and RAF Lockheed Lodestars, Fairchild PT-26 and North American T-6 Harvards. The first jets, de Havilland Vampire Mk3s, were introduced in 1948, and a further twenty-five were ordered the following year. The initial composition was changed by a Royal Commission in 1949 that proposed that there should be eight interceptor squadrons, two photo-reconnaissance squadrons, a bomber squadron and a transport squadron, each with eight aircraft. Norwegian membership of NATO ensured military aid from the United States, starting with 200 Republic F-84E

Thunderjets for the eight *KNL* fighter-bomber squadrons. In 1956, one squadron converted to RF-84F Thunderflash reconnaissance-fighters. North American F-86F and F-86K Sabres started to replace the F-84s in 1957. Transport aircraft were not neglected during this period, with Douglas C-47 and Fairchild C-119F Packet transports introduced, with CCF Norseman and de Havilland Canada DHC-3 Otters for communications duties. The first helicopters, Bell 47D/G Sioux, arrived and eight Grumman HU-16 Albatross amphibians replaced the Catalinas. Lockheed F-104G Starfighters were introduced in 1963 with twenty, and during the late 1960s, these were joined by Northrop F-5A fighter-bombers and RF-5A reconnaissance-fighters. By this time, squadron strengths were twenty aircraft for the single F-104 squadron, and sixteen for the four fighter-bomber and one reconnaissance-fighter squadrons. The 1960s saw six MR Lockheed P-3B Orions enter service, while six C-130H Hercules were also introduced, as were Saab-91 Safir and Lockheed T-33A jet trainers. The helicopter force expanded, with thirty-two Bell UH-1H Iroquois for army co-operation duties, a number of Sikorsky UH-19s and, in 1971, ten Westland Sea King Mk43 (S-61) ASW helicopters. Cessna O-1E Bird Dog and Piper L-21 AOP aircraft were also provided for duties with the Army.

Norway has taken defence more seriously than many European nations or, indeed, many Western countries with a small population, but even so, there have been substantial reductions in recent years, and in particular, personnel numbers have fallen dramatically over the past thirty years from 9,000 to 2,500. Lockheed F-16s were introduced to replace the F-104s and many of the F-5s during the 1980s, and upgraded during the 1990s, along with the surviving F-5s. The Orions were replaced by the P-3C variant in 1989, with the older P-3Bs sold to Spain with the exception of two modified as P-3Ns for coastguard duties. The UH-1Hs were replaced by Bell 412SPs for army support, but it was not until 1992 that the Cessna O-1 Bird Dogs were replaced by helicopters. Despite this, post-Cold War defence cuts also affected Norway, reducing F-16 numbers from sixty to forty-eight, and deferring replacement until 2012, although this date will not be met. A C-130H replacement was shelved, although the existing force will either have to be extensively refurbished or replaced with C-130Js.

The *KNL* has a reasonable 180 average flying hours. It has three squadrons with forty-seven upgraded F-16AMs. These may be replaced in due course by up to forty-eight F-35As. Four MR P-3C Orions are in one squadron with two P-3Ns operated for the coastguard. There are four C-130J Hercules IIs in the transport squadron, which also includes three Falcon 20Cs. An SAR squadron has twelve Sea King Mk43Bs, also used for ASW, that were upgraded between 1989 and 1995. There are eighteen utility Bell 412SPs, mainly for army support, while six Westland Lynx Mk86s are used on coastal patrol and also operate from three Fridjof Nansen-class frigates. Norway participated in the Nordic Standard Helicopter Programme, which will see eight ASW NH90TTHs replace the Lynx helicopters from 2012 onwards; further aircraft might be ordered to replace the Bell 412s. Missiles include AMRAAM and Sidewinder, while ASM includes CRV-7 and Penguin Mk-3.

OMAN

- Population: 3.4 million
- Land Area: 82,000 square miles (212,380 sq.km.)
- GDP: $54.2bn (£34.6bn), per capita $15,860 (£10,138)
- Defence Exp: $4.06bn (£2.59bn)
- Service Personnel: 42,600 active

ROYAL AIR FORCE OF OMAN

Founded: 1959

Known as Muscat and Oman until 1970, military aviation started in 1959 with the formation of the Sultan of Oman's Air Force for police duties. Initially, it was largely manned by seconded RAF personnel and by ex-RAF personnel. The original equipment included Percival Provost T52 armed trainers and DHC-2 Beaver transports, later augmented by BAC 167 Strikemaster armed jet trainers and Short Skyvan 3M transports. During the 1970s, Hawker Hunter fighters entered service, followed by Sepecat Jaguar S/B strike aircraft, BAe One-Eleven and Lockheed C-130H Hercules transports, Agusta Bell 205, 206 and Bell 214 helicopters. Boeing 747SP and McDonnell Douglas DC-8-73 airliners and two Grumman Gulfstream II corporate jets were acquired for the Royal Flight. Plans to acquire Panavia Tornado strike aircraft were abandoned in favour of BAe Hawk 200s and 100s.

Now known as the Royal Air Force of Oman, it has 5,000 personnel, almost a quarter more than in 2002. One squadron has twelve F-16C/D Bloc 50-Plus and two fighter/ground-attack squadrons each have a total of twenty-four Jaguar OS/OBs, upgraded during the late 1990s to Jaguar 97 standard. A ground-attack squadron has twelve BAe Hawk 203s, while there are also twelve Pilatus PC-9s and four Hawk 103 armed-trainers. Three transport squadrons include one with three C-130H Hercules, while a second has ten Skyvan 3Ms, with seven equipped with radar for MR, and a third squadron has three BAC One-Elevens. Two helicopter squadrons operate fifteen Super Lynx 300s for transport and SAR, while twenty NH-90s have replaced Agusta Bell AB205As, and three AB212s undertake VIP duties. There are also three Bell 206B JetRangers and five 214Bs. The Royal Flight includes two Boeing 747SPs, two Gulfstream IVs, three AS330J Pumas and three AS332C/L1 Super Pumas. In addition to the Hawks and PC-9s, training uses eight PAC Mushaks, two SF-25s and four AS202 Bravos.

Missiles include Sidewinder and AMRAAM AAM, as well as Harpoon and Maverick ASM.

PAKISTAN

- Population: 174.6 million
- Land Area: 310,403 square miles (803,944 sq.km.)
- GDP: $157bn (£100.4bn), per capita $576 (£368)
- Defence Exp: $4.11bn (£2.63bn)
- Service Personnel: 617,000 active, plus 513,000 reserves

For most of the period since independence for the Indian Sub-continent, relations between India and Pakistan have been cool, with occasional outbursts of hostilities, usually over Kashmir, but also the Rann of Kutch and on one occasion because of India's perceived support for independence for East Pakistan, now Bangladesh. Since the turn of the century, relations have become much warmer, possibly just as well since both nations now have nuclear weapons. More importantly, Pakistan has sufficient internal problems with the North-West Frontier tribal provinces being used as training grounds for Taliban terrorists forced to move from their native Afghanistan, and having links with al-Qa'eda. Pakistan ground forces have been heavily engaged in attempting to ensure that the authority of the government extends to these areas, and it is in the interests of the West that it is successful.

While Pakistan has 617,000 men in the armed forces, there are another 304,000 active in paramilitary organisations such as the Coast Guard, National Guard, Frontier Corps, Northern Light Infantry and Pakistan Rangers. This is a significant counter-insurgency force, but one must never forget the old adage that it takes ten conventional soldiers to counter each terrorist.

Doubts remain, however, that many senior members of the Pakistan military have sympathies with the Islamic extremists, and that this has hampered action against them.

Possession of nuclear weapons has changed the structure of Pakistan's defence, with a National Command Authority, NCA, created to oversee their development and deployment. The mainstay of Pakistan's nuclear deterrent lies in ballistic and tactical missile systems, but the Pakistan Air Force could also be tasked with delivery of these weapons, using either its Mirage 5 or F-16 Fighting Falcon force. Use of nuclear weapons against the extremists would be heavy-handed and cause considerable civilian casualties, so for the foreseeable future, the 'enemy' against which such weapons might be used appears to be India. That being so, it is all the more important that Pakistan has Western help in its own battle against extremists, for the prospect of the country being taken over by extreme Islamists is not too incredible, and the thought of a nuclear-armed extreme Islamic state is one that should concern strategists and policy-makers in the democracies.

PAKISTAN AIR FORCE

Founded: 1947

Indian independence in 1947 divided the country into two states, India and Pakistan, with the latter divided into two, West Pakistan and East Pakistan (now Bangladesh), and separated by 1,000 miles of Indian territory. The pre-independence Indian armed forces were also split. The Royal Pakistan Air Force inherited two former Royal Indian Air Force squadrons, one with Hawker Tempest fighters, the other with Douglas C-47 transports, supported by de Havilland Tiger Moth and North American T-6 Harvard trainers. Assistance was provided by the RAF, which seconded personnel. The RPAF built up to a planned strength of three Tempest fighter squadrons; a Handley Page Halifax bomber squadron; a C-47 transport squadron; and a communications squadron with Auster AOP5s, a C-47, Harvards, a Vickers Viking and two de Havilland Doves. In 1950, the RPAF received Hawker Fury fighters and the first of an order for sixty-two Bristol 170M Freighter transports ordered specifically for the operation of the air link between the two parts of Pakistan. The first jets, thirty-six Supermarine Attacker FB2 fighter-bombers, arrived during the early 1950s.

In 1954, Pakistan became a member of the South East Asia Treaty Organisation, SEATO, and US military aid started. Later, Pakistan joined the other regional defensive alliance, the Baghdad Pact (which later became the Central Treaty Organisation after a revolution in Iraq). Pakistan became a republic in 1956, with the RPAF dropping the 'Royal' prefix. North American F-86F Sabre jet fighters were delivered in 1956, and followed in 1958 by Martin B-57B (Canberra) jet bombers and a VIP Vickers Viscount. Lockheed T-33A jet trainers and tactical reconnaissance aircraft were also delivered, as well as Sikorsky UH-19 and Bell 47G Sioux helicopters, while Lockheed C-130B Hercules started to supplement and then replace the C-47s and Bristol 170Ms. The US supplied Lockheed F-104A Starfighter interceptors in 1962, but suspended military aid in 1965 as a result of territorial disputes between Pakistan and India over Kashmir and the Rann of Kutch, which had seen aerial combat in which the PAF's Sabres fared badly against Indian Gnats, lighter and more agile jets. The IAF's superior performance also resulted from using cannon rather than air-to-air guided missiles, as heat-seeking versions proved unreliable in tropical conditions. Pakistan turned to Communist China for armaments, starting during the late 1960s with deliveries of Shenyang F-6s (MiG-19s) and Ilyushin Il-28 bombers, and Mi-6 helicopters. These were joined by fifteen Dassault Mirage IIIE and 3IIIR fighter-bombers, while Alouette III helicopters were also obtained.

Growing tension between India and Pakistan was compounded by unrest in East Pakistan, smaller but more densely populated than West Pakistan. Indian support for East Pakistan was resented by Pakistan. On 3 December 1971, the PAF struck at ten IAF bases, including two in Kashmir, timing the attacks at dusk on a Friday, the Moslem Sabbath, when the Indians would be least likely to expect an attack. The attack was largely unsuccessful – the IAF claimed to have lost just three aircraft. The IAF had not deployed many aircraft forward close to the frontier, and those that were had been stored in concrete shelters invulnerable to anything other than a direct hit. The IAF also had the advantage of Soviet-supplied Tupolev Tu-126 AWACs aircraft. After the initial attacks, the PAF deployed relatively few of its 300 combat aircraft, and did little to support the land campaign, in a classic example of an air force without the support of an indigenous aircraft industry struggling to conserve its equipment. The war lasted until 16 December, when East Pakistan declared independence as the new state of Bangladesh.

The Mirage IIIs were followed by thirty Mirage V fighter-bombers, replacing the surviving Sabres during the early 1970s. Additional F-6s and A-5s, also a MiG-19 derivative, were obtained from China. US arms supplies resumed after the USSR invaded neighbouring Afghanistan, projecting Pakistan into the position of a Cold War front-line state. During the 1980s, Lockheed F-16 Hornets were delivered for both the interceptor and strike roles. When Soviet forces withdrew from Afghanistan, Pakistan supported Mujahideen rebels in their fight against the Afghan government. While additional F-16s were delivered in 1990, further aircraft were embargoed by the US. Despite objections from India, fifty ex-RAAF Mirage IIIs were obtained, mainly as spares for the existing Mirages. The PAF managed to grow throughout this period, maintaining its links with China and introducing the Chengdu F-7M Airguard (MiG-21) and its F-7P Skybolt derivative, while collaborating with the Chinese aircraft industry in the development of the K-8 jet trainer, which first flew in late 1990. Pakistan had already built a basic trainer, the MFI-17B Mushshak, under licence from Saab.

The 'on-off' nature of US arms supplies continued throughout the 1990s, but came to a complete stop in 1998 in protest against Pakistan testing nuclear weapons. The situation deteriorated further with Pakistani incursions into Indian Kashmir and the overthrow of the civilian government by the military in 1999. Collaboration with China continues to include a 50:50 joint project for the FC-20/J-10 fighter, on which work started in 1999.

Today, the Pakistan Air Force has some 45,000 personnel, having trebled from 15,000 in the last forty years. Flying hours were reported to be in the region of 210 – implying a high state of operational readiness – but this may have dropped in recent years. The shape of the PAF's fighter and ground-attack inventory is changing, with up to 150 JF-17s entering service and thirty-six FC-20s (J-10) on order. Currently, there are ten fighter squadrons: two have up to fifty Mirage IIIEP/ODs; one has twenty-five F-16A/B Fighting Falcons, and more of these are entering service; seven operate up to 129 F-7PG Airguard/Skybolts. In the ground-attack role there are two squadrons operating up to fifty Mirage 5PA/PA3s and one with thirteen Mirage IIIEPs, while the remaining two operate thirty-nine A-5C *Fantan*; these last three squadrons are likely to be the first to receive the JF-17 Thunder. A squadron of fifteen Mirage IIIRPs is used on reconnaissance. ELINT is provided by a squadron with just two Falcon 20Es, while there is an AEW Saab 2000. The air transport element has grown in recent years, but there are relatively few transport aircraft. The mainstay remains sixteen C-130B/E/L-100 Hercules, although the number has doubled since 2002. There are also two Y-12 transports and four CN235Ms, an An-26 *Curl*, two F-27 Friendships, of which one is operated for the Pakistan Navy, a Boeing 707, a Beech 200 and a Bonanza, as well as a VIP 737-300 and a Dassault Falcon 20E. A small tanker force consists of four Il-78s. The communications role uses some of the forty-three MFI-17B Mushshaks also used for training, and being upgraded to Super Mushshak status, while there are four Cessna 172Ns, a Citation V and a Piper Seneca on communications duties. Training uses thirty Shenyang FT-5s (MiG-17) and thirty-six K-8 Karakorums as well as the Mushshaks.

Helicopters include a squadron of four Mi-17s and another of fifteen SA-316 Alouette IIIs for SAR. The PAF operates three F-27-200MPs, three Atlantique and ten P-3C Orion aircraft on behalf of the Pakistan Navy. ASM missiles include Exocet, Maverick, AS30 and Harpoon, all Western, with AAM missiles including Sparrow, Sidewinder and Magic, as well as Harm anti-radiation missiles, with six SAM batteries operating Crotale and one with CSA-1 (SA-2). The future shape of the PAF will be interesting, as US arms supplies have resumed given the country's support for US action in Afghanistan and additional F-16s are being delivered, but on past experience Pakistan might hesitate to become dependent on a single source for arms.

PAKISTAN NAVAL AVIATION

Pakistan naval aviation initially used a handful of Sikorsky H-19 helicopters acquired during the late 1950s for transport and SAR, later augmented by Alouette IIIs for liaison and ASW duties, as well as two Fokker F27-200 Friendships for transport and maritime-reconnaissance training. Plans to acquire Kaman Seasprite helicopters during the early 1990s for operation from escort vessels were abandoned, and instead Westland Lynxes were obtained. Three P-3C Orion and four Atlantique MR aircraft have been operated by the PAF, but one of each was lost in 1999 – the Atlantique being shot down by the Indian Air Force close to the border.

No figures are available for the proportion of the 22,000 naval personnel involved with aviation. Today, PNA has ten P-3C Orions, three Fokker F-27-200MPAs and three Atlantiques for MR operated with joint PAF and PN crews, with two Britten-Norman BN2T Defenders for EEZ patrols. Three Fokker F27-200s and two F27-300s provide transport. There are six Sea King 45s and three Westland Lynx HAS3s for ASW and SAR, with the latter operating from warships with eight Alouette IIIs aboard six Tariq-class (Amazon-class) frigates acquired from the Royal Navy.

PAKISTAN ARMY AVIATION CORPS

The Pakistan Army moved into aviation with Beech L-23s and Cessna O-1E Bird Dog aircraft for AOP duties, later adding ninety Bell 47G Sioux helicopters. In contrast with most armies, Pakistan has continued to use fixed-wing aircraft in the

AOP and liaison role, including more than 100 of the licence-built MFI-17 Mushshak, also used for training. While Mi-8 helicopters were added, the mainstay of the transport element has been the Aerospatiale Puma, of which thirty-five were delivered. The PAAC received twenty Bell AH-1S Cobra anti-tank helicopters in 1984 and 1985. In recent years, Mi-17 helicopters have augmented the earlier Mi-8s, while the Bell UH-1H force has been much reduced.

The US arms embargo is largely a problem of the past with Pakistan in the front line in the war against Islamic extremists and attempting to enforce the government's authority in the tribal areas of the North-West Frontier. The combat helicopter force is relatively small, with thirty-nine AH-1F/S Cobras, with the remaining force consisting of unarmed helicopters for transport, communications and training. There are 115 MFI-17 Mushshaks in the AOP and training role. Fixed-wing transport aircraft include two Harbin Y-12s (II) delivered in 1997, while a Commander 690 and an 840 and a Cessna 421 are used on communications and survey work. Transport helicopters include thirty-one SA330J Pumas, ten Mi-8s and up to ninety-two Mi-17s, five UH-1Hs and ten 206B JetRangers, some of which are also used on training. Communications uses twenty SA316B Alouette III and twelve SA315B Lama helicopters, with twelve Bell 47Gs used on liaison and training, with the latter role also filled by ten Hughes 300Cs.

PALESTINE

- Population: 4 million
- Land Area (Gaza only): 140 square miles (363 sq.km.)
- Reliable official statistics are not available, although some sources put GDP at $6.5bn (£4.15bn), per capita $1,663 (£1,063)
- Service Personnel: 56,000 paramilitary active

PALESTINIAN DIRECTORATE OF POLICE

Palestine covers the former Gaza Strip and West Bank area, formerly part of Jordan and now known as Jericho. Gaza is controlled by Hamas, while Jericho is controlled by the Palestine Authority Administration. There are a number of paramilitary organisations, and at one time these included the Palestine Air Force, but it is no longer certain that this survives and at its peak it simply operated four Mi-8/-17 helicopters and a VIP Lockheed Jetstar. Some sources maintain that the Palestinian Directorate of Police has a small number of helicopters, but there is no evidence of this, although MANPADS are supposed to be in use for anti-aircraft protection.

PANAMA

- Population: 3.4 million
- Land Area: 28,575 square miles (74,009 sq.km.)
- GDP: $23.5bn (£15.0bn), per capita $6,993 (£4,470)
- Security Exp: $269m (£172m)
- Service Personnel: 12,000 paramilitary

PANAMA NATIONAL AIR SERVICE/*PANAMA SERVICIO AÉREO NACIONAL*

Panama has only paramilitary organisations, providing internal security and policing the approaches to the Panama Canal, as well as transport and liaison. The former Panama Air Force, *Fuerza Aérea Panama*, attempted to obtain Cessna A-37B Dragonfly armed jet trainers for COIN duties in 1983, but was disbanded after the US invasion of Panama in 1989. The *PSAN* has 400 personnel. Aircraft include a CASA CN235M for patrol and transport, two C-212-200s, three C-212-300s, and a BN2A Islander. A Piper Seneca handles communications duties. VIP aircraft include a Boeing 727 and two S-76s. There are thirteen UH-1H helicopters, of which five were purchased from Taiwan in 1997, two Bell 205s and six 212s. Training uses six Chilean-built Enaer T-35D Pillans.

PAPUA NEW GUINEA

- Population: 5.9 million
- Land Area: 183,540 square miles (461,693 sq.km.)
- GDP: $6.3bn (£4.0bn), per capita $1,076 (£687)
- Defence Exp: $50m (£32m)
- Service Personnel: 3,100 active

PAPUA NEW GUINEA DEFENCE FORCE

Founded: 1975

Papua New Guinea was administered by Australia until independence in 1975. A Papua New Guinea Defence Force was established after independence to maintain internal security, with an air element for coastal and border patrols. Four GAF Nomad Mission Masters fitted with search radar, and five ex-RAAF Douglas C-47s for patrols along the border with the Indonesian territory of Irian Jaya were soon operational. Three IAI-201 Aravas were later acquired and four UH-1Hs were provided by Australia in 1989. The PNGDF joined Australian forces in rescue missions after Papua New Guinea was hit by a tidal wave in July 1998.

Today, just 200 of the PNGDF's 3,100 personnel are involved in aviation. There is just one remaining Nomad Mission Master, but the service now has three IAI-201 Aravas, two CASA-IPTN CN235M and a CASA C-212 transports, as well as the four UH-1H Iroquois helicopters for SAR. Australia donated a Bo105 in 1999.

PARAGUAY

- Population: 7 million
- Land Area: 157,047 square miles (406,630 sq.km.)
- GDP: $14.1bn (£9.0bn), per capita $2,020 (£1,291)
- Defence Exp: $127m (£81.2m)
- Service Personnel: 10,650 active, plus 164,500 reserves

PARAGUAYAN AIR FORCE/*FUERZA AÉREA PARAGUAYA*

Paraguayan military aviation originated with Army Potez XXV reconnaissance-bombers during the late 1920s as tension arose with neighbouring Bolivia over a boundary dispute in the Gran Chaco. Aided by an Italian military aviation mission and mercenary pilots, the army's air corps, or *Fuerzas Aéreas Nacionales*, acquired Fiat CR30 fighters, Caproni Ca101 bombers and Breda Ba25 trainers during the 1930s as war broke out with Bolivia. Fiat CR32 fighters were obtained later. Paraguay won the war. In 1938, Breda Ba65 ground-attack aircraft were obtained, and, in 1939, Muniz M-9 trainers from Brazil. The late 1940s saw Beech C-45 and Douglas C-47 transports, and Vultee Valiant, Fairchild M-62, North American NA-16 and Boeing-Stearman PT-17 Kaydet trainers in service.

In 1948, Paraguay became one of the founder members of the Organisation of American States, by which time *FAN* had become the *Fuerza Aérea Paraguaya*, although still under army control.

During the 1950s and 1960s, a variety of transport types augmented the earlier aircraft, including Douglas C-54s and a Convair 240, while some North American T-6 Texan trainers were armed for COIN duties. The first helicopters were introduced; Bell 47G Sioux and Hiller UH-12Es. Although land-locked, Paraguay sits across the Paraguay River and much of the terrain is swampy, so a Grumman JRF Goose amphibian was operated for many years. Combat aircraft were acquired during the 1980s, Brazilian-built Embraer EMB326B Xavantes (Aermacchi MB326) augmenting a number of AT-33As. Enaer T-35D Pillan trainers were acquired from Chile in 1991 and 1992. The most significant development in recent

185

years has been the arrival in 1998 of Northrop F-5E Tigers donated by Taiwan, which also provided six Bell UH-1H helicopters in 2001.

The *FAP* has 600 personnel, down by almost half since 2002 as total personnel numbers in the armed forces have been reduced. The F-5E Tiger IIs have gone and the sole combat capability lies in armed-trainers, just two AT-33s, with others in store, three EMB312 T-27 Tucanos and five EMB326 Xavantes. Transport aircraft include a VIP Boeing 707, five C-212-200 Aviocars, a C-47 and a number of Beech, Cessna and Piper light aircraft. Helicopters include seven remaining UH-1B/H Iroquois with three HB350 Esquilos (Ecureuil). Training uses the Tucano and Xavante already mentioned plus seven T-35A/B Pillans.

PARAGUAYAN NAVAL AVIATION/*AVIACION DE LA ARMADA PARAGUAYA*

Paraguay's Navy handles river patrols and SAR, with 100 out of its 1,100 personnel engaged in naval aviation. Aircraft are used for communications and patrol duties, including a Cessna 210, two 310s and a 401, with helicopters including two HB350 Esquilos and a UH-12E Raven. Training uses two Cessna 150s and an OH-13H Sioux.

PERU

- Population: 29.5 million
- Land Area: 496,093 square miles (1,249,048 sq.km.)
- GDP: $138bn (£88.2bn), per capita $4,674 (£2,987)
- Defence Exp: $1.57bn (£1.0bn)
- Service Personnel: 114,000 active, plus 188,000 reserves

PERUVIAN AIR FORCE/*FUERZA AÉREA PERUANA*

Formed: 1950

Peru established army and naval air arms in 1920. The government acquired twelve Avro 504K trainers for the Army, as well as a Curtiss JN4, four Morane-Saulnier Parasols, two Spad S-7Cs and three Bristol F2B fighters, and a Blackburn Kangaroo bomber. The USA supplied two Curtiss Seagull flying boats for the new Peruvian Naval Air Service, which later received DH9 bombers and Vought UO-Q Corsair observation aircraft. The two air arms merged in 1929, forming the Peruvian Aeronautical Corps, *Cuerpo de Aeronáutica del Perú*. Initially, the *CAP* operated Nieuport 121C and Curtiss Hawk fighters, Douglas M-4 seaplane bombers, Vought UO-Q and O2U-1E Corsair and Potez 39 observation aircraft,

Boeing 40B transports, and Avro 504R Gosport, Stearman C-3R, Hanriot 240 and Morane-Saulnier MS110 trainers. New aircraft during the early 1930s included Fairey Fox II bombers and Fairey Gordon general-purpose aircraft. In 1933, the *CAP* fought in a border dispute with neighbouring Colombia.

An Italian air mission visited Peru in 1935, leading to the purchase of Caproni Ca114 fighters, Ca135 bombers, Ca111 transports, Ca310 general-purpose aircraft and Ca100 trainers. An abortive attempt was made to establish a Caproni factory in Peru for local assembly of the aircraft. The late 1930s saw Curtiss Hawk 75-A and North American NA-50A fighters, Douglas DB-8A bombers, Vultee 54 and Curtiss-Wright CW-22 trainers introduced, with some Faucett F-19 transports. During World War II, there were few new aircraft as Peru remained neutral, although the *CAP* did see action in 1941 against Ecuadorian forces in a border dispute.

Peru became a member of the Organisation of American States in 1948. US military aid included twenty Republic F-47D Thunderbolt fighter-bombers, twenty North American B-25J Mitchell bombers, and twelve Lockheed PV-2 Harpoon MR aircraft. There were also Consolidated PBY-5A Catalina amphibians, Curtiss C-46 and Douglas C-47 transports, and Beech T-11 Kansan, North American T-6 Texan, Boeing-Stearman PT-17 Kaydet and Fairchild PT-26 trainers. These were followed in 1949 by four DHC-2 Beaver transports purchased from Canada. Some pre-war aircraft remained operational until the mid-1950s.

The *CAP* became the Peruvian Air Force, *Fuerza Aérea Peruana*, in 1950. No further aircraft were introduced until 1955, when the *FAP* received its first jets, North American F-86F Sabre fighters and Lockheed T-33A trainers. These were followed in 1956 by sixteen Hawker Hunter F52 fighter-bombers and eight English Electric Canberra B8 bombers, which joined eight ex-USAF Douglas B-26 Invader bombers introduced in 1955. Additional Canberras were purchased later, as well as Lockheed F-80C Shooting Star fighters. Peru has had border disputes with its neighbours, Bolivia, Chile, Colombia and Ecuador, and anxious to avoid an arms race, the USA requested the UK to embargo BAC Lightning interceptors, but fourteen Dassault Mirage 5 fighters were ordered instead.

To avoid the arms embargo and benefit from lower Soviet prices, Peru became the first South American country to buy Soviet equipment, introducing Sukhoi Su-22 fighter-bombers, Antonov An-26 transports, and Mil Mi-6 and Mi-8 helicopters in 1976. Despite poor serviceability, a second batch of Su-22s followed in 1980. The An-26s were replaced with An-32s in 1987. A substantial increase in transport helicopters occurred during 1989–90, with thirty-two Mi-17s. Six Harbin Y-12 transports were purchased from China in 1991. The USSR's collapse brought both opportunities and problems in procurement. The opportunities included buying MiG-29s and Su-25s at an attractive price from Belarus in 1996. The problems lay in the manufacturer refusing to sell spares, as the aircraft had not been purchased from them! The issue was resolved by a further Peruvian purchase of three MiG-29SEs from the manufacturer in 1998, on condition that spares be supplied for the original aircraft, which replaced some Canberras and Su-22s. Mi-24 attack helicopters were purchased from El Salvador in 1993. Peru also purchased Dassault Mirage 2000 interceptors from France, and Embraer EMB312 Tucano trainers from Brazil. At this time, US assistance resumed to help Peru fight the Maoist Sendero Luminoso, 'Shining Path', guerrillas and drug traffickers.

Today, the *FAP* has 17,000 personnel, having almost doubled in manpower over the past forty years. It operates a wide variety of aircraft, giving logistics and maintenance problems. A fighter squadron operates eighteen MiG-29E/SE *Fulcrum* while five fighter-bomber squadrons include one with twelve Mirage M-2000P/DPs; one with ten A-37 Dragonflies; while the remaining three have seventeen Su-22 *Fitter* and eighteen Su-25 *Frogfoot*. Sixteen Mi-24 *Hind* form an attack helicopter squadron. Reconnaissance aircraft consist of a photo-survey unit with two Learjet 36As and four C-26Bs. Only one Boeing 707-320 tanker survives out of the original three, with two DC-8-62CFs. Other transports include six An-32s, two An-72s, five C-130A/L-200-20 Hercules, five DHC-6 Twin Otters, two Y-12(II)s, with four more in store, and eight PC-6s. The Presidential Flight has a Boeing 737-500, an F-28-1000 Fellowship and a Falcon 20F. There are twenty-eight Mi-8/-17, fourteen AB (Bell) 212, five 214ST, two 412 and ten Bo105C helicopters. A Super King Air and four Metro IIIs are employed on communications duties. Training uses six T-41A/D Mescaleros, eighteen EMB312 Tucanos, six Il-103s, fifteen Zlin 242Ls, seven MB339APs and three AS350B Ecureuils.

With the exception of Magic, AAM are of Russian origin and include *Atoll*, *Aphid*, *Alamo* and *Adder*, with AS-30 ASM.

PERUVIAN NAVAL AIR SERVICE/*SERVICIO AERONAVALE DE LA MARINA PERUANA*

The Peruvian Navy initially used eight Bell 47G Sioux helicopters, but the introduction of ex-Dutch guided-missile cruisers in 1973 and 1976, followed by Carvajal-class (Lupo-class) frigates during 1978–81, led to shipboard helicopters. Only one of the guided-missile cruisers remains in service, but this can carry three small helicopters or a single medium helicopter. The *Aeronavale* is also responsible for MR, but lacks sophisticated long-range aircraft. Currently, aviation accounts for 800 out of the Navy's 23,000 personnel. MR is handled by five Beech B200Ts, which can also be used in the light transport role, and an F-27 Friendship. A DHC-6 Twin Otter is operated on behalf of the coast guard. Three AB212 helicopters can be deployed aboard eight frigates on ASW and ASV duties, while there are three Agusta-built AS-61D Sea Kings for a similar role aboard the guided-missile cruiser. Two An-32 and four Mi-8 helicopters are used for transport, with a Cessna 206 and an EMB120 Brasilia for communications. Training uses two T-34C Turbo Mentors and two Bell 206 and six Enstrom F28F helicopters.

PERUVIAN ARMY AVIATION/*AVIACIÓN DEL EJÉRCITO PERUANA*

Peruvian Army Aviation is primarily for transport, with just two of its ten Agusta A109K helicopters, which could be armed, remaining. It has a small number of fixed-wing aircraft, including two Antonov An-28s and three An-32s, a Cessna 150 and a 172, and seven Cessna 206s and two 303s. There is an Mi-26 heavy-lift helicopter with another in store, as well as fourteen Mi-17s, with another eight in store, two Mi-2s and eight Enstrom F28s, the latter being used mainly for training.

PHILIPPINES

- Population: 97.9 million
- Land Area: 115,707 square miles (299,681 sq.km.)
- GDP: $164bn (£104.8bn), per capita $1,680 (£1,073)
- Defence Exp: $1.16bn (£0.74bn)
- Service Personnel: 120,000 active, plus 131,000 reserves

PHILIPPINE AIR FORCE

Founded: 1947

In 1935 the Philippine Constabulary formed an aviation branch using Stearman Model 76 AOP aircraft and Curtiss JN-4 trainers to assist ground forces in detecting bandit camps. This grew into the Philippine Army Air Force with US assistance in 1940, equipped with twelve Boeing P-26 fighters, most of which were destroyed on the ground when Japan invaded on 10 December 1941. Those Philippine personnel that could escape were absorbed into the US forces for the duration of the war.

The Philippines were liberated in 1944, and became a republic the following year. In 1947, the Philippine Air Force was established as a separate service, with North American F-51D Mustang fighter-bombers and Douglas C-47 Dakota transports, used primarily on COIN. A US defence treaty in 1951 was followed in 1955 by the Philippines becoming a member of the South East Asia Treaty Organisation, receiving further US aid. The Mustangs were replaced by the PAF's first jets, North American F-86F Sabres, in 1957. Japanese-built Beech T-34 Mentor trainers arrived in 1959. Other aircraft included Sikorsky H-19A and H-34, and Hiller FH-1100 helicopters, Lockheed T-33A jet trainers, Grumman HU-16A Albatross amphibians, and North American T-6G Texan and T-28 Trojan trainers. Northrop F-5A/Bs replaced Sabres during the late 1960s, when Fokker F-27 Troopship and NAMC YS-11 transports also arrived, and Mitsubishi-built Sikorsky S-62A helicopters were introduced for SAR.

Defence was accorded a high priority throughout the 1970s and 1980s by the unpopular Marcos regime, and when this was overthrown in 1986, severe spending constraints were introduced. The F-5A/Bs remained a viable force through the provision of ex-Taiwanese aircraft through the US Military Assistance Program in 1989. Economy measures saw the withdrawal of many older aircraft, and a number of SIAI SF260M/W armed-trainers. Funds were made available for eighteen SIAI S211 trainers to be assembled in the Philippines. Plans to replace the F-5A/B force were postponed through the turn of the century while the force shrunk to just ten aircraft from the original twenty. Plans to increase the defence budget with a fifteen-year plan starting during the mid-1990s in response to a reduction in US forces in the region and growing Chinese interest in the Spratly Islands were hampered by the Asian economic crisis of the late 1990s.

Currently, the PAF has 16,000 personnel, 7,000 higher than the figure in the late 1960s. Despite having a considerable 'wish list' of replacing F-5A/Bs with either A-4 Skyhawks from Kuwait or New Zealand, or surplus USAF F-16s as well as MR aircraft, its combat capability consists of a squadron of up to fifteen OV-10 Broncos, and MR is conducted by a single F-27MPA. Transport is provided by eight C-130B/H/K Hercules, although another six are in store, an F-27-200 and a Rockwell Turbo Commander 690A. There are also two GAF Nomad utility aircraft. The importance of COIN is emphasised by the presence of four squadrons of helicopters operating a total of sixty-seven Bell UH-1H/M Iroquois, five Sikorsky AUH-76s (S-76 gunship) and twenty MD520s, as well as an S-70 Black Hawk, an SA330L Puma and twelve AB/Bell 412s. There are eleven Cessna T-41D Mescaleros (172) also available for light reconnaissance as well as training, fifteen S-111s and forty-three SF260s.

UAV consists of just two Blue Horizon IIs, while Sidewinder AAM missiles are available.

PHILIPPINE NAVAL AVIATION

The Philippines Navy has a small force of aircraft and although no helicopters are embarked, two offshore patrol craft and two LST have helicopter decks. Current aircraft include four BN2 Defenders, two Cessna 177 Cardinals and a Cessna 152, with the latter used for training, as well as five Bo105 helicopters which are used for SAR, communications and utility duties.

PHILIPPINE ARMY AVIATION

The Philippine Army has a UH-60 VIP helicopter, plus a Beech 80, a Cessna 170 and a 172, and uses Blue Horizon UAV.

POLAND

- Population: 38.5 million
- Land Area: 120,733 square miles (311,700 sq.km.)
- GDP: $465bn (£297.2bn), per capita $12,072 (£7,716)
- Defence Exp: $8.63bn (£5.52bn)
- Service Personnel: 100,000 active, plus 223,003 reserves

POLISH AIR DEFENCE AND AVIATION FORCE

Re-formed: 1944

Poland did not become an independent nation until 1918, during the chaos of the Russian Civil War. Polish personnel had served as aviators in the Russian, German and Austrian armed forces. In 1919, a seven-squadron air corps was formed as part of the Army, with almost 100 aircraft including Spad S7C fighters, Breguet Br14A reconnaissance and Br14B2 bomber aircraft, and Salmson SAL2-A2 reconnaissance-bombers. Aircraft and personnel of volunteer flying units were absorbed, giving a wide variety of military and civil aircraft. Bristol F2B fighters, Sopwith Dolphins and Martinsyde F4s were bought from the UK, which donated war-surplus DH9 bombers and Sopwith Camel fighters. Former German and Italian aircraft were also obtained, as the new state prepared to fight the USSR for its independence. The conflict started in late 1919 and lasted for a year, with air power saving Poland from defeat.

Between the two world wars, a determined attempt was made to establish a Polish aircraft industry, with both private and state-owned concerns producing their own designs and building foreign designs, including Fokker FVII transports and Avia BH33 trainers, under licence. Significant Polish aircraft included the PZL P-1, P-6, P-7 and P-11 fighters that appeared during the 1930s, as well as P-23B reconnaissance-bombers, and immediately before the outbreak of World War II, the PZL P-37 bomber. Other aircraft included the RWD-8, RWD-14 and Lublin R-XIII AOP aircraft and PWS-2 trainers. In 1938, the Polish Air Force, *Polskie Lotnictwo Wojskowe*, became a separate air service following army reorganisation. It had fifty-five fighters and seventy-six bombers, as well as transport, liaison and training aircraft. More than 250 aircraft were attached to the Army for AOP and liaison duties. Ambitious plans for expansion of the PAF recognised that many of its aircraft were obsolete. Yet, of the many modern aircraft planned, only the PZL P-37 bomber was in service when German forces invaded on 1 September.

The overwhelming strength of the *Luftwaffe* with its modern aircraft gave the Germans aerial superiority, aided by the rapid advance of the Panzer divisions and invasion by Soviet forces on 17 September, although Poland did not finally surrender until 27 September. Polish personnel took the few surviving aircraft to Romania, where they were abandoned, and continued to France, arriving in time to

191

provide a pool of experienced pilots for the *Armée de l'Air*. As German forces advanced westwards, Polish personnel flew Morane-Saulnier MS40C and Caudron C714 fighters before the fall of France, when they fled to the UK. Polish squadrons were formed in the RAF, initially being equipped with Hawker Hurricane fighters and Fairey Battle bombers, later replaced by Supermarine Spitfires and Vickers Wellingtons respectively. One squadron flew Boulton-Paul Defiant night-fighters, replacing these with Bristol Beaufighters and then with de Havilland Mosquitoes. By the end of the war, Polish pilots had also flown North American F-51D Mustang fighters and B-25 Mitchell bombers, Handley Page Halifax and Consolidated B-24 Liberator bombers. Meanwhile, other Polish airmen had flown with the Soviet forces once Germany invaded Russia. Poland was liberated by advancing Soviet forces in 1944, but few Polish personnel serving with the RAF returned home, and of those that did, few were allowed to stay in the reformed PAF before being sent to detention camps.

The post-war Polish Air Force continued to use its wartime Soviet aircraft at first, including Polikarpov Po-2, Yakovlev Yak-1, Yak-3 and Yak-9 fighters, Ilyushin Il-2 bombers, and Yakovlev and Petlyakov Pe-2 trainers. The fighter units standardised on the Yak-9s, which were joined by Ilyushin Il-10 ground-attack aircraft and Tupolev Tu-2 bombers. Later, Lisunov Li-2 (C-47) transports and Yak-11 and Yak-18 trainers were introduced. The first jets arrived in 1950, Yak-23 fighters. Over the next few years, these were joined by MiG-15 fighters, many of which were built in Poland as the LIM-2. The Polish-designed Junak-3 trainer was introduced, with Yak-17 jet trainers, Ilyushin Il-28 jet bombers and Mil Mi-1 helicopters, produced in Poland as the SM-1. In 1957, the Yak-23s and MiG-15s started to be replaced by MiG-17 jet fighters, and in turn these gave way to first the MiG-19 and then the MiG-21, while ground-attack duties passed to the Sukhoi Su-7. These aircraft were accompanied by the inevitable conversion trainers, with the MiG-15UTI filling the role of advanced jet trainer. A substantial transport force of almost fifty aircraft developed, including Antonov An-2 and An-12, Ilyushin Il-14 and Il-18, as well as the Li-2. Mil Mi-4 and the heavy-lift Mi-6 helicopters joined the Mi-1s and SM-1s. The PAF became one of the largest and best-equipped air forces of the Warsaw Pact, second only to that of the USSR itself, due to the country's size and population, but also because dissent came much later to Polish political life than elsewhere in Eastern Europe. MiG-23 and Su-20 and Su-22 aircraft updated the combat squadrons, as well as Mi-8 and Mi-17 transport helicopters, and Mi-24 attack helicopters.

The changing political situation at the end of the 1980s forced a reappraisal of the strengths of the Polish armed forces. The last new aircraft to be delivered by the USSR were twelve MiG-29s in 1989, and further orders for this type were cancelled. The Soviet 37th Air Army moved its 350 aircraft from Polish territory in 1990, and on 1 April 1991, the Warsaw Pact ceased to exist. The PAF was merged in 1990 with the Air Defence Force, *Wojska Obrony Powietrznej Kraju*, taking its present title of Polish Air Defence and Aviation Force, or *Polska Wojska Lotnice I Obrony Powietrznej, PWLiOP*. The new service had almost 400 combat aircraft and another 250 or more transport and training aircraft. Initially, Soviet structures continued, with two squadrons of twelve to sixteen aircraft in an air defence regiment and three squadrons of nine to twelve aircraft within other air regiments. There were six air defence fighter regiments, three fighter and three fighter-bomber regiments, a reconnaissance-bomber and a tactical reconnaissance regiment. In 1995, many of the helicopter units used in support of the Army were transferred to army control, creating an Army Air Force.

Plans to acquire Western equipment were put on hold due to the economic situation, but Poland became a member of NATO in April 1999. In September 1999, the MiG-23s were withdrawn and replaced by F-16s, after consideration of F/A-18, Mirage 2000, Saab Gripen and Eurofighter Typhoon. While the decision was being made, twenty-three ex-*Luftwaffe* MiG-29s were introduced during 2002 after being upgraded in Germany to meet NATO standards and communications requirements, while the Su-22s were upgraded in Israel.

The *PWLiOP* has 17,500 personnel, having been cut from more than 43,000 in 2002. Annual flying hours have increased from around 60 to 120 hours, too low for full combat effectiveness, to between 160 and 200. The *PWLiOP* is divided into two air defence corps, North and South. Fighters include two squadrons with twenty-four MiG-29A/UB *Fulcrum* as well as three multi-role squadrons of forty-eight F-16C/D Fighting Falcon, while three ground-attack squadrons have forty-eight Sukhoi Su-22M4 *Fitter*. Transport is provided by four squadrons with five C-130Es, which have replaced five An-26s, eight C-295Ms, twenty-three M-28 Bryzas, two VIP Tu-154 *Careless* and four Yak-40 *Codling*. Helicopters include eleven Mi-8s and thirty-eight Mi-2s, as well as seventeen W-3 Sokols and a Bell 412HP. Training uses fifty-four TS-11 Iskas, six PZL SW-4s and thirty-six W-3 Sokols. AAM includes *Aphid*, *Anab*, *Archer*, Sidewinder and AMRAAM, while AS includes *Kerry* and Maverick.

POLISH NAVAL AIR FORCE

Polish naval aviation has traditionally been land-based, with a single air regiment operating helicopters and fixed-wing aircraft, but this has changed with the acquisition from the USN of two Pulawski-class (Perry-class) frigates which operate SH-2G Super Seasprite helicopters. Currently, naval aviation accounts for 1,560 out of the navy's 10,864 personnel. Shore-based MiG-21s have now been withdrawn. MR, EEZ and pollution control uses eight PZL-Mielec M-28E Bryzas, with transport by an An-2 and two An-28s. ASW is handled by four SH-2Gs and ten Mi-14PLs, with another three Mi-14PSs and seven PZL W-3RMs for SAR. Training uses eight W-3s.

POLISH ARMY AIR FORCE

Formed: 1995

Polish army aviation today dates from the decision to transfer tactical helicopters from the *PWLiOP*, starting in 1995. The initial force of Mi-24 attack helicopters was augmented by an additional eighteen machines formerly belonging to East Germany and handed over after German reunification. Today, the Polish Army Air Force has thirty-two Mi-24 *Hind* and twenty-two Mi-2 attack helicopters, thirty Mi-8/-17 and thirty-seven W-3A Sokol transport helicopters, some of which are also used for training.

PORTUGAL

- Population: 10.7 million
- Land Area: 34,831 square miles (91,945 sq.km.)
- GDP: $240bn (£153bn), per capita $22,441 (£14,344)
- Defence Exp: $2.72bn (£1.74bn)
- Service Personnel: 43,330 active, plus 210,900 reserves

PORTUGUESE AIR FORCE/*FORÇA AÉREA PORTUGUESA*

Founded: 1957

By the time that Portugal established both army and navy air arms in 1917, a flying school had been in existence for some years and officers from both services had received flying training in the UK and France. The *Arma de Aeronáutica* used Spad S7C fighters and Breguet Br14A2 bombers, while the *Aviação Maritima* introduced Fairey Campania seaplanes and Short Felixstowe F3 flying boats, and later acquired Fairey III seaplanes. By 1924 the *AdA* had twenty-five aircraft, including Martinsyde F4 Buzzard fighters, Caudron GIII and Avro 504K trainers, and its future establishment had been set at three squadrons, one each with fighters, bombers and reconnaissance aircraft. In 1927, the *AM* had three Fokker TIV reconnaissance-seaplanes, three HS2L and seven CAMS37 flying boats, and five Hanriot H41 training seaplanes. Many of the older aircraft, such as the *AdA*'s Br14A2s, soldiered on while more up-to-date equipment entered service, including sixteen licence-built Potez XXV bombers and twenty Vickers Valparaiso reconnaissance aircraft. In the late 1930s, the *AdA* received Hawker Hinds and thirty Gloster Gladiator fighters, Breda Ba65 ground-attack aircraft and Junkers Ju86K bombers, finally replacing the wartime vintage aircraft. For training, the *AdA* had twenty-two de Havilland Tiger Moths, but both services received licence-built Morane-Saulnier MS233 and Avro 626 trainers.

Portugal remained neutral throughout World War II although retaining cordial relations with the Allies, even after they occupied the Azores, with the RAF using Santa Maria as a maritime-reconnaissance base. In exchange for the use of the base *AdA* received a small number of Hawker Hurricane and Supermarine Spitfire I fighters, and to these were added Bell Airacobra and Curtiss Hawk 75A fighters, Bristol Blenheim and Consolidated B-24 Liberator bombers, Miles Master, Magister and Martinet trainers, and Airspeed Oxford trainers. The *AM* received Bristol Beaufort and Blenheim bombers, Short Sunderland flying boats, Grumman G-21 amphibians, and Lockheed Hudson bombers. Post-war, Portugal became a member of NATO, but re-equipment was slow, with only twenty Republic F-47D Thunderbolt fighter-bombers being delivered during the late 1940s, and a few Douglas C-47 and C-54 transports.

In 1952, the two air arms were amalgamated to form the autonomous Portuguese Air Force, *Força Aérea Portuguesa*. The first jets arrived in 1953, Republic F-84G Thunderjets, finally replacing the Hurricanes (used for training, making Portugal the last country to operate this aircraft) and Spitfires, with Lockheed T-33A jet trainers and DHC-1 Chipmunk basic trainers. The Royal Navy provided fifteen North American T-6 Harvard trainers as a gift. The *FAP* also introduced Boeing SB-17G Fortress and Douglas B-26 Invader bombers and Lockheed PV-2 Harpoon MR aircraft, Grumman SA-16A Albatross SAR amphibians, and Alouette II and Sikorsky H-19A Chickasaw helicopters. North American F-86F Sabre fighters later joined the F-84Gs, while Lockheed P-2E Neptunes were introduced for MR. Other acquisitions included thirty-six Fiat G91R-4 fighter-bombers and twenty-five Dornier Do27 liaison aircraft.

Despite its continued membership of NATO, the 1960s and 1970s found Portugal facing difficulty in defence procurement because of international hostility to her retaining her African colonies, where the *FAP* provided support for ground forces. The international community failed to act when India occupied Goa and Indonesia annexed East Timor. The situation eased considerably after both Angola and Mozambique were granted independence in 1975.

The loss of the colonial commitments and the general reduction in defence budgets in the 1990s has meant that the *FAP* has seen its personnel numbers fall from 17,500 to 7,100 over the past four decades. Nevertheless, re-equipment and modernisation has taken place, helped by the grant of a renewed lease of the Azores base to the United States in 1989. This resulted in the sale to Portugal of Lockheed F-16s for the interceptor and attack role, six refurbished ex-USN Lockheed P-3P Orions for MR duties in 1988, and LTV A-7 Corsair IIs in the strike role. The arrival of the Orions released Lockheed C-130H Hercules from MR for their transport role. The *Luftwaffe* transferred surplus Alphajets during the early 1990s to replace the earlier Lockheed T-33A and de Havilland Vampire jet trainers. The first F-16s arrived in 1994, and a second batch was delivered in 2001.

Today, the *FAP* has upgraded its F-16s under the European Mid-Life Upgrade Programme, replaced its armed Alphajets with the second squadron of F-16s, making a total of twenty-four F-16A/Bs. The six Lockheed P-3P Orions have also been upgraded to match those of other NATO operators. SAR and fisheries protection uses twelve Merlin and four SA330 Puma helicopters in one squadron. An army air arm has been created, and this will take over many army co-operation

195

light aircraft and helicopters. Transport uses six Lockheed C-130H /-30 Hercules, twenty-four CASA C-212-100/300 Aviocars, also used for transport, SAR, ECM and EEZ duties, and seven C295s. A Dassault Falcon 20 is used for calibration duties, with three Falcon 50s for VIP use. Training still uses sixteen TB-30 Epsilons and twenty-five Alphajets, as well as eighteen Alouette III helicopters. Armor X7 UAVs are in use.

Missiles in use include AMRAAM, Sidewinder and Sparrow AAM, with Paveway II, Maverick and Harpoon bombs and ASM.

PORTUGUESE NAVAL AVIATION/*MARINHA PORTUGUESA*

Founded: 1993

Portugal re-established a naval air arm in 1993 with the delivery of five Westland Lynx Mk95 helicopters to form the Naval Helicopter Squadron, *Esquadrilha de Helicopteros de Marinha*, to operate from three Vasco da Gama-class (MEKO 200) and two Kortenaer frigates, while seven corvettes have helicopter landing platforms. Additional Super Lynx helicopters have been introduced and the Lynxes have been upgraded to Super Lynx standard.

PORTUGUESE ARMY AVIATION

Founded: 2001

Portuguese army aviation returned in 2001 with the delivery of nine Eurocopter EC635s, and both the French and German armies provided assistance in training and integration of the helicopters into the Army, which later took over the *FAP*'s FTB337Gs. Ten NH90TTHs are entering service.

QATAR

- Population: 883,285
- Land Area: 4,000 square miles (10,360 sq.km.)
- GDP: $90bn (£57.5bn), per capita $108,138 (£69,124)
- Defence Exp: $1.75bn (£1.1bn)
- Service Personnel: 11,800 active

QATAR EMIRI AIR FORCE

Qatar has developed a small but potent air force, initially built around Alphajet armed-trainers and Dassault Mirage F1s, with Westland Commando (S-61) and Eurocopter Puma transports. The Qatar Emiri Air Force was active during the Gulf

War in 1991 for the liberation of Kuwait, in addition to providing base facilities for other Coalition air forces. The Mirage F1s were sold to Spain in 1997 following the delivery of Mirage 2000-5 multi-role fighters.

Current strength is 1,500 personnel, down almost a third since 2002. One squadron operates twelve Mirage 2000-5EDA/-5DDAs, with another operating six Alphajets. An attack helicopter squadron operates eleven SA342L Gazelles equipped with HOT anti-tank missiles and another operates in the anti-surface vessel role with eight Westland Commando Mk3s (S-61) equipped with Exocet ASuW missiles. Three Commando helicopters provide transport, with one for VIP duties alongside an Airbus A340-200, a Boeing 707-320, a 727-200, and two Falcon 900s. Other transport aircraft include four C-17s and four C-130Js.

Missiles include MICA and Magic AAM and Exocet, Apache and HOT ASM.

ROMANIA

- Population: 22.2 million
- Land Area: 91,671 square miles (237,428 sq.km.)
- GDP: $180bn (£115bn), per capita $8,087 (£5,169)
- Defence Exp: $3.39bn (£2.16bn)
- Service Personnel: 73,350 active, plus 45,000 reserves

ROMANIAN AIR FORCE/*FORŢELE AERIENE ROMÂNE*

Formed: 1945

The Romanian Army formed a Flying Corps in 1910. By late 1911, it was operating four Bleriots and four Henri Farmans, to which a small number of Bristols and Morane Type Fs were added the following year, as well as some Nieuports before the outbreak of World War I. The Central Powers soon overwhelmed the Romanian Flying Corps.

Post-war, a Directorate of Army Aviation was established, with a planned strength of three groups, each of three squadrons. The fighter group operated Spad S7C1s, the bomber group de Havilland DH9s and Breguet Br14Bs, with Breguet Br14A2 and Brandenburgs in the reconnaissance group. The three groups had a total of seventy-two aircraft. During the late 1920s, seventy Armstrong-Whitworth Siskin III and sixty Spad S61C1 fighters, 120 Potez XXV and thirty XXVII reconnaissance aircraft, as well as Breguet Br19B2 bombers and Savoia S59 flying boats, with Morane-Saulnier MS35 trainers, were obtained. Many of the Potez and Morane-Saulnier aircraft were assembled in Romania. The early 1930s saw the Romanian-designed SET XV fighter, SET VIIK reconnaissance aircraft, SET VII and SET X trainers enter service with fifty PZL P-11b fighters and twenty

Consolidated Fleet Model 10G trainers. These were followed in 1936 by Miles Hawk and Nighthawk trainers, as well as the Romanian IAR 37, 38 and 39 light bombers. Hawker Hurricane fighters and Bristol Blenheim bombers were bought from the UK in 1939.

Romania supported the Axis Powers during World War II. Germany provided Messerschmitt Bf109E and Heinkel He112B fighters, Heinkel He111H bombers and Junkers Ju87D Stuka dive-bombers, as well as Heinkel He114 reconnaissance seaplanes and Fiesler Fi156C liaison aircraft. Savoia-Marchetti SM79 bombers were built under licence and the Romanian aircraft industry provided the IAP80 fighter. Romanian forces joined Operation Barbarossa, the invasion of the Soviet Union. Later in the war, Germany provided Junkers Ju88A bombers and Henschel Hs129A-O ground-attack aircraft, but by this time the strategic situation had swung against the Axis, and in late August 1944, Romania was overrun by Soviet forces.

Post-war, the new Romanian Air Force continued to operate its wartime equipment, until the USSR supplied Yakovlev Yak-9 fighters. A 1947 peace treaty restricted Romania to 150 military aircraft and 8,000 air force personnel, but this was swept aside when the country joined the Warsaw Pact and started to receive Soviet military aid. The first jets, MiG-15 fighters, arrived in 1953 accompanied by Ilyushin Il-10 ground-attack aircraft, Lisunov Li-2 (C-47) transports, and Yak-11 and Yak-18 trainers. Within a couple of years, these aircraft were joined by MiG-17 fighters and Ilyushin Il-28 jet bombers, as well as the first helicopters, Mil Mi-4s. During the 1960s, MiG-19 and MiG-21 fighters and Antonov An-2 transports also entered service. Later aircraft included Ilyushin Il-12 and Il-14 transports, and Aero L-29 Delfin trainers. Until the late 1960s, the *FAR* had kept to its personnel limit of 8,000, but combat aircraft numbers had reached 250. Continual modernisation of the *FAR* continued into the late 1980s, with the delivery of MiG-23s and then MiG-29s, An-24, An-26 and An-30m transports, and the inevitable Mi-8 helicopters. Licence-built Britten-Norman BN2A Islanders were also introduced in the utility role, while the Alouette III helicopter was built under licence as the IAR316. A number of Chinese aircraft were put into service as well, including the H-5R (Il-28) light bomber.

The *FAR* was seriously affected by the collapse of the USSR and the end of the Warsaw Pact in April 1991. By this time, it had twelve air defence regiments flying MiG-21s, MiG-23s and MiG-29s, while two ground-attack regiments operated the Yugoslav and Romanian joint venture IAR-93 Arao. The Romanian aircraft industry provided the IAR-28M training aircraft, and for advanced training, the IAR-99 Soim to replace the L-29 Delfin. The MiG-23s were retired, with the MiG-21s and MiG-29s upgraded. Transport aircraft delivered in recent years have included ex-USAF Lockheed C-130B Hercules, and additional aircraft have been obtained since to replace the An-24 and An-26.

The *FAR* has 9,700 personnel, down from more than 18,000 since 2002, but average annual flying hours have increased from 40 to 120 over the same period.

Fighter and ground-attack aircraft consist of six squadrons operating a total of seventy-two MiG-21 Lancer A/B/Cs. There are up to 100 IAR316 Alouette IIIs on attack and utility duties, although the former probably uses just sixteen helicopters. Transport is provided by five Lockheed C-130B/H Hercules, four An-26s and an An-30. There are sixty Romanian-built IAR330H/L Puma helicopters in the tactical transport role, with seven IAR316Bs (Alouette III). Three SA365N Dauphins provide VIP transport. Training uses twenty-one IAR99 Soims, many of which have been upgraded, twelve IAK-52s and ten An-2s. Shadow 600 UAVs are operated.

Magic, Python, *Aphid*, *Archer* and *Atoll* AAMs are in service.

ROMANIAN NAVAL AVIATION

The Romanian Navy has a small number of helicopters capable of operating from three Regele Ferdinand-class (ex-UK Type 22) frigates and two of the four corvettes. The aircraft consist of two IAR330 Puma ASW helicopters and eight IAR316 Alouette IIIs, with only the latter capable of shipboard use.

RUSSIA

- Population: 140 million
- Land Area: 6,501,500 square miles (16,838,885 sq.km.)
- GDP: $1.37tr (£0.87tr), per capita $9,806 (£6,268)
- Defence Exp: $41.05bn (£26.24bn)
- Service Personnel: 1,027,000 active, plus 20,000,000 reserves

The core of the Union of Soviet Socialist Republics, on the break-up of the USSR Russia attempted to tie the other parts of the union to itself by forming the Commonwealth of Independent States, CIS, with varying degrees of success. The collapse of the USSR also brought problems for the military, with the widely spread armed forces having to be brought back to the mother country, which lacked the barracks and other facilities to accommodate them. Greater freedom and the growth of capitalism also meant that the artificial economy created by a closed society and dictatorship could no longer work, and the cost of such vast armed forces had to be reduced.

The result was that for more than a decade personnel numbers dropped and equipment serviceability also fell, with much equipment in store and that still in service becoming increasingly obsolete. In recent years, however, there has been a determined effort to rebuild the armed forces and re-equip them. This was made possible by the rapidly rising price of oil and gas, but the worldwide recession of 2008–2009 saw energy prices fall and this will have made the continued redevelopment of Russia's military might more difficult, but only for a relatively brief period. A warning to the West and to the country's neighbours lies in the fact that the Russian armed forces total 1,027,000 personnel in a country with a population of just over 140 million; a ratio of almost 0.75 per cent against ratios of around 0.33 per cent in the UK and 0.50 per cent in the United States. Many other countries in the West do far worse than the UK. The loss of its external commitments also means that the Russian forces are far less stretched than those of the United States, which has a truly global role and on which far too much now depends.

Then there are 20 million personnel in the reserve forces, while those in the West have also been reduced and in some cases have all but disappeared, including those of the British Royal Navy and Royal Air Force.

The question is whether Russia needs this massive strength because of nervousness about the intentions of others, aggravated by the trauma of the German invasion in 1941, or whether it is intended to intimidate. The old Soviet Union has gone, but relationships with neighbouring states are uneasy. For the moment, Russia is using its control of gas supplies to inflict its will on its neighbours, but it is worth remembering that expansion beyond the country's borders started long before Communism and the Revolution.

RUSSIAN MILITARY AVIATION

Formed: 1924

The structure of Russian military aviation remains unique, a continuation of the structure used in the former Soviet Union, with separate air forces for long-range aviation, frontal aviation (as opposed to the Army's own air elements), air defence and military air transport. These are in addition to the distinct elements of naval aviation, army aviation and border guard aviation.

The Imperial Russian Army had been using balloons for observation purposes since the turn of the century and a Central Aviation School was established in 1910. The Imperial Russian Navy also established a flying school later that same year. In 1912, the Imperial Government bought small quantities of British and French aircraft for the Army, and Curtiss flying boats for the Navy. Licences were obtained for production of these aircraft so that on the outbreak of war in 1914, the Imperial Russian Flying Corps had in service 244 aircraft of Bristol, Farman, Morane and Nieuport design. The Imperial Russian Navy also had a substantial number of aircraft. The Russian aircraft industry mainly comprised firms with other interests, such as the Russo-Baltic Wagon Factory, and was incapable of meeting wartime demand. Some indigenous designs were put into production in small batches during 1915, notably Igor Sikorsky's RBVZS-16, RBVZS-17 and RBVZS-20 single and twin-seat fighters. Later, other types entered production, including the Lebed-7 and Lebed-10 fighters, and Lebed-212, *Anasal* and *Anade* reconnaissance aircraft for the IRFC, while Dmitrii Grigorovich produced the M-5 and M-9 flying boats for the Navy. Most ambitious was Sikorsky's *Ilya Mourametz*, or 'Giant', a four-engined heavy bomber originally developed as an 'airbus' from his earlier twin-engined *Bolschoi*, or 'Grand', of 1913. By late 1914, sufficient of these aircraft were available to form the 'Squadron of Aerial Ships', *Eskadrilya Vozdushnykh Korablei*, in Poland, and in February 1915, one of the squadron's aircraft dropped a 600-lb bomb on German forces near Plotsk. A total of seventy-two Giants were built, and were able to slow the German advance.

While Russian aircraft production reached almost 1,800 aircraft and some 660 engines during 1916, foreign designs predominated, including Morane MB and Parasol, Nieuport 11, 17 and 21, Spad S7, Sopwith 1½-Strutter, BE2E and RE8 fighters, Voisin, DH4 and DH9 bombers, and Farman trainers. Most of the aircraft in IRFC service were French, but in 1916, Britain supplied more than 250 aircraft. War on the Eastern Front differed from that on the Western Front in having relatively little fighter activity, and the IRFC found itself mainly engaged in reconnaissance and bombing, with some close-support strafing.

Probably the most loyal of the Russian armed forces, the IRFC continued to function throughout the initial period of transition from the Tsarist regime to the Provisional Government. The rapid deterioration in the situation after the revolution in November 1917 meant that many aircraft fell into the hands of revolutionaries. In December 1917, the revolutionaries started to establish their own aircraft and balloon units, leading to the formation of the Workers' and Peasants' Red Air Fleet in May 1918, retaining the title of Red Air Fleet until 1924, although officially becoming part of the Army, *Aviadarm*, in the meantime. By the end of the revolution and the Russian Civil War, the Red Air Fleet had some 300 aircraft, and over the next few years substantial numbers of foreign aircraft were ordered, including Fokker DXIII fighters and CIV reconnaissance aircraft, and a number of Ansaldo types. DH9A bombers were copied and produced in Russia as the R1. Several German designs were produced as the Treaty of Versailles had banned military aircraft production in Germany.

The Soviet Military Aviation Forces, *Sovietskaya Voenno-Vozdushnye Sily*, were formed in 1924, absorbing the Red Air Fleet. The first post-revolution Soviet designs included the Polikarpov I-1 and I-2 and Grigorovich I-1 and I-2 fighters, the confusing designations arising because the 'I' prefix denoted a fighter design. Later it became Soviet practice to assign designers' names rather than those of the manufacturer to aircraft. Other aircraft included the Tupolev TB-1 bomber, R-3 and R-6 reconnaissance aircraft, and G-1 transport. These were followed by the Polikarpov I-3 and Tupolev I-4, Polikarpov-Grigorovich I-5, Grigorovich I-6 and DI-3 fighters; Polikarpov TB-2 bomber, R-5 reconnaissance aircraft and V-2 training aircraft; and Tupolev G-2 transport. By 1930, the *SV-VS* had 1,000 aircraft in twenty air regiments, and had experienced combat during border clashes with China in 1929.

The 1930s saw large orders for heavy bombers and the development of mass drops by paratroops, using Tupolev's G-2 transport. Heavy bombers included the Tupolev TB-2 and the Kalinin K-7, while Tupolev also designed the SCh-1 and SCh-2 ground-attack aircraft, the SB-2 light bomber and the I-14 fighter. Ilyushin designed the TB-3 bomber, and Polikarpov the I-15, I-15B and I-16 fighters.

Despite the emphasis on heavy bombers, Soviet military doctrine was similar to that of the Germans, with air power used in close support of ground forces rather than strategically, as with British and American forces.

In 1936, *SV-VS* personnel and aircraft were sent to Spain to fight on the Republican side during the Spanish Civil War, with some 1,400 of the latest Soviet fighter and bomber aircraft deployed in Spain, many of them flown by Spanish pilots. Their performance has since been claimed to have been as good as that of the German and Italian opposition flying with the Nationalist side, although the Republicans lost the war! The explanation could be that the first generation of German aircraft deployed in Spain were soon shown to be inferior, but the second generation proved to be far more effective. This period was also marked by the Soviet dictator Stalin's purges, with Andrei Tupolev imprisoned and his ANT-42 heavy bomber design, which entered service as the TB-7, later passing to Petlyakov and becoming the Pe-8, while Tupolev's light bomber design, the SB-2, became the Pe-2.

Warfare flared up in 1938 and 1939, fighting Japanese forces and eventually pushing them out of Mongolia with heavy Soviet casualties. This conflict used the same types of aircraft sent to Spain. In addition, from 1937 onwards, large numbers of Soviet aircraft were provided for Chinese Communist forces, often with 'volunteer' pilots and ground crew. The *SV-VS* had by this time grown to 6,000 aircraft, but its efficiency was affected, in line with the rest of the Soviet forces, by the regular purging of senior officers on political grounds, with subsequent imprisonment and often execution. In common with the rest of the Soviet forces, during the Communist period each unit had a political commissar as well as the more usual commanding officer. Examples of US and German aircraft were obtained during this period to gain an insight into rapidly changing aviation technology, and this was aided further in 1938 by a Russo-German agreement which provided a number of Messerschmitt, Dornier and Heinkel aircraft. The Polikarpov I-17 fighter was being introduced in limited numbers when war started in 1939, and was accompanied by ground-attack and dive-bombing aircraft, including the Polikarpov VIT-1 and VIT-2, the Ilyushin Il-2, the Archangelski Ar-2, and the Tupolev R-10 reconnaissance-bomber.

A Russo-German pact allowed Soviet forces to invade Eastern Poland in September 1939, after German forces had all but crushed Polish resistance. Later that year, Soviet forces invaded Finland, which had seized its independence during the revolution and civil war. Determined Finnish resistance and poor Soviet planning meant that the war ended in 1940 with Finland making territorial concessions but retaining her independence, despite the fact that 2,000 Soviet aircraft had been deployed against Finland's numerically inferior forces. New aircraft types continued to enter service, including Lavochkin I-22 and Yakovlev I-26 (which became the Yak-1 shortly afterwards when designations were changed to reflect the design bureau rather than the aircraft's role), Yatsenko I-28, Mikoyan I-61 (MiG-1) and I-200 (MiG-3) fighters; Sukhoi Su-1 strike aircraft; Ilyushin DB-3 (Il-4), Petlyakov Pe-2 and Pe-8, and Yermolaev Yer-2 and Yer-6 bombers; Antonov An-7 transport and SS-2 liaison aircraft.

The number of aircraft actually entering service remained small. By 1941, the Soviet armed forces had some 18,000 aircraft, but just a fifth of these could be regarded as modern. Attempts to boost production were hampered both by the Soviet system and by the very necessary movement of the key factories eastwards, out of the range of German medium bombers, in anticipation of a possible war with Germany. In June 1941, Germany launched the invasion of the Soviet Union, Operation Barbarossa, using the well-tried *blitzkrieg* strategy of combining air power with fast-moving armoured formations. Half of the USSR's aircraft were deployed in the West, in anticipation of a German attack, and against these 9,000 aircraft, the *Luftwaffe* had 1,945 aircraft, joined by another 1,000 aircraft from Germany's allies, mainly Italy, Hungary and Romania, as well as some from Finland and Croatia. The *Luftwaffe* caught most of the Soviet airfields by surprise, striking at the sixty-six airfields that accommodated 70 per cent of Soviet air power in the West, although as many as 50 per cent of the Soviet aircraft are believed to have been non-operational. One Soviet commander committed suicide after losing 600 aircraft without making any impact on the invading German forces. Individual pilots tried desperate measures, even ramming German bombers by inserting their fighter's propeller into the elevators.

Only a determined strategic heavy bombing campaign backed by strong and well-deployed fighter defences, plus tactical aircraft such as dive-bombers and tank-busting rocket-firing aircraft could have stopped the invasion. Instead, it was the severity of the Russian winter and the over-stretched logistics of the Germans that proved to be the best defence. German technical superiority meant that it took massive American and British military aid to revive the *SV-VS*. Even here, inefficiency ruled, with fighters delivered at great cost in the infamous Arctic Convoys being left idle on airfields rather than deployed quickly to operational units. The USA supplied Bell P-39N Airacobra and P-63C Kingcobra, Curtiss P-40 and Republic F-47 Thunderbolt fighter-bombers; Douglas A-20 ground-attack aircraft; North American B-25 Mitchell bombers and AT-6 trainers; Consolidated PBY-5A Catalina amphibians; and Douglas C-47 transports, which were also built in the USSR as the Lisunov Li-2. The UK provided Hawker Hurricane and Supermarine Spitfire fighters; de Havilland Mosquito fighter-bombers; and Armstrong-Whitworth Albemarle AOP aircraft. Between them, the two Allied powers supplied some 15,000 aircraft, helping the USSR to go onto the counter-attack and carry the offensive to Berlin in 1945.

It was not until late in the war that new and improved Soviet aircraft entered service. These included the Yakovlev Yak-3 and Yak-9, Lavochkin La-7 and La-9, and Mikoyan MiG-5 fighters; Tupolev Tu-2 bombers and Ilyushin Il-10 ground-attack aircraft. Despite heavy losses, the *SV-VS* had 20,000 aircraft when peace returned. Despite earlier conflict with Japan in Mongolia, the USSR did not declare war on Japan until August 1945, and the *SV-VS* played no part in Japan's defeat.

Captured German airframes and engines were shipped to the Soviet Union, and the Yakovlev Yak-15 and Mikoyan MiG-9 jet fighters were designed in great haste, and followed by the Yakovlev Yak-17. It was not until a number of Rolls-Royce

Nene and Derwent jet engines were purchased from the UK, and copied in the USSR, that the famous MiG-15 fighter could be designed, based on captured German airframe plans. Boeing B-29 Superfortresses, that had force-landed on Soviet territory in the mistaken belief that this was still friendly, were seized and copied, entering production and service as the Tupolev Tu-4. Other, less well known and less successful, aircraft of this period included the Yakovlev Yak-23 and Lavochkin La-15 jet fighters, and the Tupolev Tu-77 twin-jet attack aircraft, from which the Tu-12 and Tu-14, and eventually the Ilyushin Il-28, jet bombers were developed. The famous Antonov An-2 biplane utility transport first flew in 1947.

An opportunity arose for the Soviet Union to test its new aircraft with the outbreak of the Korean War in 1950. With covert Soviet and open Chinese support, the North Koreans had invaded South Korea. The USSR was boycotting the UN Security Council at the time, enabling the United Nations to muster an international force, but with the United States providing the largest contingent, to drive the North Korean forces back. The MiG-15 was matched against British Gloster Meteor and North American F-86 Sabre jets, as well as North American F-51D Mustangs, Fairey Fireflies and Hawker Sea Furies, with a Sea Fury accounting for a MiG-15. The MiG-15 was to survive in production as the MiG-15UTI, initially meant as a conversion trainer but in fact an advanced jet trainer, when its deficiencies led to its replacement in service by the MiG-17 from 1952 onwards. Deliveries of aircraft to satellite countries, those supposedly liberated by Soviet forces during the closing stages of the war, resumed and enabled the Russian aircraft industry to mass-produce even the most mundane designs. Some countries, notably Poland and China, were allowed to build Soviet types under licence. The less important client states, especially those in Africa, received withdrawn *SV-VS* aircraft.

A Sukhoi experimental aircraft was the first Soviet aircraft to break the sound barrier. In 1955, the supersonic MiG-19 fighter was introduced. This was joined in the *SV-VS* by the Yakovlev Yak-25 night-fighter, the Tupolev Tu-16, the world's only long-range turboprop heavy bomber, and the Tu-20 jet bomber, Ilyushin Il-12 and Il-14 transports. Helicopters were also introduced at an early date, with the Mil Mi-1 helicopter of 1950 and the Mi-4 of 1951, the Yak-4 twin-rotor helicopter of 1954, the Mi-6 heavy-lift helicopter of 1958, and later by the Mi-8 and Mi-10 helicopters. Heavy bombers included the Myasishschev Mya-4, introduced in 1955. Antonov produced the An-12 transport. The MiG-21 interceptor first entered service in 1958, at the same time as the Sukhoi Su-7 ground-attack aircraft, with both types rapidly becoming standard equipment for the air forces of the Warsaw Pact, while the Su-9 was an interceptor development of the Su-7.

The steady flow of new aircraft into service, often unannounced and with their true designations not always immediately apparent, led NATO to resurrect the system of code-names, first used during war to identify Japanese aircraft in the Pacific. Under this scheme, fighters and fighter-bombers were allocated code-names beginning with 'F'; bombers gained code-names beginning with 'B'; maritime-reconnaissance aircraft began with 'M'; transport code-names began with 'C' and helicopters with 'H'. Sub-types were given a suffix after the codename, usually indicating an improved version of an original aircraft. This meant that the MiG-15 was identified as *Fagot.* Some improved and special versions of an aircraft retained the original designation, as with the Mi-8 and the Mi-17, and the Mi-24 and Mi-35. Details of these code-names are given in NATO Designations for Russian Aircraft and Missiles, page 4.

Although there may be doubts about the serviceability of many Soviet aircraft, and especially the transports, the vast resources put into defence production at this time ensured that NATO forces were always outnumbered. In 1968, the Mach 3 MiG-23 interceptor started to enter service, at about the same time as the supersonic Tu-22 rear-engined bomber. The Soviet regime also ensured that the largest transport aircraft were provided and the helicopters with the heaviest lift, while the Mi-24 *Hind* attack helicopter set a standard for other nations to follow. There were a number of idiosyncrasies in both aircraft design and in operations. Many Soviet aircraft, including transport aircraft, retained the tail-gunner's position long after this disappeared from Western bombers. There was continued interest in the flying boat and amphibian, long after these were almost abandoned in the West. The Mi-24 differed from other attack helicopters through having space for passengers or freight, even though this made the machine heavier and larger, increasing its radar signature and reducing manoeuvrability. Operationally, under the Soviet regime the functions of the military air transport fleet and that of the state-owned airline monopoly *Aeroflot* were often overlapping, with the airline organised on military lines. The structure evolved differently from that used elsewhere, with Soviet air power divided between the air force, long-range aviation, *Dalnaya Aviatsiya,* frontal aviation, *Frontovaya Aviatsiya,* air defence aviation, and military transport aviation, *Voennaya Transportnaya Aviatsiya,* as well as the more usual separation of naval, army and paramilitary border guard aviation. These divisions survived the collapse of the Soviet Union, although in 1998 the anti-aircraft defence force, *PVO,* was merged into the air force, or *VVS.*

For the most part, the Soviets used air power sparingly in suppressing dissent in the European satellite territories, preferring to use tanks and infantry, and using air power mainly to transport troops. Soviet intervention in the internal affairs of Czechoslovakia – the so-called 'Prague Spring' – in 1968 were a case in point, with air power including the air-landing of troops at Prague, having deceived the local air traffic controllers into believing that these were civilian transport aircraft. The Cold War period saw massive elements of Soviet air power stationed at forward bases in East Germany and Czechoslovakia, and in northern Russia, with long-range aircraft used to probe Western air defences and to monitor NATO naval exercises. As the early post-war alliance with Communist China fell apart, protection of the border also fell to Soviet air power, while a presence was maintained in Siberia to balance US forces in Japan. The use of client states and support for guerrilla movements in Africa, Asia and Latin America bypassed direct confrontation between the USSR and its satellites and the network of Western alliances, of which NATO was the most important, largest and most organised, and most successful.

The big exception to these policies arose in Afghanistan in late 1979. Civil war in Afghanistan threatened to destabilise neighbouring Soviet states with substantial Moslem communities. On 24 December 1979, an advance party of the 105th Guards Airborne Division was flown into Kabul, followed by the main party over the next two days, with 280 missions by Antonov An-12 and An-22 and Ilyushin Il-76 transports to land some 5,000 troops. In the heavy fighting that followed, close air support was provided by MiG-21 and MiG-23 fighter-bombers operating from bases in the Soviet Union, while later the Mi-24 *Hind* helicopter came to prominence for the first time in frequent battles with Afghan guerrilla groups. Helicopters proved invaluable, not only in providing heavy fire-power quickly whenever and wherever it was needed, but in reducing the time and the number of Soviet casualties taken in seizing high ground. Nevertheless, even with modern weapons, Afghanistan proved to be difficult and costly to control, and eventually the USSR was forced to withdraw its forces, speeded by the collapse of the Soviet state.

Meanwhile, continued technical progress saw the Tu-95 bomber enter service, armed with the AS-15 cruise missile, and later these were joined by a very small number of Tu-160 Blackjacks. The Ilyushin Il-76 transport also boosted Soviet capability, and was joined by the giant An-124, capable of carrying loads of more than 150 tons and originally intended to transport ballistic missiles. The advent of the nuclear submarine capable of carrying ballistic or cruise missiles did not undermine the maintenance of strong bomber forces during the Soviet era. Throughout this period the USSR provided every form of offensive weaponry in contrast to almost all other states which were forced by economic conditions to choose between the means of strategic warfare.

The collapse of the Soviet Union in 1990 was a triple blow to the Soviet Air Force. It lost manpower and bases as states broke away and proclaimed independence. It lost aircraft, as many were abandoned in airfields in the newly independent countries. Finally, the economic problems that followed during the switch from a Soviet style command economy to a capitalist market economy saw funding cut. The sudden interruption to the hitherto steady supply of new and more sophisticated aircraft meant that the average age started to increase, a situation that was soon aggravated by difficulties in the supply of spare parts. Within ten years, only a fifth of the aircraft could truly be described as modern, while total flying hours fell by 90 per cent from around 2 million annually to around 200,000. Among the few bright spots in this situation was an agreement with the Ukraine that resulted in the return of eight Tu-160 bombers during 1999 and 2000, enabling the fleet to grow to just fifteen aircraft, while the Ukraine also returned a number of Tu-95s.

In recent years the Russian Air Forces, in common with the rest of the Russian armed forces, passed through a period of low operational readiness, while the period of conscription has been steadily cut and is now around twelve months. Typical of this was that Russian intervention in Kosovo in 1999 was aided by the movement of troops by air, but once in the area, logistics support fell apart in circumstances where an air re-supply operation was needed. Combat aircraft, mainly Su-24s, are believed to have been used in operations against rebel forces in Chechnya during 1999–2001. More recently, however, there has been evidence of improving operational awareness, while the economy has been boosted by growing demand and higher prices for gas and oil.

The total number of personnel for all of Russia's Air Forces is around 160,000, a 10 per cent drop since 2002, but excluding naval and army aviation. Average annual flying hours have improved from a low of just twenty hours for pilots in long-range aviation and air defence; and even the military transport units, or *VTA*, only managed forty-four hours, a point at which safety is compromised, to between 80 and 100 hours.

Long-Range Aviation Command, also known as the 37th Air Army, consists of two heavy bomber divisions with four non-strategic regiments and another four that are strategic, and monitored under the Strategic Arms Reduction Treaty, START, with 116 Tu-22M-3/MR *Backfire C*, supported by twenty Il-78/78M *Midas* tankers and thirty Tu-134 *Crusty* transports for training. This part of the Russian Air Forces has the highest annual flying hours, at 80 to 100 hours.

Air Defence Aviation was at one time split from Tactical Aviation, but has since been reformed. Tactical Aviation has a bomber division and thirteen ground-attack regiments with a total of 550 Su-24/-24M2 *Fencer*, as well as seven ground-attack regiments with 240 Su-25A/SM *Frogfoot* and one regiment with sixteen Su-34P *Fullback*. Nine fighter regiments have 188 MiG-31 *Foxhound* and another nine have 226 MiG-29 *Fulcrum*, many of which are currently being upgraded, while another six regiments have 281 Su-27/-27SM *Flanker*, again many of which are being upgraded. Reconnaissance is by four regiments with forty MiG-25R *Foxbat* and five with seventy-nine Su-24MR *Fencer*. Conversion training units use thirty MiG-25 *Foxbat*. The force also includes twenty A-50 *Mainstay* AEW aircraft and sixty Mi-8 *Hip* ECM helicopters. UAV include Pchela, Albatross and Expert. Perhaps surprising is that annual flying hours are believed to be between twenty-five and forty.

There are thirty-five SAM units with around 2,000 SA-10 *Grumble* launchers and the new SA-20 *Triumph*, the first of which were deployed around Moscow. AAM in use include *Alamo*, *Aphid* and *Archer*, with *Kedge, Kerry, Kickback, Kent* and *Kitchen* ASM.

Military Transport Aviation, the *VTA* or *Voennaya Transportnaya Aviatsiya*, also known as the 61st Air Army has been modernised in recent years, replacing many of its fleet of 250 An-12s. It also makes use of civilian aircraft as necessary. It consists of nine regiments with fifty An-12s still in service as well as 210 Il-76M/MD/MF *Candid* and a heavy-lift division with twelve An-124 *Condor* and twenty-one An-22 *Cock*.

Training uses around 1,000 aircraft, including L-39, and conversion training versions of front-line types.

INDEPENDENT NAVAL AVIATION

The Imperial Russian Navy's involvement with aviation started in 1910. In 1912, Curtiss flying boats were purchased from the United States, and by 1914 it had a force of around sixty flying boats. Grigorovich M-5 and M-9 flying boats entered service during the war, but operations were hampered by units of the Imperial Russian Navy being among the first to join the Russian Revolution in 1917.

After the revolution, naval and army aviation were merged into the air force. During the early 1920s, Tupolev TB-1P seaplane bombers entered service, and these were followed by Tupolev MDR-2 long-range flying boats and Beriev MBR-2 short-range flying boats. Shortly before the outbreak of World War II, the six-engined Tupolev MK-1 flying boat entered service. An independent naval air fleet was created on the eve of the outbreak of war. Although no aircraft carriers were obtained, shore-based fighter and bomber units existed until 1960, when these units were transferred back to air force control. The Soviet Navy was the forgotten service for much of its history, and it was not until the late 1950s and early 1960s that it was given the resources to gain a world role. Two Moskva-class helicopter cruisers were introduced in 1968 and 1969, each capable of carrying up to eighteen anti-submarine helicopters. These were followed by the first of three Kiev-class aircraft carriers in 1976, each capable of carrying up to thirty-five helicopters and V/STOL aircraft, the new Yakovlev Yak-25 *Forger A* strike aircraft. A modified Kiev-class carrier, *Admiral Goshkov*, followed, but the Yak-25s were withdrawn in 1992, leaving the carrier to operate only helicopters. The first Russian aircraft carrier with an angled flight deck, *Admiral Kuznetsov*, joined the fleet, and can operate up to eighteen Sukhoi Su-33 fighters, Russia's most modern fighter and a navalised version of the Su-27K, as well as Su-25 ground-attack aircraft, and fifteen Kamov Ka-27 ASW and Ka-31 AEW helicopters. Despite the angled flight deck, this ship still has a ski jump.

Today, naval aviation accounts for around 35,000 of the Russian Navy's estimated 142,000 personnel, with the total down by about a fifth since 2002 although that for naval aviation has remained constant. Only two carriers remain operational, but cruisers and destroyers carry helicopters, while there are extensive shore-based aircraft to support the fleet. A French-built helicopter carrier has been ordered and further examples will be built in Russia.

Currently, air units are attached to the different fleets, which are Pacific, Northern, Black Sea and Baltic, with a Caspian Sea Flotilla.

The Northern Fleet has one of the two operational aircraft carriers assigned to it, but most of its aircraft are shore-based including thirty-eight Tu-22M *Backfire C* bombers, twenty Su-27 *Flanker* and ten Su-25 *Frogfoot* fighters and anti-shipping aircraft, with seventeen Il-38 *May* and fourteen Tu-142 *Bear* on MR duties, plus two An-12 *Cub* AEW aircraft, and twenty-five An-12, An-24 and An-26 transports. Aboard the ships are forty-two Ka-27 *Helix A* and sixteen Ka-29 *Helix B*, with fifteen Mi-8 *Hip* ashore on support duties.

The Pacific Fleet also has fourteen Tu-22s, with thirty MiG-31 *Foxhound A*, while the MR force consists of twenty-four Il-38s and twelve Tu-12s, again with An-12 AEW aircraft and An-26 transports. Aboard the ships are thirty-one Ka-27 and six Ka-29 helicopters, with twenty-six Mi-8s ashore.

The Black Sea Fleet is unusual in that its main base is actually in the Ukraine, and this has been a source of friction with the Ukraine, nervous about its independence having been re-absorbed by the USSR after breaking free during the Russian Civil War. The small air element includes eighteen Su-24 fighters and fourteen Be-12 MR aircraft, four An-12 and An-26 transports. There are thirty-three Ka-27 helicopters and nine Mi-8s, of which one is a transport and the others used on SAR.

The Baltic Fleet has twenty-three Su-27s and twenty-six Su-24s, while transport includes twelve An-12, An-24 and An-26, with two An-12s on MR duties. Helicopters include eleven Mi-24 *Hind* combat helicopters, nineteen Ka-27s and eight Ka-29s aboard the ships, and seventeen Mi-8 transports.

Strangely, the Caspian Sea Flotilla has the other aircraft carrier with eighteen Su-33 *Flanker* and four Su-25s embarked, as well as fifteen Ka-27 ASW and two Ka-31 *Helix* AEW helicopters. Other helicopters are available for the destroyers and frigates.

RUSSIAN ARMY AVIATION

Russian Army Aviation developed after World War II with helicopters closely integrated with ground forces. The force has almost 1,300 helicopters available, but many more are believed to be in store. There are twenty regiments of attack helicopters, of which the most modern are 300 Mi-28N *Havoc*, but there are still around 600 Mi-24 *Hind* and eight Ka-50 *Hokum*. A large number of regiments operate in the transport, utility and support role with around 600 Mi-8/-17 *Hip*, thirty-five Mi-26 *Halo* and eight Mi-6 *Hook*.

Independent of the Army, the border guard has its own aviation element, with some 200 Ka-27, Mi-8, Mi-17 and Mi-24 helicopters, as well as SM92 surveillance aircraft and seventy transport aircraft, including An-24, An-26, An-72, Il-76, Tu-134 and Yak-40 types.

RWANDA

- Population: 10.7 million
- Land Area: 10,169 square miles (26,338 sq.km.)
- GDP: $3.9bn (£2.49bn), per capita $237 (£151.5)
- Defence Exp: $76m (£48.6m)
- Service Personnel: 33,000 active

RWANDA AIR FORCE/*FORCE AÉRIENNE RWANDAISE*

Founded: 1962

Rwanda was a German colony until 1918, and then passed into Belgian administration until 1962. The *Force Aérienne Rwandaise* was established on independence, mainly for transport and liaison duties, but later attempted to buy Aermacchi MB326 armed trainers in 1972; the order had to be cancelled because of funding difficulties. An order for just three CM170 Magisters had to be cancelled three years later. Two Rallye 235 Guerriers were fitted with machine guns for COIN operations eventually. Civil war during the early 1990s reached a peak in 1994, when almost all of the *FAR*'s aircraft were destroyed. Afterwards, two Mi-24 attack helicopters were acquired from Belarus in 1997, and in 1999 Russia provided four Mil Mi-17MD transport helicopters. There have been reports that Belarus might supply MiG-21s, but there is no training or support infrastructure and the force's role is primarily that of transport.

Today the *FAR* has an estimated 1,000 personnel. Aircraft include a transport squadron with a Boeing 707, a BN2A Islander, two An-8s and an An-2, while another squadron has up to seven Mi-24 attack helicopters and up to twelve Mi-17s.

SAUDI ARABIA

- Population: 28.7 million
- Land Area: 927,000 square miles (2,400,930 sq.km.)
- GDP: $410bn (£262bn), per capita $14,316 (£9,151)
- Defence Exp: $41.2bn (£26.3bn)
- Service Personnel: 233,500 (inc. 100,000 National Guard) active

ROYAL SAUDI AIR FORCE

Founded: 1950

Saudi Arabia came into existence with the union of the Nejd and the Hejaz in 1926. The first Saudi military aircraft were a gift from the British government to assist in suppressing dissident tribesmen. These aircraft were followed in 1931 by four ex-RAF Westland Wapiti general-purpose biplanes, operated by British aircrew. Italian military assistance was received in 1937, but World War II intervened and there was little further development for the next decade. The arrival of a British mission in 1950 saw fresh progress, organising the new Royal Saudi Air Force and providing Avro Anson light transports and de Havilland Tiger Moth trainers. Further development followed in 1952 when an American mission provided ten

Temco TE-1A Buckaroo and a number of North American T-6 trainers, as well as Douglas C-47 transports. In 1953, eighteen Saudi pilots were given further training in the UK by Airwork and Air Services Training, while a Saudi-Egyptian Defence Agreement saw a number of pilots trained in Egypt, which also supplied four de Havilland Vampire FB52 fighter-bombers before the agreement ended in 1957. The RSAF had meanwhile acquired nine Douglas B-26 Invader bombers and a number of DHC-1 Chipmunk basic trainers. During the late 1950s and early 1960s, sixteen North American F-86F Sabre fighter-bombers were introduced, accompanied by Lockheed T-33 and Beech T-34 Mentor trainers, six Fairchild C-123 Provider and four Lockheed C-130E Hercules transports. A Westland Widgeon helicopter and a Vickers Varsity provided VIP transport.

Growing tension in the Middle East, with Saudi Arabia taking a different line from Iraq, Egypt and Syria, resulted in the creation of strong Saudi forces during the 1960s. In 1966, thirty-four BAC Lightning F53 multi-role aircraft were ordered, with six Lightning T55 and twenty-four BAC167 Strikemaster trainers. This marked the start of Saudi procurement of modern aircraft. The Lightnings were followed by Northrop F-5s for attack duties, while during the late 1980s and early 1990s Saudi Arabia became one of the few export customers for the Anglo-German-Italian Panavia Tornado, buying forty-eight interdictor and twenty-four air defence versions. The mainstay of Saudi Arabia's air defences became the McDonnell Douglas F-15 Eagle, backed by five Boeing E-3A Sentry AEW aircraft and eight KE-3A tankers. Iraq's invasion of Kuwait in August 1990 placed the RSAF on a high state of alert. The RSAF played an important role in the resulting war for the liberation of Kuwait, losing at least one Tornado IDS during strikes against Iraqi positions. The USAF accelerated delivery of F-15 Eagles to the RSAF, providing an extra twenty-four aircraft as well as stationing its own F-15s in the country.

Saudi Arabia's ability to maintain its relatively high level of defence expenditure has always been dependent on the price of crude oil, and a price fall during the late 1990s saw a number of modernisation projects shelved.

The RSAF has 20,000 personnel, having quadrupled in size over the past thirty years. It has nine fighter squadrons, of which one has fifteen Tornado ADVs, five operate eighty-four F-15C/D Eagles, and another three have sixty-eight F/RF-5E Tiger IIs. The strike role is handled by six squadrons, with three operating eighty-five Tornado IDSs and three operating seventy F-15S Eagles. AEW is provided by a squadron with five E-3A Sentries, while the tanker force now includes eight Boeing KE-3A and eight Lockheed KC-130H tanker transports with six A330s entering service which will replace the KE-3As. Three transport squadrons operate thirty-eight C-130E/H/H-30; twenty-nine C-130H Hercules; as well as three L-100-30HS CASEVAC aircraft, and four CN235s. Two helicopter squadrons operate seventeen AB212s, twenty-two AB205s, thirteen AB206As and sixteen AB412s, with the latter also undertaking SAR as well as transport, and there are twelve AS532A2 Cougars for combat SAR. The Royal Flight operates two Lockheed TriStars and two McDonnell Douglas MD-11s, a Boeing 737-200 and a 757-200, two 747SPs and a 747-300 and A340-200. In addition, a VIP unit has four

VC-130H Hercules, four BAe 125-800s, two Grumman Gulfstream IIIs and two AS61As. Training uses forty-three Hawk Mk65/As, forty-five Pilatus PC-9s, a BAe Jetstream 31 and twenty MFI-17 Mushshaks, as well as a number of JetRangers and conversion trainers for the combat aircraft. There are separate air defence forces. The RSAF uses Sparrow and Sidewinder AAM, while AGM include Maverick, Sea Eagle, AS-15 and AS-30.

ROYAL SAUDI NAVY

The Royal Saudi Navy became involved in aviation following the success enjoyed by Iraq and Iran in anti-shipping operations during the war between these two countries. Seven frigates can each operate SA365SA Panthers, of which there are twenty-one, mainly shore-based, or twelve AS532AL Cougars, six of which can carry Exocet anti-shipping missiles. Four Panthers are used for SAR, but the others have AS15TT anti-shipping missiles. Up to fifty NH90 could be ordered.

ROYAL SAUDI LAND FORCES ARMY AVIATION COMMAND

The Saudi Army, or Royal Saudi Land Forces, maintains a small but up-to-date aviation element, with twelve Boeing AH-64A Apache attack helicopters purchased in 1993 upgraded to AH-64D standard, supported by twelve S-70A Desert Hawks and fifteen Bell 406CS Combat Scouts equipped with TOW anti-tank missiles. Other helicopters include twenty-two UH-60A Black Hawks, of which four are used for CASEVAC, and six AS-365N Dauphin 2s.

SENEGAL

- Population: 13.7 million
- Land Area: 76,104 square miles (197,109 sq.km.)
- GDP: $13.5bn (£8.6bn), per capita $985 (£629)
- Defence Exp: $217m (£139m)
- Service Personnel: 13,620 active

SENEGAL AIR FORCE/*ARMÉE DE L'AIR DU SENEGAL*

The *AAS* was established after independence from France in 1965, and is mainly a transport and communications operation with a very limited counter-insurgency capability. Its initial force was the standard French post-colonial package of Max Holste Broussards, an Alouette II helicopter and a Douglas C-47 transport. Despite an attempt at a pact with Gambia, which is surrounded on three sides by Senegal, in 1981, Senegal and Gambia have followed their own ways in defence, and Gambia has not established an air arm. It is not clear whether the *AAS* still operates over Gambia.

Today, the *AAS* has 770 personnel. At one time it had five CM170 Magisters and three Rallye 235A Guerrier armed trainers, but these have now gone and the sole combat capability rests with two Mi-35 helicopters. EEZ duties are handled by a C-212 and two BN2T Islanders, while transport includes a VIP Boeing 727-200 and six F-27-400M Friendships, with two Mi-17s, two Mi-2s, two Alouette IIs and a SA341 Gazelle. Two Rallye 235s and two TB30s are used for training.

SERBIA

- Population: 7.4 million
- Land Area: 34,107 square miles (88,337 sq.km.)
- GDP: $47.4bn (£30.3bn), per capita $6,426 (£4,107)
- Defence Exp: $1.06bn (£0.67bn)
- Service Personnel: 29,125 active, plus 50,171 reserves

SERBIAN AIR FORCE AND AIR DEFENCE

Founded: 2006

The Serbian Air Force has been established from what had been the core of the Federal Republic of Yugoslavia Air Force, itself the rump of the former Yugoslav Air Force, and for a while was a joint service with neighbouring Montenegro. In 2002, it lost its autonomy and became one of the nine corps within the Army. The current service was created in 2006 on the separation of Serbia and Montenegro.

Yugoslavia was part of the Austro-Hungarian Empire, but independence came with the end of World War I. When the Yugoslav Army Aviation Department was formed in 1923, it was able to use Serb, Croat and Slovene officers who had flown with the Serbian Military Air Service, formed in 1912. Early equipment included Spad S-7C1 and Dewoitine D-1C fighters, Breguet Br19A/B2 reconnaissance-bombers, and a number of Brandenburg trainers built under licence. These aircraft were followed by Avia BH33 fighters, Potez XXV reconnaissance aircraft and Hanriot A32 trainers, as well as a number of H41 seaplanes for a naval co-operation flight. In 1930, semi-autonomy was attained as the Yugoslav Army Air Corps.

The YAAC continued to grow, and by 1935 had forty-four squadrons and 440 aircraft. Modernisation started with orders for Dewoitine D500 and Hawker Fury fighters, Bristol Blenheim I, Dornier Do17K and Savoia-Marchetti SM79 bombers, with some Dornier Do22 general-purpose aircraft. By September 1939, it was introducing Hawker Hurricane, Curtiss P-40B Tomahawk and Messerschmitt Bf109E fighters, and Me108 and Westland Lysander army co-operation aircraft. Yugoslavia was occupied by German forces in 1941, following a revolution that overthrew the monarchy and installed pro-Axis government. A number of personnel managed to escape, reaching the UK where Yugoslav squadrons flew Supermarine Spitfire fighters in the RAF. The Axis Powers partitioned Yugoslavia, with Italy governing the Croatian areas and forming a Croatian Air Force with Caproni Ca310, Fiat G50 and Messerschmitt Bf109G fighters, Morane-Saulnier MS40 and Dornier Do17 bombers, and some AOP aircraft.

After the war ended in 1945, Yugoslavia became a republic, initially within the Soviet sphere of influence, and the Yugoslav Air Force, *Jugoslovensko Ratno Vazduhoplostvo*, was formed. Initially its former RAF equipment was used, but Soviet aircraft, including Yakovlev Yak-3 and Yak-9 fighters, Ilyushin Il-2 and Il-10 ground-attack aircraft, Petlyakov Pe-2 bombers, Lisunov Li-2 (C-47) transports, and UT-2 and Po-2 trainers soon arrived. There were some 400 aircraft by 1950. The country was the first to break away from Soviet influence, remaining outside the Warsaw Pact although still a Communist state, and this encouraged the US to provide aid. During the 1950s, the *JRV* equipped with 150 Republic F-47D Thunderbolt and 140 de Havilland Mosquito FB6 fighter-bombers, plus a few Mosquito night-fighters. A nationally-designed fighter, the S-49A, entered service, but was soon withdrawn. The first jets, Lockheed T-33A trainers, were introduced in 1953, and were soon joined by 200 Republic F-84G Thunderjet fighter-bombers, Westland Dragonfly (S-51) helicopters, and both Douglas C-47 and Ilyushin Il-14 transports, as well as Aero 3 trainers. Canadair Sabre Mk2 and North American F-86D Sabres were obtained during the late 1950s, and in the 1960s, these were followed by MiG-21F interceptors, reflecting a policy of buying aircraft from both sides of the Iron Curtain. Soko Jastreb armed-trainers and Galeb jet trainers entered service, with Yugoslav manufacturers developing a niche for themselves in the trainer and armed-trainer market.

The 1970s and 1980s saw the *JRV* come to operate Soviet combat aircraft and transports, with Western helicopters. This situation reflected Western concerns about selling the most advanced combat aircraft to a Communist country, and the higher cost of Western aircraft. Additional MiG-21s replaced Western aircraft from the 1950s and early 1960s, but were not joined by anything more sophisticated until 1988, when the first MiG-29A/Bs were introduced.

The collapse of the Yugoslav Federation in 1991 and 1992 saw Serbia retain a firm grip on the assets of the former *JRV*, with little equipment passing to the breakaway states. The worsening economic situation, an arms embargo and, finally, in 1999, air strikes by the Western powers, all contributed to a weakening of the new FRYAF. The MiG-29 force of sixteen aircraft was destroyed, mainly on the ground but six aircraft were shot down. It was believed that 150 other combat aircraft were destroyed, but this is now thought to have been an exaggeration, with part of the air force operational.

The Serbian Air Force and Air Defence is still being developed from the country's share of the original combined force. It has 4,155 personnel. There is a composite squadron with five MiG-29s and ten MiG-21bis, but the main combat force consists of two squadrons with thirty-three J22 Orao armed trainers, augmented by up to fourteen G-4 Super Galebs. Transport is provided by another composite squadron with eight An-26s, two Yak-40s and two Do28s. Helicopters include ten Mi-8/-17s, while there may be two Mi-24 attack helicopters. Training uses fifteen UTVA-75s.

SEYCHELLES

- Population: 82,247
- Land Area: 140 square miles (404 sq.km.)
- GDP: $930m (£594m), per capita $10,754 (£6,874)
- Defence Exp: $8.51m (£5.4m)
- Service Personnel: 200 active, plus 450 paramilitary (includes coast guard)

SEYCHELLES COAST GUARD AIR WING

The Seychelles are more than 100 small islands in the Indian Ocean, with a small army which includes a coast guard with twenty personnel in its air wing, briefly known as the Seychelles People's Air Force before the present title was adopted in the late 1990s. At one time it had a Merlin III for maritime patrols and replaced this with a Cessna Citation V, but now it has three aircraft, a Cessna F406 Caravan II, a 152 and a BN2A Islander.

SIERRA LEONE

- Population: 5.1 million
- Land Area: 27,925 square miles (72,326 sq.km.)
- GDP: $2.38bn (£1.52bn), per capita $474 (£303)
- Defence Exp: $11m (£7m)
- Service Personnel: 10,500 active

SIERRA LEONE DEFENCE FORCE

Founded: 1973

Sierra Leone in West Africa became independent of the UK in 1961. It was not until 1973 that a defence force was established with Swedish assistance, including four Saab MFI-17 Safirs, a Hughes 269C and two Hughes 369 helicopters. These aircraft were joined by a VIP Bo105C helicopter in 1976. A Safir was soon lost, while most of the other aircraft were sold due to a shortage of money and spares. The Bo105C remained, but was not joined by other aircraft until two AS355 Ecureuil helicopters were obtained in 1983. By this time, the country was embroiled in internal unrest, with the government fighting the so-called Revolutionary United Front with the assistance of a South African mercenary organisation, Executive Outcomes, which used two Mil-17 helicopters to support its operations. The government also used an Mi-8 and an Mi-24 flown by Belarussian pilots, and in 1999 acquired two more Mi-24s from the Ukraine. Successive peace agreements throughout the 1990s were broken, and both the United Nations and neighbouring African states, notably Nigeria, have been attempting to maintain the peace. British troops had to be deployed in 2000, and have since been rebuilding Sierra Leone's own defence forces with an eventual strength of up to 14,000. It is believed that an Mi-24 and an Mi-8 remain operational.

SINGAPORE

- Population: 4.7 million
- Land Area: 225 square miles (580 sq.km.)
- GDP: $170bn (£108.7bn), per capita $36,454 (£23,302)
- Defence Exp: $8.23bn (£5.26bn)
- Service Personnel: 72,500 active, plus 312,500 reserves

REPUBLIC OF SINGAPORE AIR FORCE

Founded: 1965

Singapore withdrew from the Federation of Malaysia in 1965 and immediately established its own defence forces, while continuing strong defence links with Malaysia. The embryonic air force was known initially as the Singapore Air Defence Command, and its early equipment included twenty Hawker Hunter FGA9 ground-attack aircraft and sixteen BAC167 Strikemaster armed-trainers, as well as Alouette III helicopters and Cessna 172s for AOP and liaison. These were soon joined by Northrop F-5E/F Tiger II and, later, by ex-USN Douglas A-4S Skyhawk fighter-bombers with the first forty-seven Skyhawks refurbished and upgraded in the US, but a follow-on order for eighty-six aircraft was upgraded in Singapore. A limited MR capability was developed using Fokker 50 Enforcer 2s, with AEW using four Grumman E-2C Hawkeyes. Eight Lockheed F-16A/B Hornets were ordered in 1984, but today there are more than sixty F-16C/Ds. During 1986–89, Singapore upgraded its Skyhawks to 'Super Skyhawk' status. The F-5s were upgraded to F-5S standard in 1997.

An unusual feature of the RSiAF's training is that Singapore's limited territory has forced most of its training abroad, mainly in Australia and the United States. A Skyhawk squadron was based in France until 2001. The availability of training facilities does influence aircraft orders.

A substantial helicopter force has been developed, including Boeing CH-47D Chinooks and, currently being delivered, AH-64D Longbow Apache helicopters, augmenting a substantial Super Puma, Iroquois and Fennec operation. Tanker and transport aircraft also provide a longer reach than might normally be expected of a country of this size. Initial pilot training is provided by BAe Flying Training Australia. In 2000 Singapore joined the NATO Flying Training in Canada, NFTC, programme.

The RSiAF has 13,500 personnel of whom 3,000 are conscripts, up from 6,000 in 2002 when half were conscripts. It is likely to remain a substantial and well-equipped air force given the political instability of much of the surrounding area.

In the fighter and ground-attack role, there are six squadrons. Three squadrons are equipped with sixty F-16C/D Fighting Falcons, with some of the F-16Ds having conformal fuel tanks for longer-range strike. Two squadrons have twenty-eight F-5S/T Tiger IIs, being replaced by twenty-four F-15SGs, and another operates ninety-eight RF-5Ss in the reconnaissance role. AEW is provided by four Grumman E-2C Hawkeye. There are five Fokker 50MPA Enforcer 2 aircraft for ASW and EEZ duties, with another four 50UTA and five C-130H Hercules transports, and five KC-130B/H and four KC-135R tankers. Helicopters include sixteen Boeing CH-47D/SD Chinooks, of which several are based in the USA, eighteen AS332M/UL Super Pumas and twelve AS532UL Cougars for transport and SAR, five Bell 205s and a 412. Trainers include twenty-seven S211s based in Australia, as well as F-16D conversion trainers based in the USA. Currently, the RSiAF is responsible for all service aviation, providing aircraft for naval and army use as required, with six S-70B helicopters entering service with the Republic of Singapore Navy for operations from six Formidable-class frigates.

UAVs include Hermes and Searcher. Missiles include AMRAAM, Sparrow and Sidewinder AAM, and Shrike, Hellfire, Maverick, Harpoon and Exocet ASM.

SLOVAKIA

- Population: 5.5 million
- Land Area: 18,940 square miles (49,435 sq.km.)
- GDP: $91bn (£58.2bn), per capita $16,666 (£10,653)
- Defence Exp: $1.46bn (£0.93bn)
- Service Personnel: 16,531 active

REPUBLIC OF SLOVAKIA AIR FORCE

Founded: 1993

Slovakia became an autonomous state again in 1993 when the former Czechoslovakia split in two, with the former air force as a whole divided 2:1 in favour of the Czechs, the larger and more populous, as well as more prosperous, portion of the country. The 2:1 split applied to the overall strength, but varied slightly on particular items: the MiG-29 force was divided equally while the Czechs received all MiG-23s. Since then, the Slovaks have added eight extra MiG-29s in 1995–96 and retired older aircraft. Since becoming a member of NATO, the MiG-29s and Mi-24s have been upgraded with Western equipment, including IFF. There are plans to replace the Su-25s, Su-22s, MiG-21s, L-39s and L-29s with a single armed-trainer type to cut costs, with a requirement for up to fifty aircraft, once funding permits.

Currently there are 4,190 personnel, down from 10,200 in 2002, but annual flying hours have doubled to an average of ninety, although this is still too low. The fighter units operate twenty-two MiG-29A/UBs and twenty-one MiG-21s, while three Su-22M-4/Us equip a reconnaissance unit. There are sixteen Mi-24 attack helicopters, as well as seventeen Mi-8/-17 transport helicopters and four Mi-2s for liaison duties. Transport is provided by three L-410Ms and two An-26s, likely to be replaced by two C-27 Spartans. Training is provided on seven Aero L-39C Albatros.

SLOVENIA

- Population: 2 million
- Land Area: 7,796 square miles (16,229 sq.km.)
- GDP: $53bn (£33.9bn), per capita $26,343 (£16,839)
- Defence Exp: $879m (£561.9m)
- Service Personnel: 7,200 active, plus 4,500 paramilitary and 3,800 reserves

SLOVENIAN ARMY AIR ELEMENT

Founded: 1991

Formerly part of Yugoslavia, Slovenia has escaped the upheaval and fighting that affected much of the rest of that state. The armed forces are centred on the Army, which maintains maritime and air elements, the latter accounting for around 530 of the Army's 7,200 personnel. The initial role of the air element has been communications and transport support for ground forces, using aircraft based in Slovenia at the time of independence being declared in 1991. These were augmented by three ex-US Army Pilatus PC-9 armed-trainers, and a further nine

PC-9 Mk2s were later purchased. Despite speculation over the purchase of fighters, defence expenditure was cut by almost a third between 1999 and 2000, placing expensive procurement programmes in doubt.

Today, the combat aircraft are still twelve PC-9/9Ms, also used for training, while some of the four AS532 Cougars and eight Bell 412EPs are armed. There are also three AB206 helicopters, as well as two PC-6 Turbo-Porters and an L-410 Turbolet.

SOMALI REPUBLIC

- Population: 9.8 million
- Land Area: 246,000 square miles (637,658 sq.km.)
- GDP: No reliable figures available
- Service Personnel: Nil

SOMALIAN AERONAUTICAL CORPS

Founded: 1960

The former protectorate of British Somaliland gained independence in 1960 at the same time as the Italian Trust Territory of Somalia, the two amalgamating to form the present state. The Italian-sponsored air corps provided the basis for an air arm, which developed with Soviet aid. The Somalian Aeronautical Corps reached 2,000 personnel by the mid-1970s, by which time it was operating six MiG-17 and twelve MiG-15 fighter-bombers, six Beech C-45s, a Douglas C-47 and an Antonov An-24 in the transport role, with twenty Yakovlev Yak-11, ten Piaggio P148 and six MiG-15UTI trainers. Many of these aircraft soon became unserviceable, but were replaced by at least eight MiG-21s and, during a brief flirtation with China, more than twenty Shenyang F-6s (MiG-19). During the 1980s the country was courted by the West, and received military aid in return for making airfields and ports available to US forces. Aircraft were not provided by the USA, but the SAC was able to obtain at least nine Hawker Hunters, mainly FGA76 fighter-bombers, and some Britten-Norman BN2 Islander communications aircraft from Abu Dhabi.

In 1991, a revolution split the country into fiefdoms dominated by rival clans. From late 1992 until mid-1994, the United Nations attempted to restore the peace and provide aid to ward off the effects of a severe famine, largely using forces supplied by the USA and Italy. These forces had to be withdrawn as the position within the country made a military presence virtually impossible. The country is now in a state of anarchy with upwards of ten different groups fighting so that the central government's authority carries little weight, while the country is also used as a base for pirates and militant Islamic groups. The SAC's aircraft are reported to lie abandoned at Mogadishu Airport.

SOUTH AFRICA

- Population: 49.1 million
- Land Area: 471,445 square miles (1,224,254 sq.km.)
- GDP: $283bn (£181bn), per capita $5,767 (£3,686)
- Defence Exp: $4.35bn (£2.78bn)
- Service Personnel: 62,082 active, plus 15,071 reserves

SOUTH AFRICAN AIR FORCE

Founded: 1920

The country with the longest history of military aviation in Africa, South African Army officers received flying training as early as 1913, and in 1915, the South African Aviation Corps was formed. In World War I, the Royal Flying Corps had a number of South Africans among its personnel. The SAAC's initial equipment consisted of a number of BE2As and also some ex-RNAS Henri Farmans. At first, the new force was expected to operate against German South-West Africa, but in the event there was little aerial activity; however, the SAAC did fight with some distinction against German forces in East Africa. It was disbanded in 1918.

The South African Air Force was formed in 1920 as a separate service using the 'Imperial Gift' of war-surplus aircraft: in the SAAF's case this amounted to 100 aircraft, including SE5A fighters, forty-eight DH4, DH9 and DH9A bombers, and a number of Avro 504K trainers. During the early years, the SAAF was involved in a number of police actions. From 1925, an air mail service was operated between Durban and Cape Town. Other duties included surveying and the training of Citizen Defence Force pilots. In 1929, twenty Avro Avian trainers were purchased to replace the Avro 504Ks, followed by thirty-one Westland Wapiti general-purpose biplanes, while another twenty-seven Wapitis were assembled locally. During the early 1930s, sixty Avro Tutors were built under licence in South Africa. The SAAF differed from other air forces in the British Empire by not having a 'Royal' prefix, reflecting the different constitutional arrangements and colonial history of the country. Air force ranks were based on those of the Army rather than the Royal Air Force system.

In 1936, major expansion started. Licence-production began of sixty-five Hawker Hartebeest bombers, as the Hawker Hind was known in South Africa. Aircraft supplied direct from the UK included Hawker Fury fighters, and the last biplane fighter, the Gloster Gladiator, followed by Hawker Hurricane fighters, Fairey Battle and Bristol Blenheim bombers. When war broke out in Europe in September 1939, the SAAF had 100 modern combat aircraft, and was able to put the Junkers Ju86 airliners of South African Airways into service on MR patrols.

Although not formally a member of the Empire Flying Training Scheme, SAAF training facilities expanded considerably and were placed at the disposal of the

RAF. SAAF bases were also made available for the Royal Navy's Fleet Air Arm, using them for carrier air wings in transit to Egypt and the Mediterranean after the direct route across the Mediterranean became impassable. They proved invaluable rear bases for the Fleet Air Arm once the Royal Navy returned in force to the Indian Ocean and then to the Pacific. Many South African personnel also served with the RAF. The SAAF itself operated MR over the South Atlantic and the Indian Ocean, while its fighter and bomber units operated in North Africa against Italian and German forces. SAAF units were stationed in East Africa protecting Kenya and the Sudan from Italian forces, and helped liberate Ethiopia. After taking part in the Allied invasion of Sicily, the SAAF saw further service in Europe. Inevitably, the rapidly-expanded SAAF operated a wide variety of aircraft during the war years. Fighters included Curtiss Kittyhawk, Mohawk and Tomahawk, Hawker Hurricane, Supermarine Spitfire and the North American F-51 Mustang. Bombers included Martin Maryland, Marauder and Baltimore, Vickers Wellington, Fairey Battle, Bristol Beaufort and Blenheim, Douglas Boston and the Consolidated Liberator. Vickers Warwick, Lockheed Ventura, Lodestar and Harpoon MR aircraft operated alongside Short Sunderland flying boats and Consolidated PBY-5A Catalina amphibians. Transport aircraft included the Ju86s, as well as the more usual Douglas C-47 and the C-47's angular rival, ex-civil Junkers Ju52/3Ms. Training used de Havilland Tiger Moth, Airspeed Oxford, Avro Anson, North American T-6 Harvard, Northrop Nomad, Miles Master and Hawker Audax aircraft.

Post-war, the SAAF was reorganised as a small force capable of rapid expansion from SAAF Reserve and Active Citizen Air Force personnel. Many wartime aircraft remained in service, working out their useful lives. The SAAF took part in the Berlin Airlift, helping the city to survive the Soviet siege. In 1950, the SAAF sent a small force of F-51D Mustang fighters to Korea to join the UN forces there, facing MiG-15 jet fighters. It was during the Korean War, in 1952, that the SAAF received its first jets, North American F-86 Sabres loaned by the USAF. Meanwhile, back in South Africa, de Havilland Vampire FB5 jet fighter-bombers, with Vampire T55 trainers, started to replace the Spitfires. Canadair CL-13 Sabre 6 fighters, de Havilland Dove and Heron transports, and Sikorsky S-55 helicopters were delivered in 1956. In 1957, the Short Sunderlands were replaced by Avro Shackleton MR3s for defence of the Cape shipping route, whose importance had been reinforced by the closure of the Suez Canal during the Suez Crisis of the previous year. The bomber force was modernised with English Electric Canberra B12 jet bombers. The South African Navy obtained eight Westland Wasp helicopters to operate from its frigates. US aircraft introduced during this period included seven Lockheed C-130B Hercules transports and Cessna 185 liaison aircraft for a small Army Air Corps.

Limited expansion started during the early 1960s to maintain the balance of power in Africa, where newly-independent states hostile towards South Africa were backed by the USSR. The newer members of the British Commonwealth were also hostile, and the country withdrew in 1961. A United Nations resolution against the sale of arms to South Africa specifically excluded armaments for defence against external threats, but by the mid-1960s, a new British government extended the embargo to cover all arms. Before this took effect, under the terms of the Anglo-South African Defence treaty, the Simonstown Agreement, the SAAF was able to obtain sixteen Hawker Siddeley Buccaneer low-level strike aircraft. The Buccaneers differed from the standard aircraft in having rocket-boosters to improve take-off performance from 'hot and high' airfields. The ban enabled France to become the main supplier of arms to South Africa, although more than 200 Italian Aermacchi MB326K armed-trainers were built under licence as the Impala in South Africa during the late 1960s. The Impala were used both for training and for Active Citizen Force units. French equipment introduced to SAAF service at this time included sixteen Dassault Mirage IIICZ interceptors, twenty Mirage IIIEZ fighter-bombers, four Mirage IIIRZ reconnaissance aircraft, nine C-160 Transall transports, six Alouette II and fifty Alouette III helicopters, as well as twenty SA330 Puma and sixteen Super Frelon helicopters. During the early 1970s, licence-production of Mirage III and F1 aircraft started in South Africa.

Hopes during the early 1970s that the strict British interpretation of the arms embargo would be eased proved groundless, and other nations also applied stricter controls. South African support for the regime in neighbouring Rhodesia, now Zimbabwe, aggravated these tensions. South Africa started to develop its own armaments industry. In 1986, the Cheetah development of the Mirage III first appeared, and a substantial number of the SAAF's Mirage IIIs were upgraded to this standard. The C-47s, of which there were forty, were re-engined with turboprop engines. Many older types were retired, first the Sabres replaced by Mirage F1s, and then Canberras and Buccaneers, which finally left the SAAF in 1991.

Substantial political changes were taking place inside South Africa during the early 1990s, culminating in majority rule in 1994. This result ended UN sanctions, but the SAAF was not able to modernise quickly due to financial constraints, while the USA did not relax its arms embargo until 1998. Plans to replace the Cheetahs and older Impalas with new aircraft were cut back, but the SAAF has introduced Saab Gripens and BAe Hawks, while Agusta A109s replaced the Alouette IIIs. The C-130B fleet has been renovated and upgraded. The first of an initial twelve Denel CSH-1 Roivalk attack helicopters – itself a demonstration of the capability of the South African industry – entered service in 1999.

Southern Africa remains strategically important and there is also a frequent need for humanitarian aid for the areas to the north. The SAAF remains the most capable air force on the African continent, although the period of sanctions and the financial constraints of the post-sanctions era have meant that certain significant capabilities have been lost, notably MR. Strike capability is now confined to the tactical level. The current strength of 10,653 personnel has risen from around 8,000 over the past forty years.

Today, the SAAF has introduced the first of what may be as many as twenty-six Saab JAS-39C/D Gripens, and has up to twenty-four Hawks for lead-in fighter training and light attack. Nevertheless, there are signs that financial constraints are having an impact with an order for A400 transports cancelled in late 2009. There are thirty Agusta A109 armed helicopters as well as eleven CSH-1 Rooivalks, supported by thirty-nine Oryxes (AS332 Super Puma) and eight BK117s. Transport aircraft include nine C-130B/BZ Hercules, eleven Basler Turbo-67s, and four C-212s, with eleven Cessna 208 Caravan utility aircraft, and four King Air 200/300s, a number of Cessna 185, a Falcon 50 and a Falcon 900. Training uses fifty-four PC-9 Astras. There are Seeker II UAV in service.

The SAAF operates four Westland Super Lynxes from four Valour-class frigates.

Missiles include Darter AAM, while the SAM capability has been closed.

SPAIN

- Population: 40.5 million
- Land Area: 194,945 square miles (504,747 sq.km.)
- GDP: $1.54tr (£0.98tr), per capita $38,082 (£24,342)
- Defence Exp: $11.7bn (£7.5bn)
- Service Personnel: 128,013 active, plus 80,210 paramilitary and 319,000 reserves

SPANISH AIR FORCE/*EJÉRCITO DEL AIRE DE ESPAÑA*

Founded: 1939

Spain's long history of military aviation started with a balloon company within the Spanish Army in 1896. In 1910, the balloon company saw action during the Riff uprisings in Morocco, and the following year aircraft were introduced and it became the *Aeronáutica Militar Española*, with two Henri Farman, two Maurice Farman, two Bristol and six Nieuport aircraft. Additional Maurice Farman and Nieuport aircraft were acquired, with Morane-Saulnier MS14 and Lohner biplanes, before the outbreak of World War I. A neutral country, Spain had difficulty in obtaining aircraft for the *AME* and the newly-formed *Aeronáutica Naval* during the war years, resolving this through licence-production of DH4 bombers and Morane-Saulnier Parasols. A Spanish design, the Flecha, also entered production. The *Aeronáutica Naval* managed to acquire some Curtiss F flying boats.

The return of peace meant that the two Spanish air arms were able to benefit from the release of large numbers of war-surplus aircraft by the belligerent powers, acquiring Ansaldo, Bristol F2B, Martinsyde F4A and Spad S13C1 fighters; Farman F-50, Salmson SAL2-A2, Breguet Br14A-2 and additional DH4 bombers; Macchi M9 and Savoia S16 flying boats, and a number of Caudron GIII trainers. Many of these aircraft were deployed in Morocco where there was renewed fighting in the Riff that continued until 1926. Additional aircraft entered service at this time, with Dornier Wal and Macchi M18 flying boats, Blackburn Velos seaplanes and Supermarine Scarab amphibians for the *Aeronáutica Naval*; and licence-built Breguet Br19A-2 and DH9 bombers, Fokker CIV AOP and CIII training aircraft, and Avro 504K trainers, for the *AME*. A number of Wals were also built in Spain under licence, as were Nieuport 52C1 fighters. Spain's own Loring R-1 reconnaissance aircraft entered production. Nevertheless, by the late 1920s aircraft were entering service in smaller numbers, although some Vickers Vildebeest torpedo-bombers were obtained in 1931, and the combined aircraft strength of the two services fell sharply from 700 to 300.

In 1931 the Spanish monarchy was replaced by a republic, followed by growing civil unrest within Spain itself. In 1936, the Spanish Civil War began between the Nationalists, supported by Germany and Italy, and the Republicans, supported by the USSR. The Republicans managed to obtain most of the *AME*'s aircraft, and at the outset had some 200 aircraft to the sixty or so of the Nationalists. Soviet support for the Republicans encouraged Germany and Italy to provide overt support for the Nationalists. France sold aircraft to the Republicans, including 100 Dewoitine D373, D500 and D510, Liore-Nieuport LN46 and Spad 510C fighters, and Potez 56 and Bloch MB200 bombers. Czechoslovakia supplied Letov S231 fighters and Aero 100 general-purpose aircraft. Some of the supporting nations were able to test tactics and aircraft in this theatre of war, while their 'volunteer' pilots also developed their skills. Bombing was used extensively, while fighter combat moved on from the elementary operations of 1914–18. Of significance was the use of Ju52/3M transports to move the Moroccan troops of the Spanish Foreign Legion from North Africa to Spain, giving the Nationalists an important edge. The war ended in 1939 with a Nationalist victory. The new government immediately reorganised Spanish military aviation, merging the two air arms into the Spanish Air Force, or *Ejército del Aire de España*, a separate service. Despite heavy losses during the war, the *EdA* on its creation had 1,000 aircraft available.

Spain remained neutral throughout World War II, although aligned politically with the Axis Powers. Once again, the country was effectively cut off from outside aircraft supplies, leaving the two Spanish manufacturers, CASA and Hispano-Suiza, to produce aircraft under licence. Although a few Messerschmitt Bf109F fighters, Junkers Ju88A bombers, Heinkel He114 seaplanes and Dornier Do24 flying boats were supplied from Germany, with some Fieseler Fi156 Storch general-purpose aircraft, for the most part Spain's needs were met by licence-production of German and Italian designs. These included additional Bf109Fs and He114s, 250 He111H bombers, 100 Junkers Ju52/3M transports, and Fiat CR32 fighters. Volunteer Spanish pilots flew with the *Luftwaffe* on the Russian front. Post-war, the *EdA* was to become notable for continuing the operation of many wartime aircraft, some of which remained in production until 1953, with the fitting of Rolls-Royce Merlin engines to the Bf109s, the same engines as those of their wartime rivals, the Hurricane and Spitfire!

Spain was late in joining NATO, but agreed a separate defence treaty with the United States in 1953, receiving US military aid in return for the use of Spanish air and naval bases. This resulted in the *EdA* receiving 200 North American F-86D/F Sabre fighters, fifteen Douglas C-47 transports, a number of Grumman HU-16 Albatross amphibians, Sikorsky H-19 Chickasaw and Bell 47G Sioux helicopters, thirty Lockheed T-33A and 100 North American T-6G Texan trainers. The Spanish aircraft industry produced the CASA 201B, 207 and 352L transports, and the Hispano Aviacion HA-100 and HA-200 trainers. These aircraft were followed during the early 1960s by twenty Lockheed F-104G Starfighters, enough to equip one interceptor squadron, joined later by CASA-built Northrop SF-5A/B fighter-bombers. The *EdA* also received Dassault Mirage IIIE fighters at this time. When the Spanish-American Defence Agreement was extended in 1970, Spain received McDonnell Douglas F-4 Phantom fighter-bombers and a number of other US aircraft, including Lockheed P-3 Orion MR aircraft and C-130 Hercules transports. Other aircraft included DHC-4 Caribou tactical transports. The Mirage IIIs were later joined by Mirage F1s.

The restoration of the monarchy and of democracy in Spain in 1975 brought a major change in Spain's international position, joining NATO and, later, the European Union, and also joining aircraft programmes with the country's European partners. Nevertheless, US and French designs continued to dominate the *EdA*, which introduced seventy-two licence-built EF-18A/B Hornet fighters during the 1980s, replacing the F-4 Phantom in the fighter-bomber role, while still retaining some of these aircraft for reconnaissance. The SF-5s and Mirage IIIs were modernised during the late 1980s and early 1990s. A plan to develop a Spanish attack aircraft, initially designated the AX, was abandoned during the early 1990s. Spain's hopes that participation in the Eurofighter programme with the UK, Germany and Italy would cover its future needs were dashed as the programme ran into repeated delays, necessitating the acquisition of additional F/A-18 Hornets from the USN between 1995 and 1998. Additional Mirage F1s also had to be obtained from France and Qatar in 1994 and 1997, and all the F1s have been updated to extend operational life to 2015. The first Eurofighter 2000 Typhoons have since entered service.

The *EdA* has 21,300 personnel, a fall of a third over the past forty years, partly due to the end of conscription. The *EdA* is organised into four air commands, Central, Eastern, Strait and Canary Islands, plus a Logistic Support Command.

Currently, the *EdA* has two fighter squadrons operating the Typhoon 2000 but another two are in the process of converting from the Mirage F1M, building up to a total of fifty-nine Typhoons with another thirteen on option. There are five squadrons with ninety EF-18A/Bs. The *EdA* is responsible for most shore-based maritime aviation, and has a single MR squadron with seven P-3A/B Orions, while SAR and EEZ patrols are provided by three Fokker F-27-200MPAs, normally based in Canary Islands Air Command, and there are also seven C-212s employed in this role. A Boeing 707 is used for ELINT and four Falcon 20s are on ECM duties. Tanker aircraft include two KC-135s and five KC-130H Hercules. There are seventeen Bombardier 215/415s for aerial firefighting duties. Transport aircraft include a squadron of seven C-130H/H-300 Hercules; one with twelve CN-235VIPs plus another two for VIP duties; plus another fifty-seven C-212s, some of which will be replaced by CN295 aircraft. Up to twenty-seven A400Ms are on order. Extensive SAR coverage around the long coastline of Spain and her islands is provided by five SA330H/J Pumas (known locally as the HD19) and twelve Super Pumas (HD21), and three AS332BM1 Super Puma (HT21A), and eight Sikorsky S-76A Spirits, some of which are also used for training. Two Cessna V Citations (TR20) are used for photographic surveys. Training aircraft include five Beech B55 Barons and twenty-two F33A Bonanzas, forty-six C101B Aviojets and thirty-seven Enaer T-35C Pillans, with fifteen EC120 Colibris (HE25).

AAM include AMRAAM, Sidewinder and Sparrow; ASM include Maverick, Harpoon and HARM, while there are Mistral and Skyguard/Aspide SAM batteries.

SPANISH NAVAL AIR ARM/*ARMA AÉREA DE LA ARMADA ESPAÑOLA*

Although the history of Spanish naval aviation dates from the early years of the last century, all service aviation was merged into the new *EdA* at the end of the Spanish Civil War. Spanish naval aviation reappeared with the loan by the USN of an Independence-class light aircraft carrier, the USS *Cabot*, in 1967, although the *EdA* retained shore-based MR. Renamed *Daedalo*, the carrier was sold to Spain in 1973, and at first mainly operated helicopters as an ASW carrier. The late 1970s saw the Spanish Navy become one of the first to deploy V/STOL aircraft at sea, obtaining five AV-8A Harrier and two TAV-8A trainers through the US Marine Corps. These aircraft provided the *Daedalo* with a strike capability, primarily for action against shore-targets without the fighter/anti-shipping capability of the Sea Harrier. Meanwhile, frigates capable of carrying helicopters also entered service with the Santa Maria-class (US Oliver Hazard Perry) frigates, while many older destroyers and frigates were modified for helicopters.

The *Daedalo* was replaced by a new Spanish-built aircraft carrier designed specifically for the operation of V/STOL aircraft, the *Principe de Asturias*, which usually operates with an air wing of up to eight EAV-8B-plus Harrier II/Harrier II-plus aircraft and ten helicopters. A new Juan Carlos-class LPH can also operate

Harrier II aircraft if necessary, giving the *Armada Española* a much enhanced capability.

Today, naval aviation accounts for around 814 of the Spanish Navy's 23,200 personnel, down by almost a third since 2002. There are sixteen EAV-8-plus Harrier IIs. Helicopters include twelve SH-3/3H Sea Kings with twelve SH-60B Seahawks (HS23) for ASW and ASuW operations from frigates, while ten AB212ASWs (HA18) are used on transport and SAR duties. There are three communications Cessna Citation IIs. Training uses a TAV-8B Harrier leased from the USMC and ten 500MDs.

SPANISH ARMY AVIATION/*FUÉRZA AEROMOVILES DEL EJÉRCITO DE TIERRA*

Spanish Army Aviation developed during the late 1950s and early 1960s, initially using Cessna O-1 Bird Dogs for liaison and AOP duties. These were joined by six Bell 47G Sioux helicopters, and then by twelve Agusta Bell UH-1D Iroquois. Over the past forty years, this force has grown substantially, initially as a transport and liaison force but moving into the attack role with seventy Bo105ATH/CBs, known in the *FAMET* as HA/HE15s, with twenty-seven equipped with HOT anti-tank missiles and the remainder armed with 20mm cannon. The transport capability of the force was also enhanced with CH-47C Chinook, or HT17 in Spain, heavy-lift helicopters. In recent years, the attack capability has been enhanced by AS-665 Tiger helicopters.

Today, the *FAMET* has twenty-five Tiger helicopters in the combat role, augmented by some of the twenty-six Bo105s, although others are on training duties. Transport uses eighteen CH-47D Chinooks, with thirty-two UH-1H Iroquois, four AB212 and twelve AS332 Pumas and sixteen 532UC/UL Cougars. The AB212s also have an SAR role. Training uses eleven OH-58B Kiowas.

SRI LANKA

- Population: 21.3 million
- Land Area: 24,959 square miles (65,610 sq.km.)
- GDP: $44.3bn (£28.3bn), per capita $2,076 (£1,327)
- Defence Exp: $1.57bn (£1bn)
- Service Personnel: 160,900 active, plus 5,500 reserves

SRI LANKA AIR FORCE

Founded: 1950

Although Sri Lanka, or Ceylon, was used as a major British base during wartime, its own history of military aviation dates from 1950, two years after independence from the UK, when the Royal Ceylon Air Force was formed with assistance from the RAF. After some delay, in 1953, twelve DHC-1 Chipmunk basic trainers and nine Boulton-Paul Balliol trainers arrived, followed by two Airspeed Oxford training and communications aircraft. In 1955, a de Havilland Dove light transport arrived, with Scottish Aviation Pioneers and Westland Dragonfly helicopters. Additional Doves and Pioneers were delivered in 1958, as well as eight Hunting Jet Provost armed-trainers. The RCAF operated against illegal immigrants and smugglers, and on internal security duties. The latter role soon assumed greater importance, and in 1971, Soviet personnel were seconded to the RCAF and six MiG-17 fighter-bombers were supplied to help suppress an armed rebellion.

The 'Royal' prefix was dropped when the country became a republic in 1972, although remaining within the British Commonwealth, adopting the name of Sri Lanka.

The association with the USSR was short-lived, and by the mid-1970s, the Sri Lanka Air Force had largely returned to transport and communications duties, almost allowing the combat capability to lapse. Internal security duties returned when the Tamil population in the north of the island pressed for autonomy, with a terrorist campaign by the Liberation Tigers of Tamil Eelam, LTTE. From 1983, the SLAF rebuilt its combat capability, initially using armed trainers. In 1991, two Shenyang FT-5 (MiG-17U) trainers were obtained from China with four F-7M Airguard (MiG-21) fighter-bombers – the small numbers reflecting both financial constraints and the availability of suitable personnel. Steady growth in the combat units continued throughout the 1990s, with Mi-17 transport and Mi-24 attack helicopters, while the LTTE, although lacking air support of their own, introduced surface-to-air missiles in 1995. Transport also grew in importance, with nine Y-12 transports joining the earlier small force of three HS748s. A civilian aircraft was shot down in 1998, forcing the SLAF to provide internal passenger flights, and aircraft were fitted with infra-red warning systems and chaff dispensers in response to the SAM threat. A significant increase in offensive capability occurred in 1996, with ex-Israeli Air Force Kfir C2/CT2s as well as MiG-27s acquired from the Ukraine. A number of Pucara COIN aircraft were acquired from the Argentine but were soon unserviceable, as were the F-7s, retired with wing cracks. The intensity of the conflict was such that in 2000 the SLAF lost four Mi-24/-35s, having received seven in 1996–2000. The internal security situation forced some training overseas, but new aircraft have brought this activity back to Sri Lanka.

Today, the SLAF has 28,000 personnel, having more than trebled in strength since 2002. The main offensive capability lies in eleven Kfir C2 fighter-bombers, while there are nine Chengdu F-7s and four MiG-27Ms, believed to be flown by Ukrainians. There are thirteen Mi-24/-35 attack helicopters. Ten Bell 212/412 helicopters provide transport, as do two Lockheed C-130K Hercules, seven An-32B, three Y-12, three Mi-17 and six Bell 206 helicopters. The Bell 412s also have a VIP role, as has a single Beech King Air 200. A Cessna 421 is used for survey work. Training uses eight K-8s and six CJ-6s (Yak-18).

The end of the 'Tamil Tiger' campaign in 2009 should have meant that the size of the armed forces could be reduced, but current plans are for continued expansion and a heavy military presence in those areas liberated from the Liberation Tigers of Tamil Eelam, LTTE.

A Chetak helicopter is used on naval liaison duties, but a plan to create a naval air wing within the Sri Lanka Navy appears to have been abandoned.

SUDAN

- Population: 41.1 million
- Land Area: 967,500 square miles (2,530,430 sq.km.)
- GDP: $49.6bn (£31.7bn), per capita $1,207 (£771)
- Defence Exp: no figures available
- Service Personnel: 109,300 active, plus 85,000 paramilitary reserves

SUDANESE AIR FORCE

Founded: 1959

Sudan became an independent republic in 1958 after being administered jointly by the UK and Egypt, with the Sudanese Air Force formed the following year. Initial equipment comprised four Hunting Provost T51 armed-trainers, soon joined by Hunting Pembroke and Douglas C-47 transports. During the 1960s, the SAF's role grew with the introduction of sixteen MiG-21 fighter-bombers and eight BAC Jet Provost T52 armed-trainers and five T45 trainers. Three Fokker F-27M Troopships and five Antonov An-24 transports, and Alouette III, Mi-4 and Mi-8 helicopters soon followed. Equipment originally came from a wide variety of sources, with twelve Northrop Grumman F-5E/Fs delivered during 1982–84, but more recently supplies from Western sources have been hampered by concern over the internal political and military situation and the treatment of the Christian minority in the south by the Muslim majority in the north. Donations of equipment by other Muslim states have not helped standardisation, with a Fokker F-27-100 Friendship donated by South Yemen and a DHC-5 Buffalo by Oman in 1986, and Libya donating a squadron of MiG-23s in 1988, although these appear to have suffered heavy attrition. Puma and Agusta Bell AB412 helicopters were also placed in service, while combat aircraft in recent years have come from China, with first Shenyang F-5s (MiG-17), and then F-6s (MiG-19) and F-7s (MiG-21), believed to have been paid for by Iran.

Today the SAF has 3,000 personnel. Many aircraft are believed to be unserviceable, although reports of air attacks suggest that at least some remain operational. Mainstays of the fighter and ground-attack force are up to twenty Chengdu F-7Bs (MiG-21) and three MiG-23s, with eight F-6s and three A-5s, while five F-5Ss remain. Attack helicopters include twenty Mi-24/-35s, with sixteen Mi-8/-17, four IAR330 (SA330 Puma) and three AB212 transport helicopters. Fixed-wing transport is provided by four C-130H Hercules, three

DHC-5D Buffaloes, two Y-8s, seven An-24/-26s, with at least one modified to drop bombs, and an An-74. A Dassault Falcon 50 and 20 provide VIP transport. Training uses twelve K-8s.

SURINAM

- Population: 481,267
- Land Area: 70,087 square miles (161,875 sq.km.)
- GDP: $2.3bn (£1.47bn), per capita $4,906 (£3,316)
- Defence Exp: $39m (£24.9m)
- Service Personnel: 1,840 active

SURINAM AIR FORCE

Founded: 1982

Formerly Dutch Guiana, Surinam formed a small air force in 1982 within the Army. It provides border and coastal patrols, light transport and MEDEVAC, initially using Britten-Norman BN2B Defenders and a Cessna TU206G, as well as a Bell 205 and two Alouette III helicopters. Today, it has 200 personnel. Two CASA C-212-400s were delivered in 1998 and 1999, with one used for transport and the other, equipped with radar, for offshore patrols, and these remain with a BN2 Defender, a Cessna U-206 and 182, three Dhruv ALH helicopters and a Pilatus PC-7 trainer.

SWAZILAND

- Population: 600,000
- Land Area: 6,705 square miles (17,363 sq.km.)

SWAZILAND DEFENCE FORCE AIR WING

Swaziland is an independent kingdom completely surrounded by South Africa, which donated three Alouette III helicopters in 2000 to join two IAI-201 Arava armed-transports and a Piper Cherokee training aircraft.

SWEDEN

- Population: 9.1 million
- Land Area: 173,620 square miles (449,792 sq.km.)
- GDP: $429bn (£274bn), per capita $47,367 (£30,278)
- Defence Exp: $5.61bn (£3.58bn)
- Service Personnel: 13,050 active, plus 800 paramilitary and 42,000 in voluntary military organisations, and 200,000 reserves

ROYAL SWEDISH AIR FORCE/*FLYGVAPNET*

Formed: 1926

Swedish military aviation started with the Royal Swedish Navy being presented with a Bleriot monoplane by an air-minded citizen in 1911, a year before the Swedish Army was presented with a Nieuport IVG in similar circumstances. Sweden was neutral during World War I, by which time the RSwN had added two Henri Farmans and a Donnet-Leveque flying boat, while the Army Air Corps had added three aircraft, including a Breguet. The wartime famine in aircraft for non-belligerents led to licence-production of Farman F23, Albatros CIII and Morane-Saulnier Parasol aircraft. When peace returned, the RSwN had twenty-five aircraft and the AAC around fifty. While licence-production continued post-war, including Phoenix 122 fighters and Avro 504K trainers, Swedish designs also appeared, including J23 and J24 fighters, S18, S21 and S25 reconnaissance aircraft, and O1 trainers.

In 1926, the two air arms were amalgamated into a separate service, the Royal Swedish Air Force, *Flygvapnet*. The RSwAF received some new aircraft, including Nieuport 29C-1 fighters and Fokker CV reconnaissance aircraft, but then entered a period of neglect. It was only when the shortage of aircraft became desperate that new aircraft were obtained: these included twelve Bristol Bulldog fighters, Hawker Hart light bombers and Osprey reconnaissance aircraft, with some of both types built in Sweden, ASJA J6s and ASJA RK26s, and forty de Havilland Tiger Moth trainers. There was a revival in the RSwAF's fortunes as the political situation in Europe showed the threat to national security. The service was organised into eight wings: F1, F4, F6 and F7 operating bombers; F8 fighters; F2 naval reconnaissance; F3 army reconnaissance; and F5 flying training. Some sixty Gloster Gladiator fighters were built, with forty Junkers Ju86K and 100 Douglas DB-8A bombers and forty North American NA-16-4 trainers. These aircraft were joined by sixty Republic EP-1 fighters in 1940. Before the wartime demands on the belligerent nations cut Sweden off from supplies again, seventy-two Fiat CR42 and CR60, and Reggiane Re2000 fighters were purchased, replacing the obsolescent Gladiator biplanes, and eighty Caproni Ca313 bombers. Two new fighter wings, F9 and F10, a new reconnaissance wing, F11, and a new bomber wing, F12, were quickly formed. During late 1939 and early 1940, a small but effective *Flygvapnet* unit fought alongside Finnish forces during the so-called 'Winter War' that followed an attempted Soviet invasion, but otherwise Sweden once again remained neutral. Swedish-designed aircraft included the Saab-17 light bomber, Saab-18 bomber and Saab-21 fighter, of which some 300 were delivered.

The final batch of Saab-21 fighters was not delivered until after the war ended, by which time outside supplies were resumed, with an initial fifty North American F-51D Mustang fighters followed by a further ninety. The *Flygvapnet's* first jets, seventy de Havilland Vampire F1 fighters, arrived in 1946. The Saab-21 was a twin-boom design with a single pusher propeller which meant that it could be redesigned as a jet fighter, the Saab-21R, with sixty of these delivered from 1949 onwards. New aircraft entered service in quantity, including 200 Vampire FB50 fighter-bombers and T55 trainers, sixty de Havilland Mosquito NF19 night-fighters and seventy Supermarine Spitfire PR19 reconnaissance fighters. Piston-engined aircraft remained an option at this time due to the short range of many of the early jet fighters, but the Mosquito night-fighters were replaced during the early 1950s by sixty de Havilland Venom NF51s. A new purpose-designed Swedish fighter, the Saab-29 Tunnan, or 'barrel' (an appropriate name given its shape) appeared at this time, along with the first Saab-91 Safir piston-engined trainers.

In 1956, 120 Hawker Hunter F4 jet fighters were introduced, and the Saab-32A Lansen attack aircraft entered service, as did sixteen Hunting Pembroke C52 light transports. The first helicopters, Vertol 44s, entered service in 1957. Saab-35 Drakens (Dragon) started to replace the Vampires and Saab-29s in 1959. By this time, the pattern had emerged of Sweden maintaining substantial reserve forces, a position of armed neutrality, with a core of professionals augmented by conscripts under training. Aircraft numbers were accordingly far higher than the size of the standing air force would indicate. Two other characteristics also helped the low personnel to combat aircraft ratio: the lack of a substantial MR force, reflecting Sweden's geographical position, and the small size of the air transport element, reflecting the absence of overseas commitments. Regular replacement of aircraft also became the pattern, with the Saab-37 Viggen (Viking) starting to replace the Saab-32 Lansens during the late 1960s, before replacing the Saab-35 Drakens during the 1970s. Vertol 44s were replaced during this period by Boeing-Vertol 107s, and transport capability increased with the purchase of two C-130E Hercules, augmenting a force of seven Douglas C-47s. External purchases of aircraft continued, including fifty-eight Scottish Aviation Bulldog trainers delivered during the early 1970s. An advanced trainer, the Saab 105, also entered service.

In recent years, the *Flygvapnet* has been through another upgrading exercise, replacing the Saab-37 Viggens with the JAS39 Gripen, which became operational during the late 1990s. As with the Viggen, the Gripen uses a standard airframe as the basis for attack, interceptor and reconnaissance variants, in an attempt to maintain an affordable indigenous aircraft development and production facility, although Sweden is now more active in selling combat aircraft in world markets than in the past. Saab's brief move into the airliner market, with the Saab 340 and 2000, also provided the basis for a low-cost AEW aircraft, the S100 Argus. In 1998, the helicopters of the *Flygvapnet* and the naval and army air arms were merged into a Helicopter Wing, *Helikopterflottij*. Despite remaining neutral throughout the Cold War period, Sweden has also reduced the size of its armed forces since the collapse of the Warsaw Pact.

Although the initial 1,000 personnel of the combined Helicopter Wing were on secondment from the other services, today their numbers are included in those of the air force. The original equipment included ex-*Flygvapnet* Super Pumas (HKP10 in Sweden) SAR helicopters, ex-*Marinflget*, or Naval Aviation, ASW Boeing-Vertol 107 (HKP4) Retrievers and AB206 (HKP6A/B) JetRangers, and ex-*Armeflygkar*, or Army Air Corps, AB204B (HKP3C) and 206A, Bo105CB (HKP9) and Hughes 300C (HKP5B). The Royal Swedish Navy had been operating helicopters since 1958, while the Swedish Army had started an aviation element in 1964. The *Helikopterflottij* initially maintained the thirteen squadrons provided by the constituent services, but in 1999 these were reorganised into four peacetime battalions. Standardisation started with the medium-lift helicopter chosen under the Nordic Standard Helicopter Programme, with nineteen NH90s replacing the 107s from 2004, while twenty A109s replaced the 204Bs and 206s as well as the 300Cs.

The *Flygvapnet* now has 4,300 personnel, of whom 500 are conscripts; a reduction of almost 75 per cent in personnel strength over the past forty years, and the true reduction is even greater as the strength includes *Helikopterflottij* personnel. The use of many reserve personnel may account for the relatively low average annual flying hours of between 110 and 140. A total of 204 JAS39 Gripen entered service between 2000 and 2006, with many of the earlier versions upgraded from A/B to C/D standard, undertaking interceptor, strike, reconnaissance and combat training. There are four squadrons and 165 A/B/C/D Gripens, with more than half the total in reserve. Supporting the interceptor force are six Saab S100B Argus (Saab 340) in the AEW role, with a seventh aircraft for VIP duties, and two ELINT S102Bs (Gulfstream IV), with a third aircraft as a light transport. Air transport uses eight Lockheed C-130E/H Hercules, known as the Tp84 in Sweden, one of which provides in-flight refuelling, as well as five Saab 340s. A Cessna Citation II is used for communications duties. Training uses eighty Saab 105 (Sk60) jets. AAM include Sidewinder and AMRAAM, with Maverick ASM.

The *Helikopterflottij* has twenty AW109 attack helicopters while fifteen Bo105s remain; transport being provided by nineteen NH90s and fifteen AS332 Pumas, with the latter also used for SAR.

SWITZERLAND

- Population: 7.6 million
- Land Area: 15,941 square miles (41,310 sq.km.)
- GDP: $518bn (£331bn), per capita $68,071 (£43,512)
- Defence Exp: $4.42bn (£2.82bn)
- Service Personnel: 4,059 active, plus 18,000 conscripts and 174,071 reserves

SWISS AIR FORCE/*SCHWEIZER LUFTWAFFE*

Founded: 1939

Swiss military aviation dates from the creation of an Air Troop, *Fliegertruppe*, in 1914, with eight aircraft of Aviatik, Bleriot, Henri Farman, Morane and Schneider manufacture. Although Swiss pilots flew with the French *Aviation Militaire* during World War I, Switzerland remained neutral. To overcome the shortage of aircraft from the belligerent countries, Swiss-designed aircraft were put into production, including the Haefeli DH1, DH2 and DH3 observation aircraft. In 1919 the force was reorganised and renamed as the *Militar-Flugwesen*, by which time it had 100, mainly Swiss, aircraft. Some foreign aircraft were acquired, including war-surplus Fokker DVII fighters, but the 1920s also saw continued production of Swiss aircraft, including Haefeli M7 fighters, and DH5 and M8 bombers. The late 1920s and early 1930s saw foreign designs produced under licence, including Dewoitine D9, D26 and D27 fighters, Potez 25 general-purpose aircraft and Fokker CVE reconnaissance-bombers, while the *M-F* also obtained Hawker Hind bombers and de Havilland Moth and Tiger Moth trainers.

Immediately before World War II, Potez 63 fighter-bombers were obtained, as well as ninety Messerschmitt Bf109E fighters and thirteen Bf108 liaison aircraft. Morane-Saulnier MS406C fighters and Bucker Bu131 Jungmann and Bu133 Jungmeister trainers were built under licence for what had now become the *Schweizer Luftwaffe* although still a corps of the Swiss Army. (At one time it was known as the Air Force and Anti-Aircraft Command.) At the outbreak of World War II in 1939, during which Switzerland again remained neutral although there were incursions by *Luftwaffe* aircraft, the *SF* was operating 100 fighters and 100 AOP aircraft. Additional aircraft were obtained from Germany early in the war, including Bf109Es, Fieseler Fi156 Storch AOP aircraft and Bucker Bu181 Bestmann trainers.

The first post-war aircraft were 100 North American F-51D Mustang fighter-bombers and forty T-6 Harvard trainers. In 1949 and 1950, the first jets were introduced, seventy-five de Havilland Vampire FB6 fighter-bombers, followed by a further 100 of these aircraft built in Switzerland to replace the Mustangs. The

Vampires were followed by 250 licence-built de Havilland Venom FB50 fighter-bombers, and, in 1958, 100 Hawker Hunter F58 fighters entered service. Swiss-designed aircraft continued to enter service, including Pilatus P-2 and P-3 trainers. During the late 1960s, fifty-seven Dassault Mirage IIIS fighters were built under licence for what had now become the Swiss Air Force and Anti-Aircraft Command, *Kommando Flieger und Fliegabwehrtuppen*. The Mirages were joined by Mirage IIISR reconnaissance aircraft; the *KfuF* entered the 1970s also operating Hunters and Venoms, as well as thirty Alouette II and ninety Alouette III, and twenty Bell 47G Sioux helicopters. A few Bucker trainers and three Ju52/3M transports survived from the war years.

During the 1980s, the Venoms were retired and replaced by Hunters, upgraded to carry Maverick air-to-surface missiles, while the Hunter in turn was replaced by Northrop F-5E/F Tiger IIs. Vampire trainers were replaced by the BAe Hawk F58 and T68, providing ground-attack as well as training capability. Tactical transport was improved with fifteen AS332M Super Puma helicopters delivered during the early 1990s. Pilatus PC-7 Turbo Trainers were also introduced, with PC-9s for communications duties. Despite being neutral, the ending of the Cold War saw a reduction in the armed forces starting in 1995, largely by reducing the length of training periods. New equipment continued to enter service, with the F/A-18C/D Hornet assembled in Switzerland. The Hunters retired in 1995.

The Swiss Air Force has a mobilised strength of 33,300 personnel. Flying hours are between 150–200, but closer to fifty hours for the high proportion of reservists. There are seven fighter squadrons: three operating thirty-three F/A-18C/D Hornets, and four operating fifty-seven F-5E /F Tiger IIs. Transport is provided by a squadron with fifteen Pilatus PC-6s, a Dornier Do27, a Falcon 50, a Beech 350 and a 1900. Six helicopter squadrons operate fifteen AS332M-1 Super Pumas, twelve AS532 Cougars and eighteen EC635s, which replaced the SA316 Alouette III. Training uses thirty-seven Pilatus PC-7s, six PC-21s and eleven PC-9 target tugs. The force is tactical without tankers or a strategic transport element. Some training takes place in Sweden since supersonic aircraft soon reach this small country's borders! UAVs include the Ranger, while there are Sidewinder and AMRAAM AAM.

SYRIA

- Population: 19.7 million
- Land Area: 71,210 square miles (184,434 sq.km.)
- GDP: $53.3bn (£34.1bn), per capita $2,448 (£1,568)
- Defence Exp: $1.87bn (£1.2bn)
- Service Personnel: 325,000 active, plus 314,000 reserves

SYRIAN AIR FORCE

Founded: 1946

Syria gained independence in 1943 having previously been a French-mandated territory, but foreign forces were not completely withdrawn until 1946. Despite being formed in 1946, the first Syrian Air Force aircraft did not enter service until 1949, including Fiat G46 and G59, and DHC-1 Chipmunk trainers, with Beech C-45 and Douglas C-47, as well as some French-assembled Ju52/3M transports. Ex-*Armée de l'Air* bases were taken over. Despite a British arms embargo between 1951 and 1953, the SAF managed to obtain thirty de Havilland Vampire FB52 jet fighter-bombers via Italy, although these were quickly passed on to Egypt. Syria's first combat aircraft did not enter service until 1953, with twenty-three Gloster Meteor F8 fighters, NF13 night-fighters and T7 trainers, as well as forty Supermarine Spitfire 22 fighters.

In 1955 Soviet aid began, promising twenty-five MiG-15 fighters and personnel to train Syrian air and ground crew. Of the aircraft that did arrive, all were destroyed on the ground in Israeli raids during the 1956 Suez crisis. Syria obtained some MiG-15s and sixty MiG-17 fighters later. An attempt by Egypt and Syria to form a United Arab Republic in 1958 was short-lived as Syria withdrew following a *coup d'état*. In the 1960s, the SAF received MiG-21 interceptors, Ilyushin Il-14 transports and Mil Mi-1 and Mi-4 helicopters, as well as Yak-11 and Yak-18 trainers. The Arab-Israeli War of June 1967 saw up to 75 per cent of this equipment destroyed. The USSR made good Syria's losses and then increased the SAF's strength, with ninety MiG-21 interceptors, eighty MiG-15, MiG-17 and Sukhoi Su-7B fighter-bombers, and transports, helicopters and trainers. Syria remained committed to the Soviets and continued to introduce new aircraft throughout the late 1970s and 1980s, including MiG-23, MiG-25 and, in 1989, MiG-29 interceptors, and Su-20, Su-22 and Su-24 strike aircraft. A small naval air arm developed, using SAF personnel. While transport aircraft remained an assortment of Soviet types, all in small numbers, a substantial helicopter force was established with more than 100 Mi-8 and Mi-17 transport helicopters, and thirty-six Mi-24 attack helicopters.

Syria opposed the 1990 Iraqi invasion of Kuwait and committed troops to Saudi Arabia to assist in the liberation of Kuwait, but only helicopters and transport aircraft were deployed, with no Syrian combat missions.

The Syrian Air Force has 40,000 personnel, having quadrupled in size over the past forty years. A further 60,000 personnel are in a separate Air Defence Command, which is responsible for anti-aircraft defences. Flying hours are very low, with an average of twenty per year, suggesting poor combat readiness. There are eleven fighter squadrons, with four operating forty MiG-25 *Foxbat*, four with eighty MiG-23MLDs and three with up to forty-eight MiG-29A *Fulcrum*. Another fifteen squadrons operate in the strike role, with five operating sixty-nine Su-24 *Fencer*, two with sixty MiG-23BNs, one with twenty Su-24s and seven with 159 MiG-21H *Fishbed*. Four reconnaissance squadrons operate forty MiG-21H/Js and eight MiG-25Rs. Three attack helicopter squadrons operate thirty-six Mi-24s and thirty-five SA342L Gazelles, supported by 100 Mi-8/Mi-17 transport helicopters, twenty Mi-2s and ten heavy-lift Mi-6s. Transport aircraft include six An-26s, four Ilyushin Il-76Ms and a VIP flight with six Yak-40s and two Falcon 20Fs. Training uses seventy L-39A/ZO Albatros, thirty MiG-17Fs and fifteen MiG-15UTIs, six MiG-29UBs, thirty-five MBB Flamingos and six MFI-17 Mushshaks. Missiles include *Alamo*, *Atoll*, *Acrid*, *Apex* and *Aphid* AAM, with *Kerry* and HOT ASM.

SYRIAN NAVAL AVIATION

Syrian Naval Aviation operates fourteen Mi-14 and two Ka-25 ASW helicopters, all operated from shore bases by Syrian Air Force personnel. The small force of frigates and missile craft does not include ships capable of carrying helicopters.

TAIWAN

- Population: 23 million
- Land Area: 13,890 square miles (35,975 sq.km.)
- GDP: $349bn (£223bn), per capita $15,172 (£9,698)
- Defence Exp: $9.78bn (£6.25bn)
- Service Personnel: 290,000 active, plus 1,657,000 reserves

REPUBLIC OF CHINA AIR FORCE

Founded: 1949

The late 1940s saw Communist victory in China, with Nationalist forces forced to withdraw to the offshore island of Formosa in 1949. With a number of smaller islands, this formed the basis of the new state of Taiwan, at one time known as 'Nationalist China'. Taiwan is not recognised by the Chinese People's Republic, which still claims sovereignty.

At the end of World War II, the Central Government Air Force had been reorganised and renamed the Chinese Air Force. The surviving wartime aircraft, supplied by the USA, were augmented by ex-USAF aircraft, including additional North American F-51 Mustang and Lockheed P-38 Lightning fighters, and North American B-25 Mitchell bombers. Aircraft types new to the Chinese Air Force included Republic F-47 Thunderbolt fighter-bombers and Consolidated B-24 Liberator bombers, while 250 ex-RCAF de Havilland Mosquito fighter-bombers were purchased. Many of these were lost in the withdrawal to Taiwan, where the new Chinese Nationalist Air Force started with just 160 aircraft. In 1951, US military aid started, providing Republic F-84G Thunderjets, and these were followed in 1954 by North American F-86F Sabres to replace the Thunderbolts, while the remaining Lightnings were replaced by Republic RF-84F Thunderflash reconnaissance-fighters. The next generation of aircraft into service was fifty North American F-100 Super Sabres, soon joined by Lockheed F-104A Starfighters. In 1963, Lockheed F-104G Starfighters replaced the Thunderjets and the earlier F-104As. During the late 1960s, Northrop F-5A/B fighters took over from the F-86 Sabres. A number of transport aircraft were also supplied by the USA, including Fairchild C-119 packets and C-123 Providers, as well as the necessary helicopters and trainers. Grumman S-2 Trackers were acquired for MR, and re-engined with turboprop engines in 1991, before transferring to the Navy in 2000.

Although the Cold War continues between Taiwan and Communist China, attempts by the United States to seek an accommodation with the latter have left Taiwan at the mercy of the prevailing diplomatic mood, with arms supplies fluctuating according to the state of relations between the US and China. Overall, the US has always supported Taiwanese autonomy, while the country's successful industrialisation enables it to purchase aircraft, although many countries are wary of upsetting the People's Republic. Northrop F-5E/F Tiger II interceptors were introduced during the 1980s. An Indigenous Defence Fighter (IDF) project during the 1990s led to the A-1 Ching-Kuo interceptor, although orders for this aircraft were cut from the planned 250 to 130 in 1992. IAI Kfir C7 fighters were considered at one stage. In 1991, the People's Republic acquired Su-27 interceptors from Russia, and this prompted the CAF to buy sixty Dassault Mirage 2000-5 attack aircraft and 150 Lockheed Martin F-16A/B interceptors. Older aircraft have been upgraded, and substantial numbers are held in store, for use by reservists or against attrition losses.

The CAF is organised on US lines with tactical fighter wings, composite wings and an air force academy. An F-16 conversion unit is based in the United States. It currently has 55,000 personnel, down from 68,000 in 2002, before which its size had remained steady over the previous quarter-century. The number of reservists has halved to 45,000 since 2002. Flying hours are reasonable, at around 180. Three fighter squadrons have fifty-seven Mirage 2000-5EI/5DIs, while fifteen ground attack squadrons include six with 146 F-16A/Bs, five with 128 Ching-Kuos, one with twenty-two AT-3/3B Tzu Chungs, and three with up to eighty-nine F-5E/F Tiger IIs, although some are in store. A reconnaissance squadron operates RF-16As and eight RF-5Es. Supporting the combat aircraft are six E-2C/T Hawkeye and two C-130HE AEW aircraft. Two transport squadrons have nineteen Lockheed C-130H Hercules while there are also two C-47s, ten Beech 1900s, three Fokker 50s and four VIP Boeing 727-100s and a 737-800. There are three CH-47 Chinook, seventeen Sikorsky S-70C Black Hawk and a VIP S-62A helicopters. Training uses thirty AT-3/3B Tzu Chungs, some Northrop T-38A Talons and forty-two T-34C Turbo Mentors. Missiles include Falcon, Sidewinder, Shafrir, Skysword I/II, Mica and R-550 Magic I AAM, with Maverick ASM.

REPUBLIC OF CHINA NAVAL AVIATION

The Republic of China Navy initially used aviation for communications and liaison, but it now operates ASW helicopters from frigates and destroyers, as well as having thirty-two Grumman S-2E/G Trackers for ASW and MR. There are eighteen Sikorsky S-70 Seahawk and eight MD500 helicopters. There are four Keelung-class (ex-USN Kidd) destroyers and twenty-two frigates capable of operating helicopters.

Early in 2010 it was announced that additional Seahawks would be part of an arms deal with the United States, raising objections from Communist China.

REPUBLIC OF CHINA ARMY AVIATION

Founded: 1970

The Republic of China Army became involved with aviation in 1970, with the introduction of Bell UH-1H Iroquois helicopters. It is introducing thirty AH-64D Apache attack helicopters, which are supported by sixty-two anti-tank AH-1W Super Cobras, and in the transport and utility role by nine Boeing CH-47SD Chinook and forty-four UH-1H Iroquois, with thirty-eight OH-58D Kiowas for observation and liaison. Training uses thirty TH-67 and fifteen TH-55A helicopters.

TAJIKISTAN

- Population: 7.3 million
- Land Area: 55,240 square miles (144,263 sq.km.)
- GDP: $4.6bn (£2.94bn), per capita $635 (£406)
- Defence Exp: $80m (£51m)
- Service Personnel: 8,800 active

TAJIKISTAN AIR FORCE

Tajikistan is a member of the Commonwealth of Independent States, CIS, which has provided troops and aircraft to help the Tajik government in its struggle with Muslim rebels. Forces from Russia, Kazakhstan and Uzbekistan are based in the country. The government has formed an air force which now has 1,500 personnel. Equipment includes four Mi-24 *Hind* attack helicopters, supported by twelve Mi-8/-17s and a Tu-134A *Crusty* transport.

TANZANIA

- Population: 41.1 million
- Land Area: 361,800 square miles (939,706 sq.km.)
- GDP: $21.3bn (£13.6bn), per capita $519 (£639)
- Defence Exp: $183m (£117m)
- Service Personnel: 27,000 active, plus 80,000 reserves

TANZANIAN PEOPLE'S DEFENCE FORCE AIR WING

Founded: 1964

Tanzania was formed on the federation of two former British colonies, Tanganyika and Zanzibar, in 1964. The Tanzanian People's Defence Force Air Wing came into existence with *Luftwaffe* assistance, including six Nord Noratlas transports, eight Dornier Do28 liaison and communications aircraft, and nine Piaggio P149 trainers. In 1965 the aid ended abruptly before deliveries could be completed after Tanzania recognised the German Democratic Republic and East German aid took over. As a result, the aircraft of the TPDFAW included an Antonov An-2, five DHC-3 Otters and four DHC-4 Caribou transports, and seven Piaggio P149 trainers. The Air Wing bought Chinese combat aircraft supported by Western transports and helicopters, taking first the Shenyang F-4 (MiG-17), then the Shenyang F-6 (MiG-19) and finally the Chengdu F-7 (MiG-21) into service, with a squadron of each successive type, although attrition has reduced the numbers of the older aircraft.

Training of combat pilots takes place in China after initial training in Tanzania. Today, the TPDFAW has 3,000 personnel. It has three combat squadrons operating eleven F-7s, eight F-6s and eight F-5s, as well as having two MiG-15UTIs. There are four DHC-5D Buffalo transports, three HS748s, two Y-12s and a Y-5, as well as five Cessna 310s and two 404 Titans. Four AB205B, six Bell 206 JetRanger and four ex-South African SA316 Alouette III helicopters cover transport and communications duties. Five Piper Cherokees provide basic training. There is a substantial SAM force with more than 150 SA-3, SA-6 and SA-7 launchers.

A police air wing has a Cessna 206, two Bell 206L LongRangers and a Bell 47G.

THAILAND

- Population: 66 million
- Land Area: 198,250 square miles (519,083 sq.km.)
- GDP: $259bn (£165.6bn), per capita $3,927 (£2,510)
- Defence Exp: $5.13bn (£3.28bn)
- Service Personnel: 305,860 active, plus 113,700 paramilitary and 200,000 reserves

ROYAL THAI AIR FORCE

Founded: 1937

Siamese army officers were sent to France for flying training in 1911. They returned home in 1913 with four Nieuport and four Bleriot aircraft. The then Kingdom of Siam entered World War I on the side of the Allies, sending a contingent of the new Siamese Flying Corps to Europe as part of an expeditionary force. It was not until 1918 that the first Siamese operational sorties were flown, although more than 100 Siamese officers and NCOs received flying training in France during the war. The SFC spent a short period as part of the Allied Army of Occupation in the Rhineland before returning home in 1919 with ex-wartime aircraft, including Spad SVII and SXIII and Nieuport-Delage ND29 fighters, Breguet Br14A2 and 14B2 reconnaissance-bombers, and Nieuport trainers. In 1919, the name was changed to the Royal Siamese Aeronautical Service.

In 1920, some of the Br14s were converted to operate a domestic airline, while the remainder operated reconnaissance and liaison duties. Pilots were trained in the USA, the UK, France and Italy, as well as in Siam. In 1930, twenty Avro 504 trainers were purchased, and a further fifty produced in Siam under licence. Despite evaluating new aircraft, no further orders were placed until 1934, when seventy-two Vought V935a Corsair AOP aircraft were built in Siam to replace elderly Br14s. These were followed by twelve Curtiss Hawk II and twelve Hawk III fighters, with another twenty-five Hawk IIIs assembled in Siam.

Another new name, the Royal Siamese Air Force, was adopted in 1937, changing to the Royal Thai Air Force in 1939 when the country was renamed.

Further re-equipment occurred in 1937, with six Martin 139 bombers followed in 1939 by an order for twenty-five Curtiss Hawk 75N fighters and a number of North American NA-69 bombers and, in 1940, six North American NA-68 fighters. Only the Hawks were delivered, with the later aircraft being taken off their ships at the Philippines and Hawaii respectively to be pressed into US service. Some support for Japan among certain sections of the Thai community led to the acquisition of nine Mitsubishi Ki21 bombers and nine Tachikawa Ki55 trainers. A border dispute led to an invasion of French Indo-China by Thailand in January 1941, with combat between Thai and French aircraft until Japan brokered a truce in May, after which the Vichy French allowed Japan to use bases in French Indo-China. Japan invaded Thailand in December 1941, and the Thai Government, faced with overwhelming odds, arranged a cease-fire and surrender after a few days.

Thailand was officially an ally of Japan for the remainder of the war, but RTAF personnel were confined to non-combatant roles. A number of RTAF members helped an underground movement, and flew Allied agents into and out of the country. Pro-Allied and pro-Japanese RTAF personnel were segregated. Additional Japanese aircraft were supplied, including a few Nakajima Ki27 and Ki43 fighters, Mitsubishi Ki30 bombers and Mansyu Ki79 trainers. Peace found the RTAF operating abandoned Japanese aircraft and some surviving pre-war aircraft. RAF personnel were seconded to rebuild the RTAF, providing thirty Supermarine Spitfire Mk14s and a number of Fairey Firefly FR13s. These were followed by twenty Miles Magisters, forty-two North American T-6G Texans, de Havilland Tiger Moths and DHC-1 Chipmunks for training.

US military aid started before Thailand became a founder member of the South East Asia Treaty Organisation, SEATO, in 1954. Personnel were trained in the USA. New aircraft included fifty Grumman F8F-1 and F8F-1B Bearcat fighter-bombers and additional Texan trainers. There were also Stinson L-5 Sentinels, Piper L-18 Super Cubs and Cessna O-1 Bird Dogs for AOP duties; Fairchild 24W, Cessna 170 and Beech C-45 communications aircraft; Westland S51 Dragonfly (S-51), Sikorsky S-55 and Hiller 360 helicopters; and in 1957, another seventy-five Texans. The first jet aircraft also arrived in 1957, thirty Republic F-84G Thunderjet fighter-bombers and Lockheed T-33A trainers. Thunderjets and Bearcats were replaced in 1962 by North American F-86F Sabres, with additional Sabres in 1966, when the RTAF also received its first Northrop F-5A/B fighters. Transport aircraft were also provided, with Fairchild C-123 Providers and Sikorsky CH-34C helicopters. Cessna C-37B trainers were supplied as well.

Thailand's importance grew during the 1960s and early 1970s, as a stable country in an unstable region when the Vietnam War spread into both Laos and Cambodia. COIN became important, with the RTAF becoming one of the few operators of the Fairchild OV-10C Bronco. During the 1980s, Northrop F-5E/F Tiger II and Lockheed Martin F-16A/B interceptors entered service, as well as Lockheed C-130H Hercules transports.

Over the past forty years, Thailand has placed a high priority on defence. A mixture of surplus aircraft, usually from the US, and new aircraft has produced a modern air force, although the Asian economic crisis of the late 1990s seriously affected procurement programmes, including an order for F/A-18C/D Hornets.

Over the past forty years, the RTAF has almost doubled its personnel from 25,000 to 46,000. Flying hours are relatively limited at an average of 100. Nine fighter squadrons include three with fifty F-16A/Bs, and four with forty-seven F-5A/Bs, some of which are being replaced by eight Gripen Cs. Two squadrons operate many of the forty-six L-39ZA/MP armed jet trainers, while there are also twenty-two AU-23A Peacemakers (PT-6). ELINT is provided by three Arava IAI-201s and a Learjet 35, while a Saab 340 provides AEW. Transport is provided by three squadrons, one of which has twelve C-130H/H-30 Hercules; one with three G-222s and six BAe748s; and the third with nine Basler Turbo-67s and some of the eighteen N-22B Nomads in service. The VIP Royal Flight has an Airbus A310-324, an A319CJ, two Boeing 737-400s and a 737-200, one King Air 200 and two Queen Airs, as well as three Commander 500s and three Fairchild Merlin IV, plus three AS532 Cougars, three AS332L Super Pumas and two Bell 412s. There are two helicopter squadrons, one operating twenty Bell UH-1H Iroquois and the other thirteen Bell 212s, while there are also five Bell 412s for transport and SAR. Training uses ten Alphajets, twenty-three PC-9s, twelve T-41Ds (Cessna 172), a number of L-39ZA Albatros, and twenty-nine CT-4 Airtrainers.

ROYAL THAI NAVY AIR ARM

Originally a shore-based force with MR Grumman S-2F Trackers and SAR HU-16 Albatross amphibians, it was updated during the 1970s and 1980s with Canadair CL-215 amphibians and Fokker F-27s, as well as Dornier Do228s for EEZ and anti-smuggling patrols. Several warships entered service able to carry helicopters for ASW and Bell 212s and Sikorsky S-70s were introduced. Lockheed P-3 Orions entered service during the early 1990s, providing longer-range MR. The most recent advance has been the acquisition of a small aircraft carrier built in Spain, *Chakri Naruebet*, to operate nine AV-8S Matador (Harrier) V/STOL fighters and up to six S-70 helicopters, although financial problems delayed the entry into service. Eight frigates can operate helicopters.

Today, the air arm accounts for 1,940 of the Royal Thai Navy's 44,751 personnel. There are seven AV-8Ss and two TAV-8S Matadors, while shore-based aircraft include three MR P-3A/UP-3T Orions for MR, three Fokker F-27-200/400Ms, five Do228-212s, two CL-215 amphibians and five GAF N24 Searchmasters. Helicopters include six S-70B Seahawks, two Super Lynxes, and five Bell 212ASWs for ASW, with five Bell 214STs for SAR and transport, plus five S-76Ns. Training uses four Cessna O-1 Bird Dogs and a number of TA-7s.

ROYAL THAI ARMY AIR DIVISION

Although badly affected by the financial problems of Asia in the late 1990s, the Royal Thai Army developed an air mobility brigade. An attack helicopter squadron was established in 1990 with a nucleus of eight AH-1F Cobra helicopters.

Today the service operates a C-212, two Beech 1900Cs and two King Air 200s, two Jetstream 41s and two Short 330s, as well as forty O-1 Bird Dogs. Helicopters include five AH-1F Cobras, six CH-47D Chinooks, fifty-four Bell 212s, eighty-nine UH-1Hs, ten S-70/UH-60Ls, three Bell 206s and six Mi-17s. Training is on fifteen T-41Bs, eighteen Star Rockets and forty-two Hughes 300Cs.

The Royal Thai Border Police also operate two CN235s, a Fokker 50, eight PC-6 Turbo-Porters, three Skyvans and two Short 330s as well as a number of helicopters, including twenty-seven Bell 205As, fourteen Bell 206s, twenty Bell 212s and six Bell 412s.

TOGO

- Population: 6 million
- Land Area: 22,000 square miles (54,960 sq.km.)
- GDP: $3bn (£1.91bn), per capita $498 (£318)
- Defence Exp: $67m (£42.8m)
- Service Personnel: 8,550 active

TOGO AIR FORCE/*FORCE AÉRIENNE TOGOLAISE*

Founded: 1960

Togo formed a small air force on achieving independence in 1960, and through the French Community gained the standard arms 'package' of a Douglas C-47 transport, two Max Holste 1521M Broussard communications aircraft, and an Alouette II helicopter. A 1963 agreement with France on air crew training remains. A light attack capability was acquired during the 1980s with Aerospatiale TB.30s and Embraer EMB326 Xavantes as well as armed Alphajets. The main roles are transport and liaison, with a small transport and helicopter fleet including two VIP aircraft.

Today, the *FAT* has 250 personnel. The three remaining TB30 Epsilons are relegated to training, leaving five Alphajets and four EMB326G Xavante armed trainers. One DHC-5 Buffalo remains in service, with two Beech 58 Baron, two Reims-Cessna F337, two SA318B Lama II and an Alouette III helicopters. VIP transport is provided by a Boeing 707-320B, and a Fokker F-28-3000 Fellowship but an AS332 Super Puma and an AS330 Puma are both in storage.

TONGA

- Population: 90,000
- Land Area: 270 square miles (699 sq.km.)

TONGA DEFENCE SERVICES AIR WING

Founded: 1996

The Tonga Defence Services formed an Air Wing in 1996 with a Beech G18S for EEZ patrols and SAR around this South Pacific group of 150 islands. An American Champion Citabria was acquired for training in 1999, but little is known about developments since.

TRINIDAD AND TOBAGO

- Population: 1.2 million
- Land Area: 1,980 square miles (5,128 sq.km.)
- GDP: $27.6bn (£17.6bn), per capita $22,420 (£14,331)
- Defence Exp: $158m (£101m)
- Service Personnel: 4,063 active

TRINIDAD AND TOBAGO DEFENCE FORCES AIR WING

Founded: 1977

The Trinidad and Tobago Defence Forces Air Wing is part of the Coast Guard, and has fifty personnel. It operates two C-26s (Merlin), a Cessna 310 and a 172 for training, and has just recently introduced two Piper Navajos. There are no armed aircraft, but there is an S-76 and an AS355F Ecureuil, available for SAR when needed, as well as a Westinghouse Skyship 600 airship.

TUNISIA

- Population: 10.5 million
- Land Area: 63,362 square miles (164,108 sq.km.)
- GDP: $39.9bn (£25.5bn), per capita $3,800 (£2,429)
- Defence Exp: $534m (£341.3m)
- Service Personnel: 35,800 active

WORLD AIR POWER GUIDE

REPUBLIC OF TUNISIA AIR FORCE

Founded: 1956

Tunisia became independent in 1956, and shortly afterwards formed the Republic of Tunisia Air Force with Swedish assistance. The first aircraft, fifteen Saab-19D Safir trainers, arrived in 1960, and were joined by two Alouette II helicopters in 1962. Eight Aermacchi MB326B jet trainers entered service in 1966, and were followed soon afterwards by the first combat aircraft, twelve North American F-86F Sabre fighters. These were later joined by North American T-6G Texan trainers and Dassault Flamant transports, as well as additional Alouette II helicopters. Armed versions of the MB326 were acquired later, as well as SF260 armed-trainers, but Tunisia also started to receive US military aid, with Northrop F-5E Tiger IIs for interception and reconnaissance. Agusta Bell 205 and UH-1H helicopters were also delivered, as well as two Lockheed C-130H Hercules transports. In 1989, additional F-5Es were obtained from surplus USAF stocks. In 1995, C-130Bs were provided by the USA, as well as twelve UH-1H and three HH-3 helicopters, while the Czech Republic supplied L-59T Albatros armed-trainers.

The main threat to Tunisia lies in action by Algerian rebels, possibly encouraged by Tunisia's ambitious neighbour, Libya. The RoTAF has 4,000 personnel, having grown from just 600 personnel over the past forty years. The mainstay of its fighter force are fifteen F-5E/F Tiger II fighter-bombers, augmented by sixteen MB326B/K/L and twelve L-59 Albatros armed-trainers, as well as eighteen armed SF260Ws, although mainly operated in the training role. There are six AS250 Ecureuil anti-tank helicopters, while fifteen CH/HH-3Es (SH-3) include two equipped for ASW, others on transport duties, where they operate alongside fifteen AB205As and twelve UH-1H/Ns, an AS365 Dauphin, six SA313s and three SA316 Alouette IIIs. Transport aircraft include eleven C-130B/E/H Hercules, five G222s provided by Italy in 2000, five Let-410s and a Falcon 20. Missiles include Sidewinder AAM.

TURKEY

- Population: 76.8 million
- Land Area: 301,302 square miles (780,579 sq.km.)
- GDP: $658bn (£420bn), per capita $8,561 (£5,472)
- Defence Exp: $9.9bn (£6.3bn)
- Service Personnel: 510,600 active, plus 378,700 reserves

TURKISH AIR FORCE/ *TURK HAVA KUVVETLERI*

Turkey has the longest history of military aviation in the Middle East, having ordered Bristol, DFW, Deperdussin, Mars, Nieuport and REP aircraft in 1912, to be flown by foreign pilots for the Army. In 1914, the Turkish Flying Corps was formed with German assistance. During World War I, Turkey aligned with the Central Powers, operating AEG CIV and Albatros reconnaissance aircraft, Halberstadt DII fighters and, for naval co-operation, some Gotha WD13 seaplanes, all flown by German pilots with Turkish observers.

When the war ended, the Treaty of Versailles banned the Central Powers and their allies from operating or building military aircraft. The ban was circumvented by the formation of the Turkish Air League, *Turk Hava Kurumu*, in 1925, with public subscription buying a Caudron trainer and an Ansaldo A300. The Air League received assistance from France and some pilots were trained in the UK. In late 1926, some Morane-Saulnier MS53 trainers were delivered, by which time the Turkish Air Force, *Turk Hava Kuvvetleri*, was operating as part of the Turkish Army. In 1928, Rohrbach RoIII flying boats entered service, followed by eighteen Curtiss Hawk fighters and some Fledgling trainers. Development of the *THK* accelerated during the 1930s, with twenty Breguet Br19B2 reconnaissance-bombers and six Supermarine Southampton MR flying boats. As the pool of trained pilots grew, substantial orders were placed for new aircraft in 1937, with orders for thirty each of Heinkel He111D, Bristol Blenheim I and Martin 139 bombers. There were smaller quantities of Supermarine Walrus amphibians; Avro Anson light bombers and Vultee VIIG fighter-bombers; Hanriot 182 and Westland Lysander army co-operation aircraft; Miles Hawk II and Curtiss-Wright CW22 trainers; and Focke-Wulf Fw58 Weihe communications aircraft. The following year, Hawker Hurricanes and additional Blenheims were ordered. These were joined by de Havilland Dragon and Dragon Rapide light transport and navigational training aircraft. Licence-built Gotha trainers entered service, but an order for forty Gotha G23 fighters was never fulfilled when the aircraft were diverted to Spain.

Turkey remained neutral during World War II, signing a non-aggression pact with Germany and obtaining aircraft from both sides. The new aircraft included further examples of those already in service, as well as Curtiss Tomahawk IIBs, Fairey Battles, Airspeed Oxfords, Morane-Saulnier MS406s and Focke-Wulf Fw190As. The objective was to keep Turkey well-disposed towards the combatant nations, although had it not been for the failure of Operation Barbarossa, the German invasion of the Soviet Union, Turkey could have been invaded. The RAF even supplied spares for He111s, salvaged from aircraft shot down over the UK. The Allies provided Lend-Lease equipment, including additional Hurricanes and Blenheims, Supermarine Spitfire VBs, Bristol Beaufighters and Beauforts, Martin Baltimores and Consolidated Liberators.

Post-war, the *THK* received large numbers of war-surplus aircraft, including de Havilland Mosquito FB6 fighter-bombers and T3 trainers; Republic F-47D Thunderbolt fighter-bombers; North American T-6 Harvard and Beech T-11B Kansan trainers, Douglas B-26 Invader bombers; Beech C-45 and Douglas C-47 transports. On joining the North Atlantic Treaty Organisation, NATO, in 1952, Turkey became eligible for the US Military Aid Programme, including aircraft and USAF personnel as additional instructors. The *THK* received its first jet aircraft in

1952, with the first of 300 Republic F-84G Thunderjet fighter-bombers, while twenty-four Beech T-34 Mentor trainers were assembled in Turkey. In 1953, Canadair F-86E Sabre Mk2 and Mk4 fighters were introduced, at the same time as the first of Turkey's own MKEK Ugar (Lark) basic trainers. Deliveries of US aircraft continued into the 1960s, with North American F-100C Super Sabre fighter-bombers, Convair F-102A Delta Dagger and Lockheed F-104G Starfighter interceptors, RT-33A tactical reconnaissance aircraft and T-33A trainers, and C-130E Hercules transports, as well as Cessna T-37 trainers. Other aircraft have included Dornier Do27 and Do28 communications aircraft, Piper L-18 AOP aircraft, and Northrop F-5A fighter-bombers.

Despite Turkey's important Cold War strategic position, deliveries had to be withheld on occasion to prevent tension between Turkey and neighbouring Greece flaring up into open warfare. Tension centred on territorial rights in the Aegean Sea and over the island of Cyprus, where a substantial Turkish minority objected to Greek Cypriot demands for union with Greece. These tensions erupted in a Turkish invasion of Cyprus in 1973, with the *THK* operational over the island and also landing troops. Direct aerial combat with Greek forces was avoided as Greek aircraft would have had to overfly Turkey to reach Cyprus.

NATO policy was to maintain a measure of equality in the quality of equipment supplied between Turkey and Greece. This resulted in a mixture of new aircraft and surplus aircraft supplied under MAP to other NATO countries. The original F-5A force was augmented by ex-RNethAF aircraft, and the *THK*'s F-4 Phantom force was later augmented by ex-USAF aircraft.

In 1975, the *THK* took charge of the Turkish Navy's thirty-three shore-based Grumman S-2A/E Tracker MR aircraft, but these were retired in 1993.

Following the Iraqi invasion of Kuwait in August 1990, Turkish bases were used by NATO forces, and this has continued as Operation Northern Watch, enforcing a 'no fly' zone in the north of Iraq to protect the Kurdish community.

The 1990s saw Turkey's most ambitious aircraft programme so far, building 240 Lockheed Martin F-16C/D interceptors in Turkey, but plans to buy thirty-two additional aircraft were affected by a serious economic crisis in March 2001. Meanwhile, fifty CASA CN235M medium transports were also built under licence. In addition, F-4 Phantoms and F-5A/Bs have been upgraded by Israeli Aircraft Industries, with some aircraft upgraded in Turkey using IAI-supplied kits. Current plans centre on four Boeing 737 AEW&C aircraft, although originally six were planned. Turkey is also planning to buy ten Airbus A400M transports and upgrade its C-130Es.

The *THK* has 60,000 personnel, a slight increase in numbers over the past forty years. Flying hours are around an average of 180 per year, suggesting a good state of readiness. There are fifteen fighter and ground-attack squadrons: eight operate 243 F-16C/Ds, mainly upgraded to Block 50 standard; five use 135 F-4E Phantoms, upgraded to Phantom 2020 standard; the remaining two have eighty-seven F-5A/Bs, many of which are being upgraded to become lead-in trainers. A reconnaissance squadron has thirty-five RF-4E Phantoms. A new AEW&C

squadron operates four Boeing 737s, while a tanker squadron has seven upgraded Boeing KC-135Rs. Transport is provided by five squadrons: one with thirteen Lockheed C-130B/E Hercules; one with seventeen C-160T Transalls; two with forty-one CN-235s and a VIP squadron with two CN235s, two Cessna Citation IIs and two VIIs, and three Gulfstream IVs. Ten A400 are on order. Another two CN235s are used for EW. Transport helicopters include sixty-five UH-1H Iroquois and twenty AS532AL Cougars, of which fourteen are used for CSAR. A King Air 200 is used for communications. Training uses forty SF260Ds, sixty T-37B/C Tweets, sixty-nine T-38 Talons and twenty-eight Cessna T-41D Mescaleros (172). Missiles include Sparrow, Sidewinder and AMRAAM AAM, Maverick, HARM and Popeye I ASM, as well as four squadrons with Nike Hercules and two with eighty-six Rapier SAM. UAV include Gnat 750 and Heron.

TURKISH NAVAL AVIATION

The Turkish Navy relinquished its thirty-three Grumman S-2A/E Trackers to the *THK* in 1975, but retained control over their deployment, to become a purely helicopter force. It has returned to fixed-wing flying with nine CASA CN235MPA Persuader MR aircraft built in Turkey, which have replaced the Trackers. At least seventeen frigates can operate ASW helicopters, and the Navy operates nineteen Sikorsky S-70B Seahawk ASW helicopters, with plans for an eventual force of around thirty. The S-70s are replacing many of the eleven AB212ASW and three AB204AS ASW helicopters. Training uses eight SOCATA TB20 Trinidads.

TURKISH ARMY AVIATION

Turkish Army Aviation originally developed primarily as a transport and liaison force, but in 1983, it received Bell AH-1 Cobra anti-tank helicopters with TOW missiles. A substantial number of S-70A Black Hawk helicopters brought a significant improvement in the transport capability during the 1990s, but the Army still needs a heavy-lift helicopter. The current force includes four King Air 200s for reconnaissance, and three Cessna 410 reconnaissance aircraft. Helicopters include thirty-seven Bell AH-1P/S/W Cobras in the anti-tank and attack roles. Transport is provided by fifty-six S-70A Black Hawks, but there could be as many as 200 eventually, when they will doubtless replace many of the sixty-nine AB205s and eighty-five UH-1H Iroquois. There are two AB212s and sixteen transport AS532ULs, as well as another ten on CSAR. AOP duties are handled by three OH-58B Kiowas and ninety-eight Cessna U-17Bs. Training uses twenty-four Cessna T-41D Mescaleros (172), thirty Bellanca Citabrias, and twenty-eight Hughes 300s, ten AB204Bs and twenty-two AB206B JetRangers.

A paramilitary gendarmerie, the *Turk Tandarma Teskilati, TTT*, operates under army control with a number of helicopters, primarily on transport duties, including nineteen Mi-17s and fourteen S-70As, as well as eight AB204Bs, six 205As, eight 206As and a 212, in addition to two Dornier Do28Ds.

TURKMENISTAN

- Population: 4.9 million
- Land Area: 188,400 square miles (491,072 sq.km.)
- GDP: $11.8bn (£7.5bn), per capita $1,546 (£988)
- Defence Exp: $84m (£53.7m)
- Service Personnel: 22,000 active

TURKMENISTAN AIR FORCE

Founded: 1993

Plans to establish an air force were finalised in 1993, two years after the collapse of the USSR, using equipment left behind by the Russians. Unlike many breakaway and newly independent states, Turkmenistan enjoys Russian support for its military efforts, possibly because it includes the main combat training range of the former Soviet armed forces. The core of the new air force is a composite regiment of MiG-29s and Su-17s, funded by selling more than 200 advanced combat aircraft.

The Turkmenistan Air Force currently has 3,000 personnel and is organised along Soviet lines. There are at least two fighter or ground-attack squadrons operating twenty-two MiG-29 *Fulcrum*, sixty-five Su-17 *Fitter* and two Su-25MKs, with more than forty of the latter being overhauled and which may replace some of the Su-17s. There are ten Mi-24 *Hind* attack helicopters. Supporting this force is a transport squadron with an An-26 *Curl* and eight Mi-8 *Hip*. There appear to be just two L-39 Albatros trainers in addition to the conversion trainers. Missiles include around fifty *Guideline, Goa* and *Gammon* SAM launchers.

UGANDA

- Population: 32.4 million
- Land Area: 93,981 square miles (235,887 sq.km.)
- GDP: $16.5bn (£10.5bn), per capita $525 (£335)
- Defence Exp: $243m (£155m)
- Service Personnel: 45,000 active

UGANDAN PEOPLE'S DEFENCE FORCE AIR WING

Formed: 1987

Uganda was a British colony and before independence in 1962, created a Police Air Wing, with an Army Air Wing created in 1964. At first, the Army Air Wing shared the two Westland Scout helicopters of the Police Air Wing. Israeli aid at the outset was replaced during the late 1960s by Soviet and Czechoslovak aid, with MiG-15 fighter-bombers joining the earlier twelve Potez Magister armed-trainers. MiG-17s and MiG-21s arrived later, but political unrest, worsened by an Israeli raid on Entebbe in 1976 and then an invasion by Tanzania in 1979 followed by a military coup in 1985, saw all but a few aircraft destroyed. Police and army aviation merged to create the Ugandan People's Defence Force Air Wing in 1987 with Libyan aid.

Today, the status of the force is uncertain, although a single fighter squadron operates a mix of five MiG-23 *Flogger* and six MiG-21 *Fishbed*. Just five Mi-24 *Hind* attack helicopters are operational. Transport includes two Y-12s, plus four Mi-17 *Hip* helicopters, one of which is for VIP use, three Bell 206s and two 412s. Training uses three L-39s, but the SF260W armed-trainers are thought to be unserviceable.

UKRAINE

- Population: 45.7 million
- Land Area: 231,990 square miles (582,750 sq.km.)
- GDP: $108bn (£69bn), per capita $2,369 (£1,514)
- Defence Exp: $1.41bn (£0.9bn)
- Service Personnel: 129,925 active, plus 1,000,000 reserves

MILITARY AIR FORCES

Founded: 1991

Many of the former USSR's military aircraft were stationed in the Ukraine and were taken over when the USSR broke up. Some have been sold, while others have been transferred back to Russia in order to reduce the Ukraine's debts. The Ukraine could have difficulty in sustaining strong armed forces since the country is not heavily industrialised, apart from transport aircraft manufacture. The Ukraine has restructured its armed forces in three phases, with the third and final phase due to end in 2015. The first two phases have seen the armed forces personnel reduced by almost two-thirds since 2002. Early in the restructuring the Naval Air Arm's combat fixed-wing aircraft were transferred to the Military Air Forces, *VVS*, which are organised along Soviet lines.

The *VVS* currently has 45,240 personnel, down from 96,000 in 2002, while the numbers of combat aircraft have been reduced over the same period from more than 900 to 211. Flying hours are very low, at forty to fifty annually. The service is organised into three air commands, West, South and Centre, plus a Task Force Crimea. There are seven brigades with a total of eighty MiG-29 *Fulcrum*, thirty-six Su-24M *Fencer*, thirty-six Su-25 *Frogfoot*, and thirty-six Su-27 *Flanker*, while two reconnaissance squadrons operate twenty-three Su-24MRs, but many more aircraft of these types are in storage. Transport aircraft are operated by three brigades with three An-24 *Coke*, twenty-one An-26 *Curl*, three An-30s, two Tu-134s and twenty Il-76 *Candid*. There do not seem to be any operational tanker aircraft. Helicopters include four Mi-9s, thirty-one Mi-8 *Hip* and three Mi-2s. Training uses just thirty-nine L-39 Albatros out of more than 300 operational in 2002.

AAM include *Alamo*, *Apex*, *Aphid* and *Amos*, as well as *Karen*, *Kilter*, *Kegler*, *Kingbolt*, *Kedge*, *Kent* and *Kyle* ASM.

UKRAINIAN NAVAL AIR ARM

The Ukrainian Navy was formed out of the former Soviet Black Sea fleet, but the major fleet units have been returned to Russia to reduce debts, and to meet treaty obligations to ensure that the Ukraine does not remain a nuclear power. Russia still maintains a naval base at Sevastopol, dating from Tsarist days and ownership of which has become a cause of friction between the two countries. Most fixed-wing combat aircraft have been transferred to the *VVS*, other than a varied handful of Antonov transports and ten Be-12 *Mail* flying boats. Some 2,500 naval personnel are involved in aviation out of 13,000, but only one frigate is helicopter-capable. The Ukrainian Navy has twenty-eight Ka-25 *Hormone*, two Ka27E *Helix* and forty-two Mi-14PL *Haze* helicopters in the ASW role, as well as five Mi-6 *Hook* transport helicopters.

UKRAINIAN ARMY AVIATION

The Ukrainian Army once had a substantial force of attack and assault helicopters, but most are believed to be either in storage or unserviceable. It is believed that 139 Mi-24 *Hind* and thirty-eight Mi-8 *Hip* are operational.

UNITED ARAB EMIRATES

- Population: 4.8 million
- Land Area: 32,300 square miles (82,880 sq.km.)
- GDP: $245bn (£156.6bn), per capita $51,220 (£32,740)
- Defence Exp: $15.47bn (£9.9bn)
- Service Personnel: 51,000 active

In defence, as in some other fields, the countries that make up the Emirates collaborate and yet at the same time also maintain forces of their own. The history of the present UAE armed forces dates from the merger in 1976 of the Defence Union Forces and the armed forces of the UAE, which comprises Abu Dhabi, Dubai, Ras al-Khaimah, Fujairah, Ajman, Umm al-Quwain and Sharjah. Dubai is one that maintains independent forces, but others also do so albeit to a lesser degree.

UNITED ARAB EMIRATES AIR FORCE

The United Arab Emirates Air Force is a composite force funded by the seven emirates, formerly known as the Trucial States: Abu Dhabi, Ajman, Dubai, Fujairah, Ras al-Khaimah, Sharjah and Umm al-Quwain. Several maintain separate royal flights, and most aircraft are assigned to one or the other of the two main states, Abu Dhabi and Dubai. Abu Dhabi had previously maintained an air wing, whose equipment included ten Hawker Hunter fighter-bombers and four DHC-4 Caribou transports. Pressure to operate jointly came from the UK, anxious to withdraw from the Gulf during the 1970s, but aware of the threats to these territories. Standardisation of equipment has been a priority, but both Abu Dhabi and Dubai operated Hawk armed-trainers and Puma helicopters.

There are 4,500 personnel, an increase of 500 since 2002. Flying hours are low, at an average of 110. In recent years, F-16C/D Block 60 multi-role fighters have entered service, joining an existing force of thirty Dassault Mirage 2000-9s and thirty upgraded 2000-Es, but replacing the earlier Mirage VA and MB326KD/LD armed-trainers.

Currently, there are three squadrons with fifty-five F-16E Block 60s; four squadrons with sixty-two Mirage 2000-9DAD/9RADE/M-2000DADs and two squadrons with thirty Hawk Mk63/Mk102s, the latter being ex-Abu Dhabi. A reconnaissance squadron has seven Mirage M-2000RADs. SAR is handled by three A109K2s and six AW139s. Two attack helicopter squadrons have thirty AH-64A Apaches, being upgraded to 'D' standard; ten SA342L Gazelles and six AS550 Fennec anti-tank helicopters. Other helicopters include a transport element with twelve CH-47C Chinooks, fifteen SA330 Pumas, four AS365F Dauphins, of which two are used on VIP duties with another two AW139s, nine Bell 206s, a 407, three Bell 214s and nine Bell 412s. Transport aircraft are organised in three squadrons, and operate an An-124 Condor, six C-130H/H-30 and two L-100-30 Hercules, four Il-76s, seven CN235s and a DHC-6 Twin Otter. Training aircraft include thirty PC-7 Turbo Trainers, five Hawk 61s, while the Hawk 63 can also be used, twelve Grob G115TAs and fourteen AS350B Ecureuils. Royal Flight aircraft include Abu Dhabi's two Airbus A300-620s, a Boeing 747SP, a BAe 146-100, two Beech Super King Air 350s, three Dassault Falcon 900s and two AS332L Super Pumas; Dubai's Boeing 747SP, a Grumman Gulfstream II and a IV, a Sikorsky S-76 and an AS365 Dauphin; Ras al-Khaimahs Cessna Citation I and Sharjah's Boeing 737-200.

AAM include Sidewinder, while ASM includes HOT, Hellfire, *Kent*, Black Shaheen, Hydra-70 and Hakeem.

UNITED ARAB EMIRATES NAVY

The acquisition of two ex-Royal Netherlands Navy Kortanaer-class frigates, known to the UAE as Abu Dhabi-class, has led to the development of a small air arm within the United Arab Emirates Navy, which has 2,500 personnel in total. There are seven AS332F Super Puma ASW helicopters, as well as seven AS565 Panthers and four SA316 Alouette IIIs.

UNITED KINGDOM OF GREAT BRITAIN AND NORTHERN IRELAND

- Population: 61.1 million
- Land Area: 92,000 square miles (242,880 sq.km.)
- GDP: $2.26tr (£1.45tr), per capita $37,000 (£23,651)
- Defence Exp: $62.4bn (£39.9bn)
- Service Personnel: 175,690 active, plus 199,280 reserves

The United Kingdom has long had wide-ranging, almost global, defence commitments, but from the late 1960s onwards moves were made to re-focus defence on European and the North Atlantic with successive governments intent on withdrawing from 'east of Suez'. The short-sightedness of this plan was highlighted first by the Falklands campaign which saw substantial forces deployed to the South Atlantic in 1982, and then in more recent years by first the need to join the so-called 'Coalition' to liberate Kuwait from Iraqi occupation and then by a continuing commitment east of Suez in Afghanistan and Iraq.

Despite this, Britain's armed forces have been scaled back to what many regard as dangerously low levels which place considerable strain on British service personnel, while also exposing serious deficiencies in equipment and in defence procurement. The percentage of GNP devoted to defence has halved since 1990 to 2.2 per cent, but this process started even before the collapse of the Soviet Bloc, with an illusory chase at what was seen as the average GNP for NATO nations, and was followed by the decision to reap the so-called peace dividend immediately the Soviet Union broke up. When the Labour Party returned to office in 1997, one of its early actions was to instigate a wide-ranging defence review which called for the armed forces to focus on expeditionary warfare, and be capable of sustaining two medium intensity conflicts at any one time. At no time since the publication of the report were the resources made available for the review's recommendations to be implemented, and since 2000 the reduction in force levels has continued, despite fighting two medium intensity conflicts, first in Afghanistan, which continues, and then in Iraq, from which the British have extricated themselves, possibly far too early.

The world remains an uncertain and dangerous place, possibly more so than during the days of the Cold War when the potential protagonists knew the risks and treated these seriously. Piracy has become a major issue after being ignored for far too long, and it is not just in the Red Sea that this age-old blight on legitimate trade continues to flourish. Yet the Royal Navy, once the world's largest, then one of the two largest, and then the third largest, has now fallen behind that of France, which not only pursues an east of Suez policy but which has recently opened a naval base in the Gulf.

The desire to save money rather than enhance defence capabilities has been paramount in recent years. In 2012, the Royal Air Force and the Royal Navy's Fleet Air Arm will lose their shared responsibility for search and rescue, which will be contracted out to a civil operator. The number of helicopters available for SAR around the coast of the United Kingdom will drop from thirty-eight to twenty-four, using just twelve bases and in some cases extending flying time to the scene of a rescue. The RAF's Nimrod MR4 is late into service and the numbers of aircraft have been cut, but for a short period there is the danger that there will be no shore-based MR until the MR4 enters service, hopefully in 2012. The Royal Navy has lost its Sea Harrier fighters and the fleet has no realistic air defence of its own.

ROYAL AIR FORCE

Founded: 1918

The world's first autonomous air force when formed on 1 April 1918, the Royal Air Force was the result of the merger of the Army's Royal Flying Corps and the Royal Naval Air Service. The new service was already in action, in the bomber and fighter campaigns of World War I, which still had more than seven months to run.

The UK has the longest continuous history of military aviation of any nation. The Royal Engineers had started experimenting with balloons in 1878 at Woolwich Arsenal, on the outskirts of London, using these in expeditions to Bechuanaland in 1884 and to the Sudan a year later. The official status of this unit was established in 1890, with the formation of a balloon section within the Royal Engineers and the building of balloon sheds at South Farnborough. It was not until the outbreak of the Boer War in 1899 that the balloons were used for observation and artillery control duties. In 1911, the Balloon Section became the Air Battalion.

Meanwhile, the Royal Navy had started flying in 1909.

The Royal Air Force's direct predecessor, the Royal Flying Corps, was formed in 1912, on the amalgamation of the Air Battalion, Royal Engineers, and the Royal Navy's Air Branch, but it was short-lived, with the Royal Navy withdrawing from the RFC in 1914, and establishing the Royal Naval Air Service. On the outbreak of war in August 1914, the RNAS had 100 aircraft, plus airships that were to prove useful as convoy escorts. The RFC, on the outbreak of war under the command of Lieutenant-Colonel Hugh Trenchard, had 180 aircraft. The RNAS was volunteered for the task of home air defence by the then First Lord of the Admiralty, Winston Churchill.

Procurement policies of the two services differed. The RFC largely concentrated on the products of its own Royal Aircraft Factory at Farnborough, while the RNAS bought the products of the civilian aircraft industry. Aircraft included British Avro, Bristol, Short and Sopwith designs operating alongside French Bleriot, Deperdussin and Farman products. Aircraft belonging to the RNAS bombed Zeppelin sheds at Hamburg, Cologne and Friedrichshafen, and torpedoed a Turkish warship in the Mediterranean, as well as flying reconnaissance missions from warships, including seaplane carriers, on a number of occasions, but most notably at the Battle of Jutland on 31 May 1916.

Fighter warfare evolved during 1915, and in 1916 German Fokker aircraft, fitted with synchronised machine guns, gained aerial superiority. At first, the answer lay in pusher-propeller Vickers FE2B and Royal Aircraft Factory DH2 aircraft, and later the Lewis-gun-fitted French Nieuport biplanes. It was not until the arrival of Bristol Scouts and Sopwith 1½-Strutters with synchronised Vickers machine guns that the balance was restored. The bomber evolved as a distinct aircraft type, with the appearance of the DH9 series, the Handley Page 0/400 and V/1500, and, at the end of the war, the Vickers Vimy. These aircraft were involved on raids on German lines of communication, including railway marshalling yards. During the war, bombs developed from artillery shells fitted with fins to purpose-designed ordnance of as much as 1,650lbs (750 kg), although smaller sizes were more commonly used.

During the closing months of the war, the RAF was in action on fighter, bombing and reconnaissance duties. In France, the service operated as the Independent Air Force. Aircraft from the world's first aircraft carrier, *Furious*, bombed the German airship sheds at Tondern in July 1918, destroying two airships. In 1919, a Vickers Vimy made the first non-stop transatlantic flight.

The new service was controversial, being responsible for all aspects of military aviation. In the event, a small compromise was permitted, which allowed the fleet spotting aircraft operated by battleships and cruisers to be flown by naval officers, but aircraft aboard the growing fleet of aircraft carriers were flown and maintained by the RAF with just a token number of naval personnel in these carrier flights.

At the end of the war, the RAF had some 360,000 men, 200 squadrons and 23,000 aircraft. Severe budgetary cutbacks and the desire to suppress the old inter-service RFC/RNAS rivalries led to a reduction to just twelve squadrons, with one in Germany, two in the United Kingdom, and the remaining nine in the Middle East and India. In 1923, it was decided that fifteen fighter and thirty-seven bomber squadrons should be available for home defence which, with those units based overseas, made an overall strength of seventy-four squadrons. The Depression years meant that this strength was not attained until 1936. British military expenditure during this period was also hampered by the 'Ten Year Rule', which stated that there would be ten years in which to prepare for a major conflict!

In 1924, the RAF's carrier squadrons became known as the Fleet Air Arm, which became part of the new RAF Coastal Area.

It was not until 1923 that the first post-war design was introduced, the Fairey Fawn. A succession of designs followed, all of which were biplanes and purchased in such small numbers that it was not unknown for a single type to equip just a single squadron. Boulton Paul Sidestrands, Fairey IIIKs, Handley Page Hinaidis and Hyderabads, Hawker Harts and Horsleys, and Blackburn Iris and Supermarine Southampton flying boats followed Gloster Glebes, Armstrong-Whitworth Siskins and Vickers Virginias. Despite the RAF High Speed Flight participation and successes in the Schneider Trophy contests, performance of service aircraft improved but slowly.

Operations between the wars included a number of police actions, especially in the Middle East, in Afghanistan and on the North-West frontier of India. 'Air control' was seen as a cost-effective means of maintaining the peace in territories such as Iraq, but aircraft on their own were no substitute for forces on the ground. Colonial and League of Nations policing meant that a network of well-equipped bases was established, complementing the efforts of the new national airline, the rapidly expanding Imperial Airways. The RAF also operated a number of diplomatically important air mail services in the period before the creation of Imperial Airways in 1924.

In 1936 and 1937, a major re-organisation of the RAF took place, with the division of the service into Bomber, Fighter, Coastal, Maintenance and Training Commands in 1936. The following year it was decided to return naval aviation to the Admiralty. It was decided in 1936 to increase the RAF's strength to 134 regular squadrons plus 138 Royal Auxiliary Air Force, or reserve, squadrons; a massive increase over the thirteen RAuxAF squadrons of the early 1930s. Development and production of new aircraft started, including the Armstrong-Whitworth Whitley, Fairey Battle, Bristol Blenheim and Vickers Wellington bombers, and the Hawker

Hurricane and Supermarine Spitfire fighters. It was not until 1938 that industrial capacity rather than finance became the limiting factor in re-equipping and expanding the RAF, so that by 1939, the service still had only an eighth of the manpower and two-sevenths of the equipment of the *Luftwaffe*.

On the outbreak of World War II in Europe in September 1939, the RAF's Bomber Command possessed fifty-five squadrons. Of these, five had Armstrong-Whitworth IIIs and IVs, six had Handley Page Hampdens, six Bristol Blenheim IVs, another six with Vickers Wellington Is, the most effective of these aircraft, and another ten had the ineffectual Fairey Battle light bomber. The remainder operated outdated aircraft. Coastal Command had ten squadrons of Avro Ansons, one of Lockheed Hudsons, and two of Short Sunderland flying boats, as well as six squadrons of obsolete Supermarine Stranraer and Saunders-Roe London flying boats and Vickers Vildebeest torpedo-bombers. Fighter Command had twenty-two squadrons of Hawker Hurricanes and Supermarine Spitfires, and another thirteen of the obsolete Gloster Gladiator biplanes.

Initially, twenty-seven squadrons were deployed to France, including the light bombers of the Advanced Air Striking Force, AASF, which accompanied the British Expeditionary Force, BEF. Just 600 British and French aircraft faced more than 3,000 *Luftwaffe* aircraft. The RAF could only provide token support for the battle for Norway, partly because of the limited number of suitable airfields, but did manage to send a small number of Hawker Hurricane fighters. After the massive German advances during spring 1940, the RAF withdrew many of its aircraft from France, ready for *Luftwaffe* air attacks against the British Isles, the prelude to invasion. Another twenty-seven squadrons were based around the Mediterranean, where, despite operating mainly obsolescent aircraft, they managed to maintain air supremacy and helped defeat the Italian attack on Greece in 1940, although the German attack the following year was successful.

The summer that followed saw the RAF face its most testing period in what became known as the Battle of Britain, as the *Luftwaffe* concentrated on attacking British airfields. The main phase of the battle was between 11 August and 30 September 1940. In the heat of the battle, with sometimes more than one pilot claiming an enemy aircraft as a 'kill', the initial claims for both sides were heavily exaggerated. At the time, the RAF claimed 2,698 enemy aircraft destroyed, while the *Luftwaffe* claimed 3,058, but the true figures are now estimated to be 1,733 *Luftwaffe* losses against 1,140 of the RAF. The shadow system of aircraft production, which saw, among others, much of the motor industry converted to aircraft production, meant that the supply of new fighter aircraft took second place to the shortage of air crew and especially experienced pilots. Desperate for pilots, the RAF borrowed heavily from the Fleet Air Arm and used French, Polish and other airmen who had fled their countries to continue the fight against Germany, while American volunteers eventually manned three squadrons. There were substantial numbers of personnel from the countries of the then British Empire. The 'Chain Home' network of radar stations was a secret weapon that enabled fighters to be scrambled in time to face German attacks.

The Battle of Britain was followed by a sustained heavy bombing campaign against British cities, known as the 'Blitz'. The *Luftwaffe* believed that they had destroyed the RAF, and that they could hamper its re-building by attacking factories and communications – but had the attack on the airfields been maintained, the RAF might eventually have run out of pilots. The German *blitzkrieg*, so successful when operating in support of ground forces in Poland, Norway, France and the Low Countries, was defeated by heavy anti-aircraft defences. Later, these were augmented by increasingly sophisticated night-fighters, initially the Bristol Beaufighter and then the de Havilland Mosquito. The *Luftwaffe* also suffered from the absence of an effective bomber. Severe damage, with heavy casualties, was inflicted on British cities, but the war effort was maintained, helped by the British breaking the German 'Enigma' codes and also bending the radar beams guiding German bombers.

After the fall of France and Norway, and then the entry of Italy into the war on the side of Germany in 1940, the only means of carrying the war to the enemy lay in a heavy bomber night offensive. The RAF's early experiences of offensive raids at the outset of the war had produced unsustainable casualties, due to inadequate aircraft and poor tactics. A generation of genuinely heavy bombers, the Short Stirling, Avro Lancaster and Handley Page Halifax, capable of delivering heavy warloads in excess of 8,000lbs and with a good defensive armament, meant that from 1942 onwards the balance began to tip in the RAF's favour. Navigation and bomb-aiming improved with aids better suited to fast-moving aircraft, and assisted by the creation of crack Pathfinder units to mark targets in advance of the main bomber force. No less important, the forces deployed against each target increased, attaining a critical mass that overwhelmed defending fighters and anti-aircraft defences. This led to the famous 'Thousand Bomber Raids', with the first on the night of 30–31 May 1942, against Cologne, with 1,050 aircraft deployed including Hampdens and Whitleys brought out of their working retirement on training and other duties. Losses on that raid amounted to forty aircraft, or 3.8 per cent, far below that on many less successful operations in which a third or more of the aircraft deployed had been lost. With the longer ranges of the new aircraft, targets in Italy could also be reached.

One of the most famous bomber operations was against the Moehne and Eder dams in May 1943, by Lancasters of 617 Squadron, using specially designed bouncing bombs. The squadron was later to become a specialist in precision bombing using ever larger weapons, including the 12,000lb (5,400kg) 'Tallboy' and 22,000lb (10,000kg) 'Grand Slam' against a variety of targets, including the battleship *Tirpitz*, the V-weapons sites, and railway junctions.

The entry of the USA saw RAF night raids and USAAF day raids alternate. Many American aircraft entered RAF service, with the Douglas C-47, known to the RAF as the Dakota, compensating for the absence of a good British transport aircraft. Transport Command was formed in March 1943, providing specialised transport units to support paratroop operations, such as that at Arnhem, and to provide assault and supply operations in many theatres of war, including towing Hamilcar and Horsa gliders used in the Normandy invasion and the crossing of the Rhine.

There had been little that the RAF could do in the Far East against the Japanese assaults of late 1941 and early 1942. In the Mediterranean, the RAF was outmatched in the defence of Malta, for example, at one time left to just three Gloster Sea Gladiator biplanes. Nevertheless, as the war progressed, tank-busting Hurricane fighters were put into service in North Africa and then later, after D-Day, the Hurricane's successor in this role, the Hawker Typhoon, undertook the same task in France.

The RAF played a major role in the fight against the German U-boat campaign against Allied shipping around the British Isles, in the Atlantic, and on the Russian convoys. Shore-based MR aircraft, notably the Short Sunderland flying boat and the Wellington bomber, maintained patrols. Later, American Consolidated Catalina flying boats, Lockheed Hudson and Consolidated Liberator bombers provided a welcome increase in Coastal Command's capability. A novel RAF unit formed for convoy protection was the Merchant Service Fighter Unit, operating Hurricanes catapulted from merchant vessels to protect them from German aircraft, with the pilot having to ditch his aircraft or bail out afterwards. This unit was made redundant by the advent of the escort carrier.

At the end of the war, the RAF was operating Hawker Typhoons and Tempests, Lockheed P-61 Lightnings and North American P-47 Mustangs, and Avro Lincoln heavy bombers in addition to many of the aircraft types introduced earlier. While the *Luftwaffe* had received Messerschmitt Me262 jet fighters during the war, Hitler's insistence that the aircraft be used as a bomber meant that the RAF was first to operate jet fighters when the Gloster Meteor entered service in 1944.

Peace in Europe found the RAF with 1.1 million personnel and 487 squadrons, of which about 100 came from the Dominions and the 'free' air forces, and 9,200 aircraft. Heavy post-war commitments meant that the service could not immediately reduce its numbers to the peacetime target of 300,000.

During the war, the RAF had shown that its strengths lay in the projection of strategic air power, with the capability of providing strong tactical support when called upon to do so.

Post-war, the RAF was heavily involved in the Berlin Airlift, which started in June 1948 when the USSR blockaded West Berlin, cutting access between the city and the British, French and American zones of occupation in what later became West Germany. A massive RAF and USAF transport operation was augmented by chartered civilian aircraft, moving food, clothing and fuel, including coal, to the beleaguered city. Upwards of 100 RAF transport aircraft were used in this operation.

Despite receiving new aircraft, such as the de Havilland Venom fighter and the world's first operational purpose-built jet bomber, the English Electric Canberra in

1950, the RAF was fully stretched with colonial-style police operations. The first of these were in Palestine, under a former League of Nations mandate, and in Malaya, while also faced with the need to countenance the growing threat from the USSR. Only limited RAF participation was possible in the Korean War because of other demands, leaving the main British contribution to United Nations air power over Korea to the Royal Navy's Fleet Air Arm.

The United Kingdom was a founder member of the North Atlantic Treaty Organisation, NATO, with much of the RAF's front-line strength in the United Kingdom, West Germany, Gibraltar, Malta and Cyprus assigned to NATO. The UK was also involved in many other regional defensive alliances, including the South East Asia Treaty Organisation and Baghdad Pact.

During the 1950s, the RAF received a considerable quantity of new aircraft of advanced performance for the period. These included the de Havilland Comet C2 jet transport and the Bristol Britannia turboprop transport, before which it gained the unwieldy Blackburn Beverley, a heavy transport with a good short field performance. The Westland Whirlwind helicopter was soon replaced by the Wessex, like its predecessor a licence-built development of the Sikorsky original, while a heavier helicopter for the RAF at this time was the twin-rotor Bristol Belvedere. The RAF's first swept-wing fighter, the Supermarine Swift, entered service in 1954, only to be withdrawn because of technical failings, but by contrast the Hawker Hunter which followed was a great success both for the RAF and for its manufacturer, as was the English Electric Canberra light jet bomber. Gloster Javelin delta-wing interceptors were also introduced, while the famous trio of 'V' bombers to carry Britain's nuclear deterrent, the Vickers Valiant, Avro Vulcan and Handley Page Victor, became operational. A major step forward in the RAF's strategic capabilities came with conversion of the Vickers Valiants as a substantial fleet of tanker aircraft for in-flight refuelling.

The most significant action for the RAF during the 1950s was the Anglo-French operation against Egypt after that country nationalised the Suez Canal in 1956. RAF, Fleet Air Arm, *Armée de l'Air* and *Marine Nationale* aircraft participated, with the RAF using British bases in Cyprus. Although a tactical success, the operation ended abruptly due to international pressure.

During the early 1960s, the RAF supported ground forces in a successful campaign against Indonesian attempts to undermine the newly-created Federation of Malaysia.

The period saw further sharp reductions in the RAF's strength, from 250,000 during the early 1950s to 110,000 two decades later. These reductions were accompanied by the disappearance of the RAuxAF, officially deemed to be impractical given the growing complexity of modern aircraft, but units of which had been called up for up to three months at a time to maintain RAF strength. A significant element in cutting RAF numbers was the ending of conscription during the early 1960s, making the United Kingdom the first major power to have all-regular armed forces. Further reductions in both personnel numbers and aircraft

261

were to follow throughout the next three decades, with the exception of a small and short-lived reversal of the process during the early 1980s.

New aircraft continued to enter service, reflecting the rapid developments taking place. The English Electric Lightning interceptor was followed by the McDonnell Douglas F-4K Phantom, itself accompanied by the Short Belfast, Vickers VC10, Lockheed C-130K Hercules and Hawker Siddeley Andover transports. After a number of policy changes and aborted projects, the Hawker Siddeley Buccaneer became the mainstay of the bomber force. Both the Phantom and Buccaneer forces received additional aircraft with the transfer of former Fleet Air Arm aircraft when the Royal Navy retired the last of its conventional aircraft carriers. Two strike aircraft types entered service during the 1970s, the Anglo-French Sepecat Jaguar and the Hawker Siddeley Harrier, a vertical take-off aircraft and the first of its type to enter operational service with any air force. Aerospatiale-Westland Puma helicopters replaced the remaining Whirlwinds, while later Westland Sea Kings (S-61) replaced the Wessex in the SAR role. For MR, Hawker Siddeley Nimrods, developed from the Comet airliner, replaced elderly Avro Shackletons. By the 1970s, with a few exceptions, procurement projects were made in collaboration with one or more European partners, or purchased from the United States. A multi-role combat aircraft project led to the Anglo-German-Italian Panavia Tornado, available in interceptor and interdictor variants.

The slimming down of the RAF led to the disappearance of its 1936 command structure, which had been augmented by the post-war creation of area commands such as RAF Germany and RAF Middle East. Two new commands were created during the late 1960s, Strike Command and Support Command, but these disappeared as manpower, and aircraft numbers, continued to fall. New aircraft that entered service during this period included the Boeing E-3A Sentry AWACS aircraft, with the RAF having its own fleet of seven aircraft in two squadrons.

Major operations for the RAF and for the British armed forces as a whole continued to arise. RAF units were deployed to Zambia for a short period after Rhodesia's Unilateral Declaration of Independence during the late 1960s, while units were deployed to Belize after a threat to the country's sovereignty. The RAF also became increasingly involved in providing humanitarian aid, most notably with air drops of supplies by Hercules aircraft during the Ethiopian famine of 1985. It provided tactical transport for the British Army in Northern Ireland from the late 1960s to the end of the century, mainly using Wessex and then Puma and Chinook helicopters. Other conflicts involving the RAF included the Falklands campaign in 1982 and a major effort during the Gulf War of 1991, followed in 1999 by operations against Serbian forces and installations in the former Yugoslavia in support of the Kosovo Albanian population. At a lower level of intensity, British forces also had to intervene in Sierra Leone to maintain stability in the country.

The Falklands campaign to recover the islands after an Argentine invasion was spearheaded by the Royal Navy's aircraft carriers, but RAF transports dropped mail and supplies to the Task Force. On the outbreak of hostilities, a solitary Vulcan made the first bombing raid, attacking the airfield at Port Stanley, although little damage was caused. RAF helicopters and Harrier GR3 ground-attack aircraft were also deployed aboard ships to assist in the recovery of the islands. Many Boeing Chinook helicopters were lost when the merchant ship carrying them was sunk. After the surrender of the Argentine forces, the RAF secured an air supply service linking the United Kingdom and the Falklands. Initially RAF Phantoms were deployed on the islands as a deterrent against any further interference by Argentine forces.

The Gulf War followed the invasion of Kuwait by Iraqi forces in 1990. Under the auspices of the United Nations, a 'Coalition' of countries under American command provided air, ground and naval forces for the liberation of Kuwait the following year. The RAF moved British troops to Kuwait during the military build-up to the conflict, although a shortage of heavy-lift capability following the sale of the Belfast fleet meant that much heavy equipment had to be moved in chartered aircraft. The main role played by the RAF's interdictor Tornado squadrons once hostilities started with an intensive air campaign against Iraq on 17 January 1991, was to attack Iraqi airfields, flying at low level with their runway denial weapons. The RAF's Tornado GR1 aircraft suffered a number of casualties, casting doubt on the effectiveness of the technique in combat conditions.

The air campaign over the former Yugoslavia started on the night of 24/25 March 1999, and continued until June. For this, the RAF mainly operated its Harrier GR7 force, augmented by a number of Tornados, deployed to the Italian Air Force's Gioia del Colle base in southern Italy.

In 2001, the RAF closed its last base in Germany. The service remains in demand for many 'out of area' operations. The RAF's tankers provided air-to-air refuelling for the USN and USMC during operations over Afghanistan during the winter of 2001/2 as the USN used a similar in-flight refuelling system to that of the RAF. Squadrons were deployed to Italy to monitor a peace agreement in Bosnia, part of the former Yugoslavia, and Turkey, to monitor a 'no fly' zone in northern Iraq, and before the mainly Anglo-American invasion of Iraq in spring 2003.

Additional Chinooks and the new Agusta-Westland Merlin helicopter have been introduced, with additional Merlins diverted from Denmark to fulfil an urgent operational requirement for more helicopters while more Chinook helicopters are on order. A variant of the Airbus A330 wide-bodied airliner is on order to replace the VC10 and ex-British Airways Lockheed TriStar 500 transport and in-flight refuelling fleets, while a trio of leased Boeing C-17 Globemaster II transports provided a heavy-lift capability once more, and this force has now been increased to six aircraft, all now owned rather than leased. Improving battlefield intelligence has been aided by the introduction of the Bombardier Sentinel aircraft, with five operated by a single squadron which is currently mainly deployed in Afghanistan.

Given the importance of strike aircraft, it seems strange that during the Afghan conflict the RAF lost its three squadrons of Sepecat Jaguar aircraft, shortly after they had received an extensive modernisation and life extension.

Some restructuring of British service aviation has seen the Royal Navy and Royal Air Force Harrier units combined into a Joint Force Harrier, tasked with operating from the Invincible-class carriers but with the loss of the Fleet Air Arm's Sea Harrier fighters, while a Joint Helicopter Command has grouped together RAF and RN transport helicopters.

Despite the heavy demands on the service, over the past seven years strength has fallen from 53,950 personnel to 33,480, with just 140 active reservists; a force that is spread thinly. The service has been organised into RAF Expeditionary Air Wings operating from its main bases, with the aim of being able to deploy effective units quickly and efficiently. There are five strike squadrons with Panavia Tornado GR4 and two with the GR4A. There is a flight of Tornado F-3 interceptors based in the Falklands, while in the UK, two squadrons of F-3s have been replaced by the new Eurofighter Typhoon and another two are converting. The RAF and Fleet Air Arm Joint Force Harrier operates three squadrons of GR7 and GR9, some of which have been deployed to Afghanistan, but have now been replaced by a Typhoon unit. Two squadrons operate seven Boeing E-3A Sentry AWACs aircraft. There are now just two MR squadrons with fourteen BAe Nimrod MR2s, being progressively re-winged and re-engined to MR4 standard, by which time numbers will be reduced to nine, or perhaps twelve aircraft at most, in 2012. An ELINT squadron operates the Nimrod R-1, while five Global Express ASTOR aircraft entered service in 2009. Transport and tanker aircraft include a squadron with Lockheed TriStar K-1/KC-2A transport tankers, while there is still a squadron of BAe VC-10C1K/K-3s, although these aircraft are due to be replaced by fourteen A330 tankers. Four squadrons operate twenty-six Lockheed C-130Ks and twenty-five C-130J Hercules/Hercules IIs, with the former likely to be replaced by twenty-two Airbus A400Ms by around 2014, while a fifth operates six C-17 Globemaster IIs. A Joint Helicopter Command controls three squadrons operating forty-seven CH-47 Chinook HC2/HC2A/HC3 heavy-lift helicopters, two squadrons with thirty Merlin HC3s, and two with Puma HC1s, while another three squadrons operate Sea King HAR-3s, and two SAR squadrons with twenty-five Westland Sea King HAR3/3As, although SAR duties will be contracted out to a commercial helicopter operator using Sikorsky S-92 machines starting in 2012. There are two BAe 146 CC2s and two Eurocopter Twin Squirrel HCC1s on VIP duties. Training uses a wide range of aircraft, including ninety-nine contractor-owned Grob 115 Tutor trainers and eighteen Slingsby T67M-2 Fireflies; with further training on seventy-three Shorts-built Tucano T1s and ninety-seven BAe Hawk T1/T1As, the latter also equipping the famous 'Red Arrows' aerobatic team. Other training uses ten Dominie (BAe125) and eleven Jetstream T1 (Jetstream 31) navigational trainers, thirty-eight Squirrel HT1/HT2 and nine Griffin HT1 helicopter trainers. Missiles include Sidewinder, Sky Flash AMRAAM, ASRAAM AAM, with Maverick and AGM-84D Harpoon ASM, and ALARM ARM, with Rapier SAM.

FLEET AIR ARM

Founded: 1924, passed into Admiralty control in 1939

Although the Royal Navy's Fleet Air Arm in its present form dates from 1939, as a result of the 1937 decision to return control of naval aviation to the Admiralty, the history of British naval aviation is far longer. In 1909, the sum of £35,000 (US $175,000 at the then rate of exchange) for an airship was included in the Naval Estimates. The pre-World War I Royal Navy made considerable strides, experimenting with take-offs from ships and converting warships as seaplane tenders. In 1912, the air branches of both the British Army and the Royal Navy were merged to form the Royal Flying Corps, but this was reversed with the formation of the Royal Naval Air Service in 1914. During the war, the Admiralty volunteered the RNAS to provide fighter protection for English cities threatened by German Zeppelin and, later, heavy bomber attack. The need to provide protection for convoys led to innovation, with landplane fighters flown off lighters towed at high speed behind destroyers, and seaplanes launched on trolleys from primitive flight decks on seaplane carriers. In 1916, fleet spotting by aircraft preceded the clash of the British and German fleets at the Battle of Jutland. The greatest innovation of all was the series of piecemeal conversions of the battlecruiser, HMS *Furious*, which led to the appearance of the first aircraft carrier. The first air strike from the sea was from *Furious* against the German balloon sheds at Tondern in July 1918. By this time, however, the RNAS had lost its independence, merged in April with the RFC to form the Royal Air Force.

In 1924, the Fleet Air Arm first came into existence as the carrier-borne arm of the Royal Air Force, although a number of naval aviators served with the fleet, mainly flying seaplanes and amphibians from battleships and cruisers. By 1924, there were four aircraft carriers, *Furious*, *Argus*, *Eagle* and, the first designed as such from the keel upwards, *Hermes*. These were later joined by *Courageous* and *Glorious*, converted battlecruisers that had been sisters of *Furious*. A seaplane carrier, *Ark Royal*, was later renamed *Pegasus* to release her name for a new aircraft carrier. The Fleet Air Arm between the wars suffered from neglect of its aircraft needs as the RAF struggled within a restricted defence budget to maintain its strategic forces. The late 1930s saw a plan to build a new carrier force, largely to replace the earlier ships. The first of these, HMS *Ark Royal*, joined the fleet before the outbreak of war. The next four were fast, armoured carriers, with two more ordered as the threat of war loomed, and with improvements included in the later ships. The new ships were *Illustrious*, *Victorious*, *Formidable*, *Indomitable*, *Implacable* and *Indefatigable*. Instead of replacing the earlier vessels they became additions to the fleet. These were to prove to be the most successful carriers during World War II.

While the Royal Navy found itself with new aircraft carriers, its aircraft on the outbreak of war were obsolescent, at best. Mainstay of the strike squadrons was the Fairey Swordfish, a biplane with open cockpits for its three-man crew and barely able to maintain 100 knots! There were no high performance fighters, with Gloster

Sea Gladiator biplanes and the monoplane Fairey Fulmar in service. Poor tactics saw the loss of *Courageous* during the first month of war. The Fleet Air Arm proved its worth during the Norwegian campaign of spring 1940. It made the first sinking of an operational warship, when shore-based aircraft attacked and sunk the German cruiser *Koenigsberg*. During the withdrawal from Norway, *Glorious* was lost to shellfire from two German battlecruisers.

One of the Fleet Air Arm's greatest achievements was the crippling of the Italian Fleet at Taranto in November 1940, using just twenty-one Fairey Swordfish flown from *Illustrious*, with only two aircraft lost in the night attack. Carrier-borne aircraft were also used against the Vichy French fleet in North and West Africa, at the Battle of Matapan and in the pursuit and sinking of the battleship *Bismarck*, as well as in many attacks against the *Tirpitz* in a Norwegian fjord. The Fleet Air Arm operated aircraft from MAC-ships, merchant aircraft carriers, converted tankers and grain ships, before receiving large numbers of escort carriers, mainly from the United States, for protection of convoys in the Arctic, North Atlantic and across the Bay of Biscay. Malta convoys were covered by large attack carriers, which also flew off aircraft to the beleaguered island. After early losses in the Indian Ocean, including *Hermes*, the Royal Navy returned to the East as the war progressed, operating against targets in the Dutch East Indies, and then working with the USN in the Pacific as the war edged towards Japan. New aircraft provided improved performance, including the Hawker Sea Hurricane and Supermarine Seafire, although the mainstay of the FAA was to prove to be the American Vought Corsair, Grumman Wildcat and Hellcat fighters, and Avenger bombers.

As the war ended, the Royal Navy had fifty-two aircraft carriers of all kinds. The US-owned escort carriers were returned. Older ships were scrapped or went into reserve. New light carriers of the Colossus-class and their derivatives were to prove both effective and also attractive to other navies, including the Dutch, French, Brazilian, Argentinian, Indian, Canadian and Australian. Successful deck-landing trials were carried out with the de Havilland Sea Hornet, a fast long-range twin-engined fighter, and with the de Havilland Sea Vampire jet fighter, the first jet to land on a ship, HMS *Ocean*, and with a Sikorsky R-4 helicopter onto the battleship *Vanguard*. The Royal Navy was to the forefront with the invention, but not always the introduction, of aids to modern naval aviation such as the steam catapult, mirror landing aids and the angled flight deck: the latter allowing take-offs and landings at the same time, and enabling landing aircraft to fly round safely if a first attempt did not succeed. Equipment also improved, with Douglas AD-4W Skyraider anti-submarine and, later, airborne-early warning aircraft; Westland Dragonfly (S-51) helicopters for communications and rescue, and de Havilland Sea Venom, Supermarine Attacker and Armstrong-Whitworth Sea Hawk fighters.

Heavy commitments prevented the RAF from participating fully in the Korean War, leaving British air operations to the RN. The FAA supported ground forces using the carriers *Triumph*, *Theseus*, *Glory* and *Ocean*, with one of the latter's piston-engined Hawker Sea Furies shooting down a jet MiG-15.

In 1956, the first rotary-wing assault on an enemy coast came with Royal Marines being landed during the Suez campaign from *Theseus* and *Ocean* using a combination of Westland Whirlwind (S-55) and Bristol Sycamore helicopters. Throughout the post-war period, the carrier force also supported ground forces in the so-called 'bush fire' wars. This included bandit activity in Malaya during the early 1950s, a mutiny by troops in East Africa in 1961 and the confrontation between Indonesia and the newly-independent state of Malaysia during the early 1960s. Later, the carriers were to be involved enforcing sanctions against Rhodesia after that country's unilateral declaration of independence.

New ships entered service, including Britain's two largest carriers, *Ark Royal* and *Eagle*, and the ships of the Centaur-class, *Albion, Bulwark, Centaur* and, last in service, *Hermes*. Aircraft for these ships included Fairey Gannet turboprop anti-submarine, AEW and COD aircraft, Westland Wessex (S-58) and Sea King (S-61) helicopters, Supermarine Scimitar and de Havilland Sea Vixen fighters, Blackburn Buccaneer bombers, and McDonnell Douglas F-4K Phantom fighters. Among the first to send helicopters to sea aboard destroyers and frigates, the Fleet Air Arm operated first Westland Wasp and then Westland Lynx helicopters, with all destroyers and frigates built after 1960 either converted to carry helicopters or, in the majority of cases, designed for helicopters. The fleet supply train, the Royal Fleet Auxiliary, RFA, also introduced ships capable of operating helicopters.

The middle and late 1960s saw a steady reduction in the strength of the Royal Navy, and especially of the Fleet Air Arm, with plans to end fixed-wing naval aviation completely. The creation of the 'Harrier Carrier' concept saved the Fleet Air Arm, although to overcome political opposition at first these had to be described as 'through deck cruisers'. The first, *Invincible*, entered service in 1980, although a defence review shortly afterwards proposed selling the ship. The Royal Navy selected the BAe Sea Harrier for its new carriers, a true fighter with secondary air-to-surface attack capability rather than the Harrier or AV-8 series favoured by the USMC.

British fixed-wing naval aviation, and the three-carrier force, was finally saved when the Argentine invaded the Falkland Islands, 1,000 miles off the coast of Argentina and 8,000 miles from the UK, in spring 1982. HMS *Invincible* and *Hermes* were despatched with a task force for the recovery of the islands, despite being heavily outnumbered by Argentine air force and naval aircraft, with just twenty Sea Harriers. The successful outcome of this campaign, despite losing two destroyers and two frigates to Argentine air attack, would not have been possible without carrier-borne aircraft, with the islands beyond the realistic operating range of shore-based aircraft, which required up to eleven tanker aircraft for each sortie! The loss of so many ships highlighted the lack of carrier-borne AEW, rectified afterwards with the conversion of a number of Sea King helicopters.

Post-Falklands, the Fleet Air Arm has suffered from further cuts, with one of its three carriers now in a low state of reserve, although a helicopter carrier, *Ocean*, has been introduced that can also act as an aircraft transport, while helicopters can be operated from the new landing ships' dock, *Albion* and *Bulwark*, which replaced *Intrepid* and *Fearless*, that had accompanied the Falklands Task Force. Both *Ocean* and *Illustrious* operated in the Arabian Sea as part of the Coalition fleet in

Operation Enduring Freedom during winter 2001/2. The Sea Harrier was withdrawn in 2004 with the RAF-RN Joint Force Harrier operating just three squadrons of Harrier GR7s and GR9s. This marks the end of fighter cover for the Fleet until the introduction of the F-35B around 2016. The anti-submarine carrier and RFA helicopter is now the Merlin with forty-four aircraft in four squadrons. The frigate and destroyer force has been reduced to just twenty-five ships, against more than sixty in the late 1960s. A new generation of Westland Super Lynx helicopters is being deployed aboard the frigates and destroyers. The Government has announced that two 65,000-tonne aircraft carriers will be introduced from 2016.

Today, the Fleet Air Arm has 6,200 of the Royal Navy's 30,690 personnel. It is part of the three-squadron Joint Force Harrier, with a squadron of GR9s earmarked for operations from the two aircraft carriers, *Ark Royal* and *Illustrious*. Front-line RN squadrons are numbered in the 8XX series, and support squadrons in the 7XX series to distinguish them from RAF squadrons. There are seven Harrier T7s for conversion training. A typical carrier air wing is a squadron of eight Sea Harriers and nine Merlin helicopters, with three Sea Kings for AEW. The Sea King HC4s have been redeployed into the Joint Helicopter Force. Six Lynx AH7s and nine Gazelle AH1s operate in a Royal Marine helicopter squadron, with RM air crew. Two SAR flights at Culdrose in Cornwall and Prestwick in Scotland operate Sea King helicopters, although their duties will be contracted out to a commercial helicopter operator using Sikorsky S-92s on behalf of the Maritime Safety Agency's Coast Guard in 2012. Training uses Beech King Air E350s, as well as twelve BAe Hawks and five Grob 115 Herons.

ARMY AIR CORPS

Founded: 1957

Although the British Army's experience of military aeronautics dates from 1878, this eventually led to the creation of the Royal Air Force. Present-day Army aviation dates from the formation of the Glider Pilot Regiment during the height of World War II, with its members taking part in a number of major operations, including the invasions of Sicily, Normandy and the Rhine Crossing, using troop-carrying Horsa and Hamilcar gliders. The Glider Pilot Regiment disbanded – the shortest-lived British Army corps – towards the end of the war. During the war, most flying in support of ground forces was conducted by the RAF, but twelve AOP squadrons were operated with mixed Army and RAF personnel. Some of these units survived the war years using Auster AOP aircraft. The eventual formation of the Army Air Corps in 1957 saw the gradual disappearance of RAF personnel.

During the 1960s, the emphasis moved from light aircraft to helicopters, including the Saro Skeeter and licence-built Westland 47G Sioux helicopters, with the Westland Scout introducing anti-tank operations with SS11 wire-controlled missiles. Units were established in the UK and to support the British Army of the Rhine, BAOR, in Germany, as well as in Hong Kong, the Persian Gulf and in Malaysia: the last two being run down first as British forces were gradually reduced and then withdrawn from East of Suez. During the early 1970s, Westland Lynxes replaced the Scout helicopters, while Westland-Sud SA340 Gazelles replaced the Sioux. At one time, DHC-2 Beaver utility transports were operated.

In 2001, the Army Air Corps started to introduce Westland-built versions of the AH-64 Longbow Apache, with the Lynx moving into the combat scout and light transport roles, but may be upgraded with mast-mounted sights and uprated engines. The sole fixed-wing aircraft are now five Britten-Norman BN2 Islander AL1s on communications duties. Heavier battlefield transport lies with the RAF's Pumas, originally to be replaced by Merlins but retained as operations in Afghanistan have highlighted an acute shortage of helicopters, and heavy-lift Boeing CH-47D Chinooks. A recent reorganisation has retained individual Army, RAF and Fleet Air Arm helicopter squadrons, but placed these in a Joint Helicopter Command, in which the AAC is 16 Air Assault Brigade.

Today, the AAC has sixty-seven WAH-64 Longbow Apache attack helicopters; 130 Westland AH7/AH9 Lynx helicopters, capable of anti-tank operations, but used mainly in combat scout and utility roles; 119 Gazelle AH1s on reconnaissance duties, four Agusta A109As on special forces duties, and three Bell 212 utility helicopters, plus the five Islanders. Many of the Lynxes are being replaced by thirty-four Lynx Wildcats.

UNITED STATES OF AMERICA

- Population: 304.1 million
- Land Area: 3,775,602 square miles (9,363,169 sq.km.)
- GDP: $13.8tr (£8.8tr), per capita $45,161 (£28,867)
- Defence Exp: $693.2bn (£443.1bn)
- Service Personnel: 1,539,587, plus 979,378 reserves

Today more than ever, the United States armed forces are the keystone which carries the burden of the defence of the world's democracies. While US defence expenditure has fallen as a proportion of GDP since the end of the Cold War, it is nevertheless much higher than that of any of its NATO allies, and three times that of its next-door neighbour Canada. Active manpower has actually increased over the past six years, although there has been a marked fall in the strength of the reserves.

During the Cold War, there was the growing realisation that the United States carried more than its fair burden of Europe's defence, but the situation is even worse today and the threat is far more complex. The cynics will suggest that the United States was protecting its overseas investments by contributing so much towards the defence of Europe and its allies in the Pacific and Middle East, and there is no doubt that these countries were also major markets for US exports, but such an imbalance is unhealthy and also leaves the rest of the world's democracies weak and vulnerable to a change in US policy.

One weakness in US defence planning and procurement is the power which Congress has over the annual budget, tackling this in detail unmatched anywhere else. There is no doubt that this is democracy in action, but it does interfere with longer-term planning and also makes true participation by the US in collaborative defence projects far more difficult than it should be. The situation is made worse by protectionist attitudes and also by attempts by members of both houses of Congress to protect industry and employment in their own constituencies.

Even within the US armed forces all is not as it might be. The division of maritime responsibilities between the United States Navy and the United States Coast Guard Service, which has been transferred from its original peacetime home in the Department of Transportation to the Department for Homeland Security, cannot but disguise the fact that the USCG is operating an increasingly elderly fleet of ships and aircraft. While it is understandable that the emphasis is on power projection to strengthen the resolve of wavering allies and support those too small to maintain well-balanced armed forces, the need to protect US interests in its own territorial waters remains important, not least with the modern challenges of illegal immigration, arms and drugs-smuggling and terrorism. It is also easy to forget that when war actually does break out, control of the USCG is passed to the USN, and surely the service does not want to find that its wartime expansion involves obsolescent ships and aircraft?

There have been moves to bring the air operations of the USN and the United States Marine Corps more closely in line in recent years, and this is to be welcomed, but it must not be an excuse to hamper USMC air operations in favour of the USN. The USMC has an important assault role, and needs to remain carrier-capable, not repeating the mistakes made at one period of the Pacific War when USMC pilots were not given carrier training. It is also helpful that it retains the capability of conducting aerial reconnaissance and providing fighter defence for the troops on the beaches.

UNITED STATES AIR FORCE

Founded: 1947

As early as the 1860s, balloons were used for observation duties during the American Civil War, but the history of American military aviation really dates from 1892, when a balloon section was formed within the United States Army during the Spanish-American War, and saw service in Cuba during 1898. The US

Army supported several unsuccessful attempts at building heavier-than-air aircraft after the turn of the century, notably the so-called 'Aerodrome A' of Professor Samuel Langley, but rejected offers of the Wright brothers' aircraft. In 1907, the Aeronautical Division of the Signals Corps was formed to take responsibility for military aviation, and in 1908, this finally purchased a Wright aircraft and a dirigible. By 1911, four Wright aircraft and two Curtiss aircraft were being operated on training, reconnaissance and experimental duties. Ten Curtiss aircraft flown during the war with Mexico in 1913 failed to impress, and just five aircraft were operational in 1914 when World War I broke out in Europe. In 1916, the Aeronautical Division became a separate corps, the US Army Aviation Section. The USA entered the war in 1917 against the Central Powers with 200 aircraft, mainly of Curtiss, Martin, Standard and Wright manufacture.

Ambitious wartime expansion schemes were initially heavily dependent on European designs, many built under licence, including Bristol F2B and Spad fighters, as well as DH4 bombers. European aircraft included almost 5,000 French-built aircraft, mainly of Breguet, Caudron, Farman, Morane-Saulnier, Nieuport, Spad and Voisin manufacture, with 300 British-built Avro, Airco, Bristol and Sopwith aircraft. Rapid expansion was helped by the return of Americans who had volunteered to fly with other Allied air arms before US entry into the war, especially in the French *Escadrille Lafayette*. Orders for 61,000 aircraft had to be cancelled within days of the Armistice in 1918.

Post-war, the strength of the US Army Aviation Section was established at 2,500 aircraft in 1920, but few new aircraft were forthcoming. Attempts to form an autonomous air service on the lines of the RAF failed, with the resulting court-martial of Brigadier-General William Mitchell as the result of intense inter-service rivalry with the USN, and the unauthorised bombing of the surrendered German battleship *Ostfriesland*. There was a further change of name to the US Army Air Service in 1920, with an initial strength of twenty-seven combat squadrons each of eighteen aircraft. The USAAS did grow, reaching forty-eight squadrons by 1923, although based on the obsolescent DH4B bomber. To maintain morale and gain publicity, long-distance flights were made. In 1926 the name changed again, to the US Army Air Corps. New aircraft started to enter service during the late 1920s, including Curtiss PW-8, P-1 and P-2 fighters and O-1 AOP aircraft; Martin MB-2 bombers; Douglas O-2 AOP aircraft, Consolidated PT-1 trainers and Douglas C-1 transports. The USAAC was hit by the Depression, curtailing a planned programme of expansion originally set to begin in 1926. Even so, a number of new aircraft continued to arrive, including 150 Curtiss A-3 Falcon ground-attack aircraft; twelve Curtiss B-2, and numbers of Keystone LB-5, LB-6, LB-7, B-3, B-4 and B-6 bombers; twelve Fokker FVIII and thirteen Ford C-3 and C-4 trimotor transports. Berliner-Joyce P-16, Boeing P-12B and Curtiss P-16 fighters followed these. Four balloon squadrons were retained for AOP duties. The USAAC also undertook internal air mail services at this time.

The USAAC steadily achieved the size and equipment quality of the air arm of a major power. As the economic situation improved, it grew more rapidly. New aircraft included the Boeing P-26A fighter of 1933, and the advanced Martin B-10 bomber of 1934. A series of accidents in 1935 led to an investigation, which gave the USAAC a greater degree of autonomy, although still remaining part of the US Army. The year also saw the first of fifty Consolidated P-30A and seventy-six Seversky P-35 fighters, thirty-two Martin B-12 and 250 Northrop A-17 bombers and attack aircraft respectively, and ninety Douglas O-46A AOP aircraft. In 1936, the Douglas B-18, the bomber version of the C-47 transport, entered service. Both the USAAC and the USN undertook maritime-reconnaissance, leading to renewed inter-service rivalry.

Following the Munich crisis of 1938, expansion accelerated, especially once war broke out in Europe in September 1939. Aircraft ordered included Boeing P-26, Seversky P-35 and Curtiss P-36 fighters; Northrop A-17 and Curtiss A-12 and A-18 ground-attack aircraft; Martin B-10 and B-12, Boeing B-17 Fortress and Douglas B-18 bombers; Douglas OA-3 and OA-4, Sikorsky OA-8 and Grumman OA-9 amphibians; Bellanca C-27, Douglas C-33, C-39 and C-47, Lockheed C-36 and C-40 transports; North American O-47 and Douglas O-38, O-43 and O-46 AOP aircraft; and North American AT-6 and BT-9, Consolidated PT-11, Stearman PT-13 and Seversky BT-8 trainers.

Many of the aircraft that followed benefited considerably from experience gained from service with the British and French armed forces before full-scale entry into US service. These included Lockheed P-38 Lightning fighters, and Consolidated B-24 Liberator, Martin B-26 Marauder and North American B-25 Mitchell bombers. In contrast to the Germans and Russians, who used air power tactically in close support of ground forces, the USAAC planned to use air power strategically. In 1941, the USAAC became the United States Army Air Force, with four main constituent air forces, the 1st and 2nd in the north, and the 3rd and 4th in the south, deployed primarily for the defence of the United States. When the Japanese attacked Pearl Harbor on 7 December 1941, bringing the USA into World War II, the USAAF had some 2,500 aircraft in service. The latest were Bell P-39, Curtiss P-40, Republic P-43 and P-47 Thunderbolt, and North American F-51 Mustang fighters; Douglas A-20 Boston and A-24, and Curtiss A-25 attack aircraft; Beech C-45, Curtiss C-46 Commando and Douglas C-47, C-53, C-54 and Lockheed C-59 and C-60 transports. Most aircraft were still based in the continental United States; many in Hawaii and the Philippines were destroyed on the ground by air attack.

US entry into the war was provoked by the Japanese attack, although the USA had given strong support to the United Kingdom, doing almost everything short of an open declaration of war. Debate raged over whether the war against Japan should be given priority, defeating that country first before settling matters in Europe, but the USA decided to wage war on two fronts, helping to shorten hostilities. The USAAF would have had great difficulty in taking the offensive against Japan until bases had been secured within range of the Japanese islands, but a major propaganda coup that influenced future Japanese strategy was Operation Shangri-La, a raid on Japanese cities, including Tokyo, using sixteen North American B-25 Mitchell bombers flown off the USS *Hornet* on 18 April 1942, proving that Japan was not immune from aerial attack.

From 1942 onwards, the USAAF operated in increasing strength in North Africa, and from the British Isles over Europe, from Hawaii and Midway over the Pacific, and from India over China and Burma. Few combat aircraft remained in the United States. New concepts included 'shuttle bombing' in which bombers engaged targets deep inside enemy-occupied Europe and then flew on to bases in North Africa. The USAAF co-operated with the RAF, notably in the 'Point Blank' initiative, with the RAF attacking targets by night with a follow-up USAAF raid by day. Day bombing demanded heavier defensive armament on bombers, and the use of long-range escort fighters, such as the Lockheed P-38 Lightning and the North American F-51 Mustang. Bomber wing commanders devised elaborate formations so that aircraft could provide covering fire for each other. Air transport grew in importance, with the USAAF flying supplies 'over the hump' to China. Individual aircraft types were deployed according to their strengths, with the long range of the Liberator proving useful for raids on Burma from India, and for raids on the Romanian oil refineries from bases in England and North Africa.

The USAAF took part in the invasion of Sicily in 1943, and the following year in the Normandy landings, as well as in landings in Italy and the South of France. Troop-carrying aircraft and gliders were used. Among the significant new aircraft was the Northrop P-61 Black Widow night-fighter and the Boeing B-29 Superfortress bomber. B-29s destroyed Japan's major cities in incendiary raids, before effectively ending the war in the Pacific by dropping the first atomic bombs on Hiroshima and Nagasaki on 6 and 9 August 1945.

The USAAF ended the war as the world's strongest air force, with some 60,000 aircraft and more than 2,250,000 personnel. In 1946, it was reorganised into Air Defence, Tactical and Strategic Air Commands, with support commands including Air Material, Air Proving, Air Training and Air Transport. The long-cherished ambition to become a separate service was realised in 1947 when the USAAF became the United States Air Force. One Army tradition remained with the main reserve units organised on a state-by-state basis as the Air National Guard. That same year the USAF received its first jet fighter, the Lockheed F-80 Shooting Star, and the switch from 'P', 'Pursuit', to 'F', 'Fighter' designations. When the US services eventually standardised aircraft designations, it was the straightforward USAF system that was adopted. The Republic F-84 Thunderjet joined the F-80s in 1948. New aircraft entering service in 1947 included the Boeing B-50, a higher-powered version of the Superfortress, and the six-engined Convair B-36, designed to bomb Germany from bases in the United States had the UK fallen. Many B-29s were converted to the tanker role for in-flight refuelling.

Structural changes also occurred. In 1948, Air Transport Command absorbed the Naval Air Transport Service to create the Military Air Transport Service, MATS, which assumed responsibility for all US strategic air transport requirements, leaving the other services with a small number of tactical transport aircraft, just as Tactical Air Command also operated some theatre transport and communications aircraft.

The return of peace was short-lived. An indication of what was to come had already occurred in 1944, when three USAAF B-29s had forced-landed on 'friendly' Soviet territory, but had been confiscated and copied, producing the Tupolev Tu-4. Germany was already being divided into zones, with the American, British and French zones eventually to become West Germany, but leaving the capital, Berlin, also divided into zones, well inside what was to become East Germany. In 1948, anxious to take power over the entire city, the USSR blockaded the land routes into Berlin from the West, forcing the Allied Powers to mount an airlift to keep the West Berliners supplied. Everything, including coal, had to be moved by air. USAF Fairchild C-82, Douglas C-47 and C-54 transports played their part in moving supplies, but the demand was such that military transports had to be augmented by chartered civilian aircraft. New aircraft continued to arrive, including the North American F-86 Sabre and Lockheed F-94 Starfire fighters, and North American B-45 Tornado light jet bombers. Progressive modernisation proved its worth when war broke out in Korea in 1950 after Communist North Korea mounted a surprise invasion of South Korea. USAF forces in Okinawa and Japan were well placed to counter the invading forces: their location was an advantage as bases in South Korea were overrun. Air National Guard units were mobilised during the Korean War, doubling the USAF's strength. During the conflict, Allied forces fought under the auspices of the United Nations. The Korean War lasted until 1953, and saw some of the first jet fighter battles, with Lockheed F-80C Shooting Stars matched against MiG-15s.

In 1951, the first heavy jet bomber, the Boeing B-47 Stratojet, entered service with Strategic Air Command. Heavier still was the eight-engined Boeing B-52 Stratofortress that replaced the Convair B-36s in 1955, starting a long period of service that has lasted into the current century with the aircraft modified as a cruise missile platform. Other aircraft included the Martin B-57, a licence-built English Electric Canberra light jet bomber. Reflecting the rapid change from piston to jet technology for combat aircraft and the growing importance of airborne radar, new aircraft included the North American F-100 Super Sabre, McDonnell F-101 Voodoo and Northrop F-89 Scorpion fighters and interceptors; Douglas B-66 Destroyer bombers; Lockheed RC-121 Constellation AEW aircraft; Douglas C-124 Globemaster and C-133 Liftmaster, and Fairchild C-119 Packet and C-123 Provider transports. These were followed during the late 1950s by Convair F-102 Delta Dagger and F-106 Delta Dart, and Lockheed F-104 Starfighter interceptors; the Convair B-58 Hustler supersonic jet bomber and the Lockheed C-130 Hercules transport, still in production today. Vertol and Sikorsky helicopters were also introduced during the decade. The need for global reach was reflected not only in increasingly heavy transport aircraft, but in a large fleet of several hundred Boeing KC-135 tankers.

The formation of the North Atlantic Treaty Organisation, NATO, in 1952, and the Organisation of American States, also led to the USAF assisting many allied air forces, while retired aircraft were welcomed by the poorer member states. The United States was the major partner in NATO, and its regional counterparts, the South East Asia Treaty Organisation, SEATO, and the Baghdad Pact, later the Central Treaty Organisation, or CENTO, in the Middle East.

Most of the late 1950s influx of aircraft continued in production into the 1960s, often suitably upgraded. The 1960s saw the Northrop F-5 tactical fighter-bomber, mainly exported under the US Military Aid Programme, MAP; the McDonnell F-4 Phantom II fighter-bomber; the LTV A-7 attack aircraft; the General Dynamics F-111 variable-geometry strike aircraft; and the Lockheed C-141 Starlifter and C-5 Galaxy transports. The F-4 and A-7 were naval aircraft that for the first time offered a performance that earned them a place in the USAF inventory. The decade was marked by a growing US involvement in Vietnam, where the USAF was involved in countering Communist infiltration, bombing build-ups of Communist Viet Cong guerrillas in the South and also attacking targets in the North. The B-52 heavy bombers were deployed, although subsequently proving vulnerable to the North Vietnamese firing SAM missiles in salvoes to ensure a hit. More successful was the F-4 operating with 'smart' bombs, able to make a precision strike against a target while enabling the aircraft to remain clear of the worst AA and SAM fire. Political uncertainties in the US meant that the intensity of the campaign varied, and eventually South Vietnam was overrun.

Changing technology has been constantly adopted by the USAF. The B-52 bomber fleet was upgraded and refurbished, gaining a new lease of life as cruise missile platforms. The ability to deploy cruise missiles proved an effective counter to a new generation of Soviet intermediate-range missiles deployed in Europe. Aircraft such as the F-111 could exploit the gap below the scope of ground-based radar. To counter such techniques Airborne Early Warning and Control, AEW&C, aircraft, the Boeing E-3 Sentry, based on the Boeing 707 airliner, were introduced. New aircraft reflected these developments, with the supersonic North American Rockwell B-1 bomber, and then the 'stealthy' Northrop Grumman B-2, with a much reduced radar signature. Post-WWII, the USAF deployed worldwide, especially in the UK, West Germany, Spain, Turkey and Japan. The Military Air Transport Service became Military Airlift Command, MAC, and by the late 1980s this had acquired a daily airlift capability of 66 million ton/miles per day. One of MAC's duties became SAR, and also Combat Search and Rescue, CSAR, using Sikorsky S-65 helicopters, and on many occasions during the Vietnam War downed US pilots were rescued from behind enemy lines.

New aircraft entering service during the final years of the Cold War included the McDonnell Douglas F-15 Eagle air superiority fighter and the Lockheed Martin F-16 fighter-bomber.

The collapse of the Soviet Union and of the Warsaw Pact meant a reassessment of priorities. This was further emphasised in August 1990, when Iraq invaded Kuwait. The immediate priority was to protect Saudi Arabia and that country's vital oilfields – Operation Desert Shield, followed later by Operation Desert Storm, the liberation of Kuwait. The forward basing of so many units in Europe meant that the USAF was able, within two months, to base the equivalent of five tactical fighter wings in the Middle East, drawing aircraft from across the USA and Europe. Other aircraft operated from bases in Turkey and in the Indian Ocean. MAC was the main force in moving not only the heavy equipment and many of the personnel of the US armed forces, but those of its allies, known as the Coalition, as well. The bulk of the Coalition air offensive was provided by the USAF, mainly using F-111s and cruise-missile- carrying B-52Gs, while the new Lockheed F-117 Nighthawk stealth fighter was also deployed for the first time. The F-117 units, deployed against important targets and using laser weapons, achieved success rates as high as 80 to 85 per cent, compared to the 30 to 35 per cent managed by advanced weaponry in Vietnam.

The success of the Coalition Forces was limited by the decision not to occupy Iraq, which would have needed a fresh mandate from the United Nations, which had sanctioned the liberation of Kuwait in the first UN operation since the Korean War. Nevertheless, the USAF remained in the region and in Turkey, enforcing two 'no fly' zones, one in the south of Iraq and the other in the north.

The collapse of the Warsaw Pact and the experience of the Gulf War led to a reassessment of the USAF's structure. Strategic Air Command was disbanded and its aircraft, personnel and bases reallocated, mainly to Tactical Air Command, which in turn became Air Combat Command, ACC, from January 1992. On the same date, the SAC tanker fleet passed to a new Air Mobility Command, AMC, which absorbed MAC. The USAF changed to provide rapid response service, rather than massive retaliation. Air mobility was aided by the new Boeing C-17 Globemaster II heavy-lifter.

Post-Cold War reductions in the strength of the USAF were used to enhance the equipment of the USAF Reserve and the Air National Guard (ANG) units, with the latter now manning the majority of the USAF's air defence squadrons. During operations over Kosovo during 1999, Operation Allied Force, the USAF performed a far larger number of the missions than the entire effort of the other allies. Aircraft deployed during these operations included the Lockheed F-117A, Lockheed Martin F-16 Falcon and the Fairchild OA-10 Thunderbolt or Warthog. The B-52 force was deployed again during 2001/2 in operations against the al-Qa'eda network, mainly operating from Gan in the Indian Ocean. The use of US-based B-2s to attack targets in Serbia demonstrated the ability to attack over extremely long distances. In 1999 further changes were made, introducing the Aerospace Expeditionary Force, AEF, concept, with almost the entire USAF, Reserves and ANG organised into ten AEFs, each with between 10,000 and 15,000 personnel and up to 200 aircraft, including air superiority, air-to-ground precision strike and air mobility capability. Each AEF will be on call for ninety days every fifteen months, with at least two AEFs on call at any one time, so that the USAF should be able to handle two medium intensity conflicts at any one time.

The USAF has 340,530 personnel, a reduction of more than 50 per cent over the past thirty years. There are also 191,038 personnel in the Air Force Reserve and Air National Guard units. Flying hours are an average of 212 annually. In the tactical role, there are fifty-two fighter squadrons, each with between twelve and twenty-four aircraft. The fighter squadrons include fourteen with F-15s and six with F-15Es, with a total of 615 aircraft, two with fifty-two F-117s, twenty-three with 782 F-16C/Ds, and seven with around sixty-three A-10s and forty-five OA-10s. In due course, 339 F-22 Raptors will replace some of the aircraft in this force, followed later by USAF versions of the Joint Strike Fighter, JSF, the F-35A. Support aircraft include three reconnaissance squadrons with thirty-one U-2s and eighteen RC-135U/V/Ws, as well as AEW&C with six squadrons, including one for training, with thirty-two E-3s, while another two squadrons have twenty-two EC-130s for AEW and ELINT. Other support aircraft include forward air control with about sixty-three A-10As and forty-five OA-10As in seven squadrons. Bomber aircraft include seventy-three B1B Lancers and twenty-one B-2A Spirits, as well as eighty-five B-52H Superfortresses. There are twenty-eight transport squadrons, of which seventeen are strategic, with five, including a training squadron, equipped with eighty C-5A/B Galaxies; three with C-17 Globemaster IIs, still in course of delivery with 120 planned; and these will eventually replace many of the aircraft in the nine squadrons (with two on training duties) of 139 C-141 Starlifters. The other eleven transport squadrons provide tactical airlift, equipped with 191 Lockheed C-130 Hercules, including the new C-130J Hercules II. There are also twenty-three tanker squadrons, of which nineteen operate 255 KC-135s, including a training squadron, while the other four have fifty-nine KC-10A Extender (DC-10) tankers. Eight squadrons operate forty-seven HH-60 helicopters and twenty-four HC-130N/P Hercules on CSAR. Three squadrons operate twenty-nine MEDEVAC C-9A (DC-9) Nightingales. Weather reconnaissance uses three WC-135s. There are a number of miscellaneous aircraft in the transport role, with small numbers of C-9s, C-12s, C-20s, C-21s, C-135s and VC-137s; a number of these operate VIP flights. Training aircraft include a squadron of F-16s in the aggressor role, as well as another thirty-five squadrons with conversion trainer variants of front-line aircraft. Basic and intermediate training is provided by contractors using Slingsby T-3 Fireflies and Boeing T-1 Jayhawks (BAe Hawk) as well as T-37 Tweets.

There are thirty-five wings in the Air Force Reserve. These include a bomber squadron with nine B-52Hs. Seven fighter squadrons include four equipped with seventy-one F-16C/Ds and three with seventy-four A/OA-10 Thunderbolt IIs. There are nineteen transport squadrons, with the seven strategic squadrons including two operating thirty-two C-5As and five with forty-eight C-141 Starlifters. The eleven tactical transport squadrons operate 130 Lockheed C-130E/H Hercules, while there is a weather-reconnaissance squadron with WC-135E/H. Seven tanker squadrons operate seventy-two KC-135E/Rs. SAR squadrons number three, with twenty-three HH-60 Black Hawks and seven HC-130 Hercules. There are another twenty-six squadrons without aircraft but with the personnel to augment the operations of active squadrons of transport and training aircraft in an emergency.

The Air National Guard has ten air defence fighter squadrons, with F-15s and F-16s, as well as forty-one in the fighter/strike role, of which thirty-two have F-16s and five F-15A/Bs, and six have A-10/AO-10s, with a total of 116 F-15, 600 F-16 and 100 A-10/AO-10 aircraft. There are twenty-seven transport squadrons, of which three are strategic, with one operating thirteen C-5As and two eighteen C-141s. The twenty-four tactical squadrons, including a training unit, have 225 C-130E/H Hercules. There are twenty-three tanker squadrons with 224 KC-135E/Rs. A special operations squadron has eight EC-130E EW Hercules. Three SAR squadrons have seventeen HH-60 Black Hawks and thirteen CH-130 Hercules. There are seven training squadrons.

The USAF is a major user of UAV, including a wing of six squadrons operating Predator and Global Hawk.

Missiles used include the AAM Sidewinder, Sparrow and AMRAAM, and the ASM Maverick, Harm, Harpoon and ALCM and JSOW.

UNITED STATES NAVY AND UNITED STATES MARINE CORPS

Founded: 1911

Although US naval aviation dates officially from 1911, in November 1910 a young naval officer, Lt. Eugene Ely had taken off from a wooden platform built over the forecastle of the light cruiser USS *Birmingham* while the ship lay at anchor. Ely repeated this exploit the following January, flying from a platform constructed over the stern of the cruiser USS *Pennsylvania*, but on this occasion flew his aircraft onto the ship first! Later that year, the first funds were voted for naval aircraft, a Wright and two Curtiss machines. Within two years, the United States Navy was operating eight aircraft, including some seaplanes operating from warships in a brief war with Mexico. It was to take the First World War to achieve rapid expansion, growing from just twenty-one aircraft in 1917 to the USN and USMC having more than 1,000 seaplanes and flying boats as well as 250 landplanes at the end of the war. At first, many personnel had served with other Allied air arms before the USN was ready to mount anti-submarine patrols. In 1918, Curtiss R-6 and HS2-2 seaplanes and H-16 flying boats predominated, but there were substantial numbers of aircraft built by the new Naval Aircraft Factory plus foreign imports.

Post-war organisation and equipment of the USN and USMC was based on Curtiss JN-4H, Martin MBY and Thomas-Morse M1 aircraft, Curtiss N-9 and R-6, and Boeing CL-4 seaplanes, with Curtiss NC series, licence-built Felixstowe F5 and F5L, Aeromarine 40L and HS2L flying boats, and Airco DH4Bs. Sopwith Scouts were flown from platforms constructed over the turrets of battleships. An attempt to fly across the Atlantic with four Curtiss NC flying boats in 1919 resulted in one of the aircraft, NC-4, reaching Lisbon via the Azores in late May, before flying on to reach England.

In 1921, the USN formed its Bureau of Aeronautics, known as BuAer, to advise the Chief of Naval Operations on all aspects of aeronautical affairs. In 1922, a collier was converted to become the USN's first aircraft carrier, the USS *Langley*; a Vought VE-7-SF biplane was first to take off. Some 300 new aircraft entered service in the early 1920s, including the Naval Aircraft Factory PT, Davis-Douglas, Martin MO-1 observation floatplanes and Vought OU-1 biplanes, while aircraft were bought from many European manufacturers for evaluation. The USN also experimented with airships. A number of landplanes, including DH4s, Douglas DTs and Vought UO-1s, were fitted with arrester hooks so that they could operate from the *Langley*. USN aviation was organised into Pacific Fleet and Atlantic Fleet air forces, each with six squadrons for bombing, torpedo-bombing, reconnaissance and fighter duties, while the USMC had four air squadrons. Large-scale exercises included torpedo attacks against battleship targets.

Squadron designations were specified from 1922 as VF, fighters; VO, observation; VS, scout; VT, torpedo and bombing; ZK, balloons and kites. The letters were followed by the pennant number of the warship.

In 1922, the Washington Naval Treaty limited the tonnage of all classes of warships in the major navies. The USN had too high a tonnage of battlecruisers, so converted the USS *Saratoga* and *Lexington* into aircraft carriers. These were the largest warships of their kind when they entered service in the late 1920s. The USN also had twenty battleships each capable of carrying three catapult-launched seaplanes, and ten cruisers, each capable of launching two seaplanes.

Progress continued throughout the 1930s, with a fourth aircraft carrier, *Ranger*, joining the fleet in 1934, the first US carrier to be designed as such but she was too small for successful operation in the heavy swell found in the open waters of the Pacific. New aircraft included Boeing F3B-1 and F4B-2 fighters, Chance-Vought O2U-4 and O3U-2 Corsair and Curtiss O2C-1/2 observation aircraft, and Consolidated P2Y-1 flying boats, followed by Curtiss Helldiver dive-bombers and Sparrowhawk fighters, some of which operated from airships as an experiment. The USMC also started carrier operations at this time. Two new aircraft carriers, *Yorktown* and *Enterprise*, entered service during the late 1930s, and were followed by a smaller ship, *Wasp*. As the end of the decade approached, the USN and USMC were building up towards a total of 3,000 aircraft, with many of the new arrivals in the early 1940s destined to achieve fame during World War II. New aircraft entering service included Brewster F2A, Vought F4U Corsair II, and Grumman F4F Wildcat and F6F Hellcat fighters; Douglas SBD Dauntless and Curtiss-Wright SB2C Helldiver dive-bombers; Douglas TBD Devastator and, later, Grumman TBF Avenger torpedo-bombers; Vought Sikorsky OS2U Kingfisher observation aircraft; Consolidated PBY-1 Catalina and Martin PBM Mariner flying boats. The Catalina was also available as an amphibian. In 1941, a new carrier, USS *Hornet*, entered service. Recognising the threat of a major war, eleven aircraft carriers were already under construction at the time of the Japanese attack on Pearl Harbor in 1941. Plans had also been made for the conversion of merchant vessels to provide low cost escort carriers, known to the USN as CVEs, with more than 100 built for both the USN and for loan to the Royal Navy. Between the escort carriers and the large, fast attack carriers, lay the Independence-class light carriers, CVLs, converted from Cleveland-class light cruisers.

The Japanese attack on the US Pacific Fleet base at Pearl Harbor on 7 December 1941 was designed to cripple the US fleet, destroy the base and give the Imperial Japanese Navy twelve months as undisputed masters of the Pacific. It failed. The base was never rendered unusable, and the carriers were safely at sea at the time. The USN then started the task of taking the war across the Pacific and back to Japan, which had quickly established a vast Asian empire in the weeks following Pearl Harbor. In April 1942, *Hornet* flew off sixteen North American B-25 Mitchell medium bombers of the USAF to attack Japanese cities, including Tokyo, and while the raid achieved little in tactical terms, it changed Japanese strategy. The following month, the two navies engaged in the first carrier-to-carrier battle, the Battle of the Coral Sea, in which the USN lost *Lexington* and saw *Yorktown* badly damaged, but sunk the Japanese *Shoho* and severely damaged *Shokaku*. In June, just six months after Pearl Harbor, the US Pacific Fleet sank four Japanese carriers in just one day at the Battle of Midway, putting paid to Japanese ambitions. The vast distances of the Pacific meant that, at first, only carrier-borne aircraft could take the lead in the war. A pattern emerged of the USN supporting invasions, with the USMC and US Army taking islands, and then the USAF would move in and use the bases once the islands were secured. At the same time, the USN had to protect shipping in the Atlantic and also took part in the invasion of North Africa before which the *Wasp* twice ferried aircraft to Malta. Escort carriers and longer-range MR aircraft, notably the Consolidated Liberator, helped to secure the North Atlantic for merchant shipping, while the United States Coast Guard Service came under USN control for the duration of the war. As the Battle of the Atlantic against German submarines was gradually won, escort carriers started to find new roles, often carrying aircraft for ground-attack duties and covering invasion forces, and as aircraft transports. Japanese resistance was strong, and included the use of *Kamikaze* suicide aircraft against warships, including aircraft carriers, as well as bombing and torpedo-bombing attack. Significant milestones in the Pacific included the campaign known to the USN as the 'Great Marianas Turkey Shoot', in the Battle of the Philippine Sea in June 1944, with 300 out of 365 Japanese aircraft based on Truk, an island in the Marianas, shot down. This was followed in October by the Battle of Leyte Gulf, the biggest naval battle in history. Japan lost three battleships, four aircraft carriers, six heavy cruisers, four light cruisers and many smaller vessels; many of them to aerial attack, while the USN lost just one light carrier, *Princeton*.

Post-war, the USN was the largest navy, with 41,000 aircraft, twenty large aircraft carriers and five under construction, eight light carriers and sixty-nine escort carriers. Most of the large carriers were new ships of the Essex-class, although *Enterprise* and *Ranger* survived the war. In addition to the Liberators, land-based aircraft in USN service included the Boeing B-17 Fortress and B-29 Superfortress, North American B-25 Mitchell and Lockheed PV-1 and PV-2 Harpoon bombers, as well as many transport aircraft, including Curtiss C-46s and Douglas C-47s, and Lockheed Lodestars. The war had seen rivalry with the USAAF over allocations of long-range aircraft.

The USN started to reduce to thirteen large aircraft carriers. Aircraft numbers were rapidly reduced to around 10,000, but many of these were to be new aircraft, of types developed too late to play a part in the conflict. These included Grumman F7F Tigercat, F8F Bearcat and Ryan FR-1 Fireball fighters, the latter being a compound aircraft with a main piston engine and an auxiliary jet engine to boost performance. These were accompanied by Douglas A-1 Skyraider and Martin AM-1 Marauder strike aircraft, while Lockheed P2V Neptunes gradually took over MR from Harpoons, Fortresses, Superfortresses and Liberators. The USN introduced its first jets, the McDonnell FH-1 Phantoms. New transport aircraft included Fairchild C-119 Packets, Douglas C-54 Skymasters and transport versions of the Grumman VF-1 Albatross amphibian and Martin JRM Mars flying boat, both of which also operated SAR. While awaiting sufficient carrier-capable jets, land-based Lockheed F-80 Shooting Stars were introduced to increase the pool of experienced jet pilots. Starting in 1948, all but the smaller aircraft in the Naval Air Transport Service were transferred to the USAF's Military Air Transport Service. The transfer was not one-way, as the USN finally gained control of all MR operations, ending yet another instance of duplication.

Carrier-capable jets eventually started to arrive in large numbers including the Grumman F9F Panther, McDonnell F2H Banshee, and Chance-Vought F6U Pirate, and the naval version of the Sabre, the North American FJ-1 Fury. Sikorsky S-51 helicopters were introduced for plane-guard and communications duties, relieving destroyers of this role. The plane-guard helicopter was more efficient than a destroyer, and eliminated the risk of a destroyer being sunk while stopping to pick up survivors.

The Korean War meant the recall of many reservists to increase USN and USMC manpower. The two air arms provided support for ground forces. On 20 September 1951, twelve USN Sikorsky S-55 helicopters lifted a company of 228 fully-equipped US Marines to the top of a strategically important 3,000-ft high hilltop in central Korea. The helicopters then carried nine tons of food to the Marines, before completing the operation by laying a field telephone system back to headquarters. The entire exercise took four hours, instead of the two days it would have taken on foot. Within a month, 1,000 combat-equipped Marines were moved in full view of enemy forces, taking six hours and fifteen minutes, twenty-five minutes less than the Marines had planned. USMC helicopters joined those of the US Army in such roles as CASEVAC, and survival rates of seriously wounded soldiers were vastly improved because of the helicopters' speed and through being spared a bumpy ambulance journey over rough terrain.

During the Korean War, the North American AJ-1 Savage attack aircraft started to enter USN service, a piston-engined design, but the first USN aircraft designed to deliver an atomic bomb. Variants of existing aircraft included airborne-early-warning versions of the Skyraider. Other new aircraft included the Douglas F4D-1 Skyray, Grumman F9F-6 Cougar and McDonnell F3H Demon fighters, as well as land-based Lockheed R70s, MR versions of the Super Constellation, and Fairchild R49. North American T-28B Trojan and Lockheed TV-2 jet trainers were also introduced.

New aircraft carriers also appeared, with the USS *Forrestal* and a new *Saratoga*. American carriers were the first to incorporate the many advances pioneered by the Royal Navy. Aircraft of the late 1950s included Chance-Vought F8U-1 Crusader and F7U-3M Cutlass fighters and the Douglas A3D-2 Skywarrior carrier-borne bomber. Grumman produced the S2F-1 Tracker anti-submarine aircraft, which was joined aboard the carriers by its derivatives, the WF-2 Tracer AEW aircraft and the Trader carrier onboard-delivery, COD, aircraft. Traditionally, carrier-borne aircraft have had an inferior performance to their land-based counterparts, usually due to the extra weight of folding wings, arrester hooks and strengthened undercarriages, but during the 1960s the McDonnell F-4 Phantom II and the LTV A-7 Corsair II proved a match for their shore-based counterparts, and both entered service with the USAF and many other air forces. Other aircraft at this time included the Douglas A-4 Skyhawk attack aircraft, the LTV F-8 Crusader, and larger and more capable AEW and COD aircraft in the Grumman E-2A Hawkeye and C-2A Greyhound. Many Crusaders found a role as reconnaissance aircraft, as did North American RA-5C Vigilante reconnaissance versions of the USN A-5 bomber. In the late 1960s, Grumman A-6 Intruders started to replace many of the Skyhawks, while the Lockheed S-3 Viking replaced the Trackers. Shore-based MR was not neglected, with the Lockheed P-3 Orion developed from the Electra airliner. Throughout this period, new and more capable helicopters entered service, including the Sikorsky S-58, followed by the SH-3A Sea King and, for small ships, the Kaman HH-43 Husky and the UH-2 Seasprite, also often based aboard carriers for plane-guard duties. The SH-3A was primarily an anti-submarine helicopter, but a number were converted for mine-sweeping duties until it was felt that a larger helicopter would be more suitable, resulting in a variant of the Sikorsky CH-53 Sea Stallion. Sea Stallions and Boeing Vertol CH-46 Sea Knights were also used by the USMC.

The 1960s were marked by the Vietnam War, and the USN normally kept at least one aircraft carrier off the coast of Vietnam, joining the USAF in attacks on key targets and also dropping mines in the main North Vietnamese port of Haiphong. US withdrawal from the war, leaving South Vietnamese forces to face the Viet

Cong, was not matched by a reciprocal fall in support by the Soviet Union, which continued to ensure ample supplies for the North Vietnamese. In the closing stages of the war, in April 1975, nine aircraft carriers from the US Seventh Fleet converged on South Vietnam to evacuate US nationals and prominent South Vietnamese political and military personnel, landing 7,000 US Marines. During this time, the fleet had risen to fifteen large attack carriers, CVA, as well as seven older carriers of the Essex-class redesignated anti-submarine carriers, CVS.

The advent of assault ships of the Iwo Jima-class and Tarawa-class, capable of carrying large numbers of aircraft, was enhanced by the introduction of the vertical take-off strike aircraft, with the USMC introducing the McDonnell Douglas AV-8A Harrier, a licence-built Hawker Siddeley Harrier, into service. The advent of the nuclear age came in 1961 with the USS *Enterprise*, with a range of some ten years between refuelling, and capable of carrying up to eighty-four large aircraft. This also led to reclassifications, and with the gradual retirement of the anti-submarine carriers, the USN began to classify all carriers as CV or CVN in the case of nuclear-powered ships. Nuclear-powered aircraft carriers also allowed larger and more capable aircraft to be operated, such as the variable-geometry Grumman F-14A Tomcat. These were augmented by the McDonnell Douglas (now Boeing) F-18, later F/A-18, Hornet, a highly capable fighter-bomber which also proved much less expensive than the Tomcat, and eventually replaced the A-6 Intruder in the strike role as well as taking over from the F-4s. The USN has now standardised on nuclear-powered carriers, mainly of the Nimitz-class, first commissioned in 1975.

The USN and USMC have been active in the UN operations in Somalia and in the Gulf War, with the first carrier reaching the area within days of the Iraqi invasion of Kuwait. The USN eventually had six aircraft carriers involved in the Gulf War, about half the number in active service at the time. Oldest of these ships was *Midway*, and the newest, *Theodore Roosevelt*, one of two Nimitz-class ships present. The operation marked the final combat sorties for the A-7 and A-6 strike aircraft, and the first for the F-14 Tomcat and the F/A-18 Hornet, which undertook bombing and fighter sorties. On one occasion, an F/A-18 carrying four 2,000lb bombs shot down an Iraqi aircraft en route to the target, without shedding its warload. Operations over Afghanistan during the winter of 2001/2002 saw aircraft operating from carriers, including the *John C Stennis* and the *Theodore Roosevelt*.

Today, there is growing integration of USN and USMC aviation. The USN has 339,453 personnel, of whom 98,588 are involved with aviation, an increase over the past six years. There are 186,661 personnel in the USMC, with 34,700 involved with aviation. The USN was historically divided into just two fleets, the Atlantic and the Pacific, but for practical purposes, there are five fleets. These are: Second Fleet, Atlantic; Third Fleet, Pacific; Fifth Fleet, Indian Ocean; Sixth Fleet, Mediterranean; Seventh Fleet, West Pacific. There are eleven aircraft carriers, all nuclear-powered with one in refit at any one time. There are five aircraft carriers in reserve at sixty to ninety days' notice. Eleven naval air wings exist, of which one is in reserve, while the average air wing includes nine squadrons, usually two with twelve F/A-18Cs each, one each with twelve F/A-18Es and twelve F/A-18Fs, one with eight S-3B Vikings, one with six SH-60 Seahawk helicopters, one with four EA-6Bs, and one with four E-2C Hawkeye AEW aircraft. The USN's twenty-two cruisers can each accommodate two SH-60B helicopters, as can fifty-two Arleigh Burke-class destroyers, while the twenty-two remaining Oliver Hazard Perry-class frigates can take two helicopters. The USMC AV-8B Harrier IIs operate from seven Wasp-class LHDs, each of which can take five AV-8Bs and up to forty-eight helicopters; three Tarawa-class LHAs, each of which can take six, plus up to twenty-one helicopters; while six Austin-class LPDs can take six helicopters each, and the first three of a planned nine San Antonio-class LPDs can accommodate either a Sikorsky CH-53E Sea Stallion heavy-lift helicopter, a Bell-Boeing MV-22B Osprey or two Boeing CH-46 Sea Knight medium-lift helicopters. The USMC uses F/A-18A/B/C/D Hornets, with the -18D for forward air control and reconnaissance as well as training, as well as AV-8B Harriers. There are also KC-130/130J Hercules tankers, with the C-130J force likely to reach fifty. The USMC CH-46E Sea Knight and CH-53D helicopters will eventually be replaced by 425 MV22B Osprey tilt-rotors, while the USN will receive forty-eight HV-22Bs. Bell has been upgrading 100 UH-1N Iroquois and 180 AH-1W Super Cobra helicopters, making these the UH-1Y and AH-1Z respectively.

USMC aircraft are included in the USN figures, which include 381 F-14A/B/D Tomcats, 545 F/A-18E/F Hornets, which are replacing the 320 F/A-18A/B and some of the 483 F/A-18C/D Hornets, and likely to be joined in due course by carrier versions of the F-35, the F-35C. There are 200 USMC AV-8B+ Harrier IIs in the attack role, and nineteen TAV-8B conversion trainers, likely to be replaced in due course by STOVL variants of the F-35, the F-35B. A substantial number of A-6 derivatives survive, including sixteen EA-6A Mercury long-endurance communications relay aircraft, now being upgraded, and 128 ECM EA-6B Prowlers. Carrier-borne aircraft also include ninety-four AEW E-2C Hawkeyes, 137 S-3A/B Viking ASW aircraft, and thirty-eight COD C-2A Greyhounds. There are 161 Lockheed P-3C Orions for MR with twelve squadrons. Another ten EP-3E/J Orions operate on ELINT, with twelve RP-3A/TP-3As on survey and training duties, while there are fourteen UP-3A/B/VP-3A Orions on VIP duties. MCM helicopters include nineteen Sikorsky RH-53D Sea Dragons and twenty-five Kaman SH-2F/G Seasprites. ASW helicopters include 220 Sikorsky SH-60B/F Seahawks, with eight MH-60R Strike Hawk multi-mission and eighty-five MH-60S Knight Hawk multi-mission support helicopters. There are fifty-two Sikorsky SH-3D/G/H Sea Kings, which join the Seahawks on SAR. CSAR is provided by twenty-four HH-60H Black Hawks, while the 150 UH/HH-1N Iroquois can also be used for SAR. Other utility helicopters include fifty-eight UH-46D/HH-46D Sea Knights, forty-seven UH-3A/H Sea Kings, and forty-five MH-53E Sea Stallions. Apart from the Super Cobra helicopters and Osprey tilt-rotors already mentioned, transport helicopters include 269 CH-46D/E Sea Knights and 226 CH-53D/E Sea Stallions. Fixed-wing transport aircraft include six Boeing C-40A Clippers (737-700), seventeen C-130F/T and seven C-130F/R Hercules, thirty-three KC-130J tankers and three DC-130A Hercules on drone control duties, and twenty-nine C-

9B Nightingale MEDEVAC aircraft. Apart from the special Orions, VIP aircraft include eight VH-60N and fifteen VH-3A/D Sea King helicopters. Communications aircraft are fixed-wing, with ten CT-39E/G Sabreliners, two C-26A Metro IIIs and seven C-20D/Gs (Gulfstream). Training uses thirty-eight F-5E/F Tiger IIs in an aggressor squadron. There are 319 T-34C Turbo Mentor, 339 T-6A Texan II, 170 T-45A Goshawk (BAe Hawk), fifty-seven T-44A Pegasus, and a small number of Sabreliner and TC-18F trainers, as well as ninety-four TH-57B/C Sea Ranger (206) helicopters.

The USN is evaluating UAV, including Global Hawk, Fire Scout and Pioneer.

Missiles include Sparrow, Sidewinder, AMRAAM and RAM AAM, with Maverick, Harpoon, Hellfire, SLAM, Penguin, HARM and JSOW ASM.

The P-3C Orion is to be replaced during the present decade by a new maritime-reconnaissance aircraft, the P-8, based on the Boeing 737.

UNITED STATES COAST GUARD AVIATION

Founded: 1916

While many air arms had earlier experience of balloons, the United States Coast Guard can lay claim to the longest connection with aviation, as Surfman Daniels took a photograph of the first flight by the Wright brothers on 17 December 1903. Yet it was not until 1915 that two young officers experimented with a Curtiss F flying boat for USCG operations before being trained by the USN at Pensacola in 1916. Later that year, Congress authorised the USCG to establish ten air stations, but did not provide additional funding. To gain experience, a number of USCG personnel flew with the USN during World War I. The USCG is counted as part of the US armed forces, but is under USN control only in wartime, otherwise being administered by the Department of Homeland Security after many years under the Department of Transportation. Post-war, Read's Curtiss NC-4 was piloted on its historic transatlantic flight by a member of the USCG.

In 1920, the USCG took over an abandoned USN air station and borrowed USN Curtiss H2-2L flying boats, but continued lack of funding led to the station closing in 1921. Prohibition ensured that Congress funded USCG aviation once the service proved that it could prevent the smuggling of whisky. Congress allowed $152,000 (£38,000 at the then rate of exchange) for three Loening OL-5 amphibians and two Chance Vought UO4s, flown from two bases in New England. SAR was not overlooked, and the USCG set about buying 'flying lifeboats', aircraft that could land in the open sea to rescue survivors. The first of these were two Douglas RD-2 Dolphins and the specially designed General Aviation Flying Life Boat PJ-15, which operated successfully throughout the early 1930s. The early 1930s also saw Grumman JF-2 amphibians operated from the larger cutters, against opium smugglers off the West Coast and on fisheries protection off Alaska, as well as later on 'plane guard' duties for experiments in transatlantic air services. By 1936, the USCG had six air stations and forty-two aircraft. Before US entry into World War

II, the USCG surveyed the coast of Greenland, looking for sites for airfields, using a cutter carrying a Curtiss SOC-4.

US entry into World War II saw the USCG capturing German weather stations in northern Greenland, which were providing forecasts for U-boats. Rescue missions were flown over the sea and the ice cap to aid downed Allied aviators and survivors from ships sunk by U-boats. During the war, USCG aircraft located 1,000 survivors and directed rescuers to the scene, and rescued 100 survivors by landing aircraft in the open sea, on one occasion having to taxi ashore as the weight of those rescued meant that the aircraft could not take off. A dedicated Air Sea Squadron was formed in 1943 in California with Consolidated PBY-5A Catalina amphibians. By 1945, SAR accounted for 165 aircraft and nine air stations, and the USCG responded to 686 plane crashes. Offensive operations were also mounted, with the USCG's first U-boat sunk in the Gulf of Mexico in 1942.

In 1943, the USCG was given responsibility for helicopter trials to assess its role for ASW. It used a merchant vessel, SS *Daghestan*, and a joint trials unit was set up with the Royal Navy after the USCG had trained British pilots, using Sikorsky HNS-1 and HOS-1 helicopters. Evaluation also took place aboard the USCGV *Cobb*, a converted passenger steamer. The first rescue by a helicopter came in 1945, after two ski-equipped aircraft ran into difficulties attempting to rescue the crew of an RCAF aircraft crashed in Labrador. A USCG HNS-1 was dismantled and loaded into a Douglas C-54 transport, flown to Goose Bay, reassembled, and flown 185 miles to the crash site via a refuelling stop.

Post-war, the USCG returned to civil control. It replaced the PBY-5As with Martin PBM-5Gs. A new task was the International Ice Patrol, which started in 1946 and continues, using Lockheed HC-130 Hercules with side-looking airborne radar. Civil disasters saw more than 300 people being rescued by helicopter during floods in 1955. Some 7,000 USCG personnel served in Vietnam, including aviators flying with USAF SAR units. In 1980, the USCG rescued many refugees fleeing from Cuba. Successive generations of helicopters displaced the flying boat and amphibian, using Sikorsky HH-52 (S-63) helicopters, and later replacing these with Eurocopter HH-65 Dolphins (Dauphin).

The USCG has 40,698 uniformed personnel, of which 7,960 are directly involved in aviation. It operates twenty-six HU-25A/B/C/D Guardians (Falcon 20), its first jet, with another twenty-nine on support duties or in storage, as a rapid response multi-mission aircraft, but many of these will be replaced by deliveries of the HC-144A as the CN235-200 is known to the USCG. There are also twenty-two HC-130H Hercules, with five more in storage, for SAR, while another six C-130Js provide transport, with a C-37, two CN235-200s and a C-143A Challenger. The main strength lies in its helicopter force, with thirty-five Sikorsky HH-60J Jayhawks, with another seven in storage, and ninety-five HH-65A Dolphins (Dauphin). There are eight MH-68A (A109E) utility helicopters. There are twelve cutters of the Hamilton-class able to operate HH-60 or HH-65 helicopters, while the latter can also operate from the thirteen Bear-class cutters. The sixteen Reliance-class also have helicopter decks.

UAV are not operational at present, but trials are proceeding.

UNITED STATES ARMY

Founded: 1942

Despite the USAAF, and before it the USAAC, being under overall US Army command until it became the autonomous USAF in 1947, from 1938 onwards its aircraft had not been under the control of local Army commanders, leaving the service free to pursue its strategic goals. This situation meant that Army aviation had to be re-invented, to provide the vital AOP and liaison duties. In 1942, shortly after the United States entered World War II, the US Army purchased some 3,000 light aircraft, including Piper L-4 Cub, Stinson L-5 Sentinel, Vultee-Stinson L-1, Taylorcraft L-2 and Aeronca L-3 AOP aircraft. Additional deliveries of these aircraft and subsequent models meant that US Army aviation reached 4,500 aircraft by the end of the war.

The war was notable for the appearance of the first helicopters, with the Sikorsky R-4, R-5 and R-6 entering service. Post-war, Sikorsky H-5, H-19 Chickasaw and Bell H-13 Sioux helicopters were used extensively by the US Army. New fixed-wing aircraft continued to enter service, with the most numerous the Cessna L-19 Bird Dog, which first appeared in 1950. By the mid-1950s, the US Army had about 1,800 fixed-wing aircraft, and about the same number of helicopters, operating AOP, liaison, communications, light transport and reconnaissance duties. The end of the decade saw Sikorsky H-34A Choctaws and H-37As, Vertol H-21Cs and H-25Cs, Hiller H-23Cs and still more Bell H-13s enter service. Other new aircraft included 370 DHC-2 Beavers, known to the US Army as the L-21, U-1A Otter (DHC-3) light transports, and Beech L-23s (Twin Bonanza), Cessna L-27As (310) and Aero L-26 Commanders for communications duties. The 1960s saw Bell UH-1A Iroquois and Boeing Vertol CH-47 Chinook helicopters enter service. Originally, the first early UH-1s had been designated HU (helicopter utility), giving rise to the nickname 'Huey'.

The Vietnam War saw the development of armed helicopters, both to suppress enemy fire and to act as gunship escorts for troop-carrying helicopters. US Army personnel in Vietnam were initially 'advisors' but eventually took over the full duties of soldiers. In 1970, these actions spread into Cambodia. The growing importance of the helicopter, its improved performance and the demands of the conflict in Vietnam meant that by 1970, the US Army was the world's largest helicopter operator, with 10,000 machines. It also had a number of fixed-wing attack aircraft, including the North American Rockwell OV-1D and RV-1D Mohawk COIN aircraft. After Vietnam, the main theatre for helicopter deployment was in Europe, where the US Army faced possible conflict with Warsaw Pact forces in Germany. New helicopters continued to enter service, with new and more powerful versions of the Chinook joined, at the lower end of the scale, by the Bell OH-58 Kiowa (206), and attack helicopters such as the Bell AH-1 Cobra. The next major US Army involvement using helicopters was the Gulf War, with several hundred machines deployed in Operation Desert Storm, liberating Kuwait. This was the debut for the McDonnell Douglas (now Boeing) AH-64 Apache, with the air war started by the US Army, whose Apache helicopters destroyed a key Iraqi air defence radar installation, easing the way for the massive air strikes by the air forces and the USN and USMC.

During the late 1980s and 1990s, almost all new aircraft were helicopters, but a few fixed-wing types continued to enter service, notably C-23 Sherpa (Short 330) light transport. By the mid-1990s, more than a third of the helicopter force, by this time down to around 8,000 aircraft, consisted of armed combat helicopters. The most significant new arrival during this period was the Sikorsky UH-60 Black Hawk utility helicopter. A squadron of AH-64 Apache helicopters was deployed during operations in Kosovo, but not used.

The US Army is in the midst of a programme of change, having retired its AH-1F Cobras in 2001 and the remaining OH-58 Kiowas and UH-1 Iroquois in 2004. In their place, more than 1,200 Boeing-Sikorsky RAH-66 Comanches entered service, while the UH-60s and the AH-64s have upgraded and many of the latter fitted with the Longbow radar system. The active units of the US Army are backed up by the state-orientated National Guard and by the US Army Reserve, with strong aviation elements in both.

The United States Army has some 4,000 helicopters, of which 1,500 are armed, a 60 per cent cut in numbers over the past thirty years, but there have been significant improvements in aircraft capability. Mainstays of the present force are 634 AH-64A/D Apache anti-tank and attack helicopters with 375 OH-58A/C/D Kiowas. Special operations use sixty-four MH-47E/H Chinooks and sixty MH-60K/L Black Hawks/Pave Hawks, with thirty-six AH-6/MH-6 Little Birds (MD 500/530). Observation is provided by 463 OH-58A/C Kiowas, and SAR by fifteen HH-60L Black Hawks. Supporting this force are 374 CH-47D/F Chinooks, 1,484 UH-60A/L/M Black Hawks, four UH-60Q Black Hawks and 447 UH-1H/V Iroquois.

SIGINT is gathered by fixed-wing aircraft, fory-nine RC-12D/H/P Guardrails (King Air), as well as special versions of the Dash-7, with eighteen RC-7s in service. Reconnaissance is by sixty O-2 Skymasters. The C-23A/B/C Sherpas are being replaced at present by up to fifty-four C-27 Spartans, while there are more than 100 King Airs of various models, ten C-26 Metros, three Beech 1900s and three C-212s. Training uses 179 TH-67 Creeks (206), plus a small number of T-34s and King Air RC-12Ds.

Increasing use is being made of unmanned aerial vehicles, UAVs, of which the most numerous is the RQ-7A Shadow, but which also includes Buster, Hunter, I-Gnat, Desert Hawk, Exdrone, Sky Warrior and Raven, which have developed from providing reconnaissance to having an attack capability.

The National Guard with 351,350 reservists, of which 67,048 are active, and the US Army Reserve with 195,000 reservists of which 27,069 are active, operate AH-1E/F/G/P/S Cobra attack helicopters, as well as AH-64 Apaches in the same role. Heavy lift is provided by CH-47 Chinooks. There are UH-1H/V Iroquois utility helicopters and UH-60A/L Black Hawks, as well as a handful of these on ECM duties. Scout helicopters include OH-58A/C Kiowas and OH-58D Kiowa Warriors. There are C-12 Huron and C-23B Sherpa fixed-wing aircraft, as well as a handful of C-20 Gulfstream and C-21 Learjet communications aircraft.

URUGUAY

- Population: 3.5 million
- Land Area: 72,172 square miles (186,925 sq.km.)
- GDP: $34.9bn (£22.3bn), per capita $9,982 (£6,380)
- Defence: $373m (£238m)
- Service Personnel: 24,621 active

URUGUAYAN AIR FORCE/*FUERZA AÉREA URUGUAYA*

Founded: 1952

Uruguay established a Department of Military Aviation in 1916, and despite being neutral during World War I, managed to obtain three Morane-Saulnier aircraft. In 1919, four Avro 504Ks were obtained, followed in 1921 by six Nieuport trainers. The new air arm became the *Aeronáutica Militar*. During the 1920s and 1930s, it operated Ansaldo A300 fighters, Potez XXVA and Romeo Ro37 reconnaissance-bombers, a Farman F190 and Stinson AOP aircraft, a Breguet Br19 transport, Waco D-7 general-purpose aircraft, and de Havilland Moth trainers.

Officially the country remained neutral during World War II, but in return for putting bases at the disposal of US forces, military aid was provided. This included Grumman F6F-5 Hellcat fighter-bombers, TBM-1C Avenger torpedo-bombers and J4F-1 Gosling amphibians; North American B-25J Mitchell bombers, AT-6 and SNJ-4 trainers; Chance-Vought OSU-3 Kingfisher AOP aircraft; Curtiss C-46 Commando and Douglas C-47 transports; Beech T-11B, Fairchild PT-23A and PT-26 trainers. In 1948, Uruguay became a member of the Organisation of American States. North American F-51D Mustang fighter-bombers entered service during the late 1940s, and were not replaced until the early 1960s, when Lockheed F-80C Shooting Star jet fighter-bombers and T-33A jet trainers arrived. The *FAU* title was adopted in 1952.

During the 1980s, the *FAU* gave up fast jet fighters and concentrated on aircraft suitable for COIN operations, transport and communications. The Shooting Stars were not replaced with fighter-bombers, but with additional AT-33A armed-trainers, and then these were joined by ex-USAF Cessna A-37Bs in 1989. It also became one of the few export customers for the Argentinian FAMA IA-58 Pucara light attack aircraft. Despite reports over the past decade that fighter-bombers have been sought, possibly A-4 Skyhawks or Northrop Grumman F-5E Tiger IIs, the service has continued to concentrate on counter-insurgency as before.

Today, the *FAU* has 3,000 personnel. There are two combat squadrons, with one operating ten A-37B Dragonflies and the other five FAMA IA-58 Pucaras on COIN duties. The second squadron may be strengthened in an emergency by five armed PC-7U Turbo Trainers. A helicopter squadron operates four Bell 212s and six UH-1Hs. A transport squadron operates three C-212 Aviocars on transport and SAR, two EMB110C Bandeirantes and an EMB120 Brasilia, three Lockheed C-130Bs, a Cessna 206 and a VIP Cessna 310. The *FAU* also operates an airline, a common practice in Latin America, *Transporte Aéreo Militar Uruguayo*, with C-212s. A number of Cessna and Piper aircraft are employed on liaison duties. Apart from the PC-7Us, trainers include twelve T-34A/B Mentors and five T-41D Mescaleros (172).

URUGUAYAN NAVAL AVIATION/*COMANDO DE AVIACIÓN NAVAL/ARMADA DE URUGUAY*

Founded: 1920

Originally founded in 1920, the *Aviación Naval* remains a shore-based force. At one time Grumman F6F-5 Hellcat fighters were operated, as well as Martin PBM-5 Mariner flying boats. Today, 211 of the Navy's 3,603 personnel are involved with aviation, and in recent years two Uruguay-class frigates have been introduced, bought from Portugal, that can each carry a single helicopter. Grumman S-2A/G Tracker ASW aircraft have been withdrawn and the service is now limited to EEZ patrols maintained by a Beech 200T and two Jetstream T2s. A squadron operates three ex-RAF Wessex (S-58) helicopters on transport and SAR, supported by six Bo105Ms which also operate from the two frigates. Communications duties are carried out by two T-34B Turbo Mentors, which also have a training role.

UZBEKISTAN

- Population: 27.6 million
- Land Area: 173,546 square miles (412,250 sq.km.)
- GDP: $27.4bn (£17.5bn), per capita $991 (£633)
- Defence Exp: $94m (£60m)
- Service Personnel: 67,000 active

UZBEKISTAN AIR FORCE

Founded: 1991

On the break-up of the Soviet Union, the Uzbekistan Air Force was formed, declaring that 300 former Soviet aircraft on its territory belonged to it, although of these just 200 were believed to be operational. The situation was complicated by the country becoming a member of the CIS, and with CIS, mainly Russian, units based in the country. Reports indicate that the UAF may have provided forces, including combat helicopters, to fight anti-government rebels in Tajikistan.

Today, the UAF has around 17,000 personnel. It is still organised on Soviet lines, with seven air regiments. A fighter regiment has thirty MiG-29/29UBs and twenty-five Su-27/27UBs. Two fighter-bomber regiments include one with twenty Su-25/25BMs and twenty-six Su-17MZ/UMZs, with the other having twenty-three Su-24s and eleven Su-24MPs. There is a helicopter regiment with twenty-nine Mi-24 attack helicopters, some of which are being upgraded, and fifty-two Mi-8 transports, twenty-six Mi-6s and an Mi-26, as well as two Mi-6AY command posts. A transport unit has two Il-76s, a Tu-134 and an An-24. Training uses fourteen L-39 Albatros, of which nine are believed to be in store. A relatively wide selection of missiles is in use, including *Alamo, Archer* and *Aphid* AAM, with *Karen, Kilter, Kegler, Kerry* and *Kyle* ASM, *Kilter* and *Kegler* ARM and around forty-five *Guideline, Goa* and *Gammon* SAM launchers.

VENEZUELA

- Population: 26.8 million
- Land Area: 352,143 square miles (912,050 sq.km.)
- GDP: $355bn (£227bn), per capita $13,228 (£8,455)
- Defence Exp: $3.2bn (£2.05bn)
- Service Personnel: 115,000 active, plus 8,000 reserves

VENEZUELAN AIR FORCE/*FUERZA AÉREA VENEZOLANA*

Venezuela formed a Military Aviation Service in 1920, with a total of sixteen Caudron GIIIs and GIVs and Farman F40s delivered the following year. Shortly afterwards, the Venezuelan Navy also formed an air arm. Little further progress was made until 1936, when the Military Air Service became the Military Aviation Regiment and a number of aircraft were bought from Consolidated. This was the start of a modest expansion programme, with the MAR starting a domestic airline, and following a visit by an Italian aviation mission, introducing Fiat CR32 fighters and BR20 bombers.

During World War II, Venezuela allowed the US to use its bases. In return, military aid was provided, including Republic F-47D Thunderbolt fighters, North American B-25J Mitchell bombers and NA-16-3 and T-6, and Beech T-7 and T-11B trainers. Post-war, Venezuela became a member of the Organisation of American States, and the two air arms merged to form the *Fuerza Aérea Venezolana*. Modernisation occurred during the early 1950s with de Havilland Vampire FB5 jet fighter-bombers and English Electric B2 jet bombers, followed in 1955 by de Havilland Venom FB4 fighter-bombers and Vampire T55 trainers, North American F-86F Sabre fighter-bombers, additional Canberras and Fairchild C-123B Provider transports. In 1967, ex-*Luftwaffe* F-86K Sabres were obtained. The 1970s saw Canadair-built CF-5A/D fighter-bombers and Dassault Mirage IIIs and Vs introduced, as well as Fairchild OV-10E Bronco COIN aircraft, with a second batch later purchased from the USAF as Canberra replacements. The *FAV* also introduced the Lockheed Martin F-16A/B interceptor during the 1980s. The Mirage IIIs and Vs were upgraded to Mirage 50 standard in 1988, with additional Mirage 50s obtained from Dassault. Ex-Netherlands NF-5As boosted the CF-5A/D force in 1992.

Sometimes styled as *Aviación Militar Bolivariana*, today the *FAV* has 11,500 personnel, an increase from 7,000 in 2002, although that lower figure was a reduction of almost a quarter over the previous thirty years. Average annual flying hours are around 155. There is little standardisation, but the introduction of F-16s has increased the force's capability. The air force is organised in groups, with one operating twenty-one F-16A/B Fighting Falcons; another with twenty-four Su-30MKVs; another with thirteen Mirage 50V/DVs and a fourth with ten remaining VF-5A/NF-5Bs, while there is a COIN group with eight OV-10A/E Broncos and three AT-27s. AEW is provided by two Dassault Falcon 20DCs and two C-26Bs. TCSAR is provided by two AS532 Cougars. Other helicopters include twelve Bell UH-1B/Hs and two Bell 412s, as well as ten AS532 Cougars, of which two are reserved for VIP duties, and six AS332B Super Pumas. Transport uses six C-130H Hercules, three Short 330/360s and a G222, as well as two Boeing 707 tankers. VIP aircraft in the Presidential Flight include an An A319CJ, a 737-200, a Gulfstream II and III, a Learjet 24D and a Bell 412. Cessna and Beech light aircraft are used for liaison and communications, including nine Cessna 182s, a Citation I and II, seven Beech Queen Airs and five Super King Air 200s. Training uses fifteen EMB312s and twelve SF260Es.

The present regime is strongly anti-American and growing links with China suggest that Chinese aircraft might be obtained in the future. In 2006, as a protest to the refusal of the United States to provide new equipment, the government even threatened to sell the *FAV*s F-16s to Iran. Future supplies of arms and spares from the United States certainly cannot be relied upon, given the present state of relations. Missiles include Magic and Sidewinder AAM, Exocet ASM and a small number of Roland and RBS-70 SAM launchers.

VENEZUELAN NAVAL AVIATION/*AVIACIÓN DE LA MARINA VENEZOLANA*

Naval aviation accounts for 500 of the Venezuelan Navy's 14,300 personnel, with six Mariscal Sucre-class (Lupo) frigates capable of operating a Bell 212 ASW helicopter. Seven AB 212ASWs are operated, with four HeliDyne modified Bell 412s for SAR and six Mi-17 transports. There are seven C-212-200AS/400 Aviocar transports, with some having radar for SAR. Two King Airs and a Turbo Commander 695 provide transport and communications, with a VIP King Air E90 and Super King Air 200.

VENEZUELAN ARMY AND NATIONAL GUARD/*EJÉRCITO VENEZOLANA E GUARDIA NACIONAL*

The Venezuelan Army and National Guard use aviation on transport, communications and liaison with a wide variety of aircraft. The Army's transport aircraft include four IAI-201 Aravas, twelve M28 Skytrucks and a number of Beech and Cessna light aircraft. The helicopter force has grown considerably since 2002 and represents Venezuela's new political orientation with twelve Mi-35M2 *Hind* attack helicopters and twenty Mi-17 *Hip* and three Mi-26s, as well as two AS-61As (S-61), ten Bell 412EPs and two 412SPs, four 206B JetRangers and a 206L Longranger, while four UH-1H Iroquois and three Bell 205As are in store at present. Two Cessna 182s are used for training.

VIETNAM

- Population: 88.6 million
- Land Area: 129,607 square miles (335,724 sq.km.)
- GDP: $94bn (£60bn), per capita $1,061 (£678)
- Defence Exp: $2.8bn (£1.79bn)
- Service Personnel: 455,000 active, plus reserves of upwards of 5 million

VIETNAM PEOPLE'S ARMY AIR FORCE

Vietnam was reunited in 1976, after being divided following the French withdrawal from Indo-China in 1954. For most of this period, the Communist North backed by the Soviet Union was infiltrating forces into the pro-Western South, backed by the United States. Both countries maintained air forces, although the main thrust of their defences lay in their ground forces. During the Vietnam War, the Vietnamese Air Force operated US-supplied Northrop F-5A/B, Cessna A-37B armed-trainers

293

and at one time a substantial number of Douglas A-1 Skyraiders, all on tactical fighter-bomber duties, as well as many helicopters and C-47 transports. The Vietnamese People's Air Force operated Soviet-supplied MiG-15, MiG-17 and MiG-21s, and a few Ilyushin Il-28 bombers. The VAF provided COIN operations and tactical support for ground forces, while the VPAAF concentrated on air defence.

After reunification, Vietnam used its armed forces to influence events in Cambodia, and then to back Communist guerrillas in Thailand. Abandoned US-built aircraft were used until spares became a problem. Supplies of Soviet equipment continued at first, including forty MiG-23ML interceptors and thirty Mi-24 attack helicopters, with Soviet and North Korean advisors. In 1994 and 1997, Su-27 interceptors entered service, replacing the older MiGs, but new aircraft deliveries have been much reduced since the collapse of the USSR. Vietnam is still committed to Russian aircraft, and has obtained spares for its MiG-21s, which were upgraded by MAPO/Sokol in 2000–2002. Despite expectations that the VPAAF would be reduced in size now Vietnam has to pay for its equipment, small numbers of new aircraft have been delivered recently.

Personnel strength has remained unchanged since 2002 at 30,000. The VPAAF is organised along Russian lines, with regiments instead of squadrons. Seven fighter regiments operate 140 MiG-21bis *Fishbed* L/N, while two regiments operate fifty-three Su-22M-3/-4 *Fitter*, seven Su-27SK *Flanker* and nineteen Su-30MKKs. The VPAAF also provides aircraft for the People's Navy, and SAR and ASW are provided by a regiment with three Ka-25 *Hormone* and ten Ka-28 *Helix*, while four Be-12 *Mail* flying boats are used by an MR regiment. An attack helicopter regiment is equipped with twenty-six Mi-24 *Hind*. Three transport regiments operate twelve An-2 *Colt*, twelve An-26 *Curl*, four VIP Yak-40 *Codling*, thirty Mi-8/-17 *Hip*, four Mi-6 *Hook* and two Ka-32s, while as many as twelve UH-1H Iroquois may still be operational. Training uses ten BT-6s (Yak-18) and eighteen L-39 Albatros as well as MiG and Sukhoi conversion trainers.

Missiles include *Alamo*, Adder, *Atoll* and *Aphid* AAM, with *Kedge*, *Krypton*, *Kazoo* and *Kyle* ASM.

YEMEN

- Population: 22.9 million
- Land Area: 184,289 square miles (486,524 sq.km.)
- GDP: $22.9bn (£14.6bn), per capita $1,000 (£639)
- Defence Exp: $1.55bn (£0.99bn)
- Service Personnel: 66,700 active

YEMEN AIR FORCE

Founded: 1990

Until 1990, Yemen was divided into North and South, although unification had been agreed in 1981. South Yemen had gained independence in 1967, having previously been a British protectorate. North Yemen, normally referred to as Yemen, had been independent, ruled by an Imam, and had formed a small Yemen Air Force during the mid-1950s, initially as a transport force until Ilyushin Il-10 ground-attack aircraft were supplied by the USSR, with helicopters and trainers, and Soviet and Egyptian pilots. The country was invaded by Egyptian troops during the mid-1960s, and the Imam overthrown.

After unification in 1981, the service became the Unified Yemen Air Force and mainly used Russian equipment, while Yemen's failure to support action against Iraq following the invasion of Kuwait led to Saudi Arabia cutting off aid, while the collapse of the Soviet Union ended military aid. A civil war in 1994 also undermined the economy and the efficiency of the armed forces. Probably only 50 per cent of its aircraft are operational, with many destroyed in the civil war or grounded by a shortage of spares. The situation of the country is critical as it suffers from Islamic terrorists, mainly al-Qa'eda, developing bases within its territory and while Western help has been given to the government, there is strong local resistance to any foreign military presence.

Today the YAF has 3,000 personnel. Four squadrons are equipped with ten F-5E Tiger IIs, eighteen MiG-29SMT/UBTs and thirty-three Su-20s. It seems likely that the MiG-29 force will be increased, perhaps by as many as another thirty-two aircraft. A transport squadron operates four C-130H Hercules, four Il-14s, three Il-76s, five An-26s and two An-2s. Helicopters include eight Mi-35 attack helicopters, twelve Mi-8s and two Bell 212s. Training aircraft include twelve Z-242s and twelve L-39Cs, as well as F-5B, MiG and Sukhoi conversion trainers. Missiles consist of *Atoll* and Sidewinder AAM.

ZAMBIA

- Population: 11.9 million
- Land Area: 288,130 square miles (752,618 sq.km.)
- GDP: $12.8bn (£8.2bn), per capita $1,078 (£689)
- Defence Exp: $229m (£146m)
- Service Personnel: 15,100

ZAMBIA AIR FORCE AND AIR DEFENCE COMMAND

Founded: 1964

Formerly the British colony of Northern Rhodesia, Zambia became independent in 1964 and immediately established the Zambia Air Force. The first aircraft were four Douglas C-47 and two Hunting Pembroke C1 transports. Initially the RAF provided assistance, but an Italian military mission later helped. The Pembrokes were replaced by six DHC-2 Beavers, while the C-47s were replaced by four DHC-5 Caribous. Agusta Bell 205 helicopters were introduced, with a mixture of DHC-1 Chipmunk and eight Scottish Aviation Bulldog trainers. In 1971 the first jets arrived, two Soko Galeb and four Soko Jastreb armed-trainers, while eight Savoia-Marchetti SF260 trainers also entered service.

In 1980, the USSR supplied sixteen MiG-21 interceptors, while twelve Shenyang F-6s (MiG-19) fighter-bombers were bought from China. Antonov and Yakovlev transports were added, with Mi-8 helicopters, but the ZAF also continued to buy Western equipment, Aermacchi MB326 and Saab MFI-17 armed-trainers, and a substantial number of AB205 and 212 helicopters. Poor serviceability soon reduced the operational numbers, with the MiG-21 force also depleted by serviceability from sixteen to seven by 1996, while the MB326s fell from eighteen to six between 1990 and 1996. Many of the MiG-21s were upgraded in Israel during 1999 and 2000.

Today, the ZAF has 1,600 personnel, the same as in 2002. Many aircraft are believed to be unserviceable. One squadron operates eight F-6s (MiG-19) and another has eight MiG-21MFs. Transport is provided by a squadron with four An-26s, four DHC-5 Buffaloes, nine Y-12s, an HS748 and five Do28s, as well as two VIP Yak-40s. There are four Mi-8, six UH-1H and five Bell 47G helicopters. Training uses conversion trainers for the combat aircraft, eight K-8s and five SF-260TPs.

ZIMBABWE

- Population: 11.4 million
- Land Area: 150,333 square miles (389,329 sq.km.)
- Accurate figures are not available for GDP and defence expenditure
- Service Personnel: 29,000

AIR FORCE OF ZIMBABWE

Founded: 1980

Originally the British colony of Southern Rhodesia, the country's history of military aviation dates from the late 1930s, when the colony's government offered three squadrons for the Royal Air Force. This was in addition to playing a major role in the planned Empire Air Training Scheme, training pilots for the British and other air forces. The RAF's three Rhodesian squadrons operated Supermarine Spitfire and Hawker Typhoon fighters, as well as Avro Lancaster bombers. Flying training took most of the available personnel, with the country offering excellent flying conditions so that courses took less time, in skies free from the threat of enemy aircraft. Post-war, the Southern Rhodesian Air Force was formed, with a single squadron initially for transport and communications, with Douglas C-47s, an Avro Anson and a de Havilland Rapide, as well as de Havilland Leopard Moths. The SRAF acquired combat aircraft in 1951, with eleven Supermarine Spitfire 22 fighters, replacing these in 1953 with de Havilland Vampire FB9 fighter-bombers, with Vampire T11 trainers following in 1955. In 1958, a bomber squadron was formed with English Electric Canberra jets, and Hawker Hunters joined the Vampires shortly afterwards.

Southern Rhodesia was part of the Federation of Rhodesia and Nyasaland (now Malawi) from 1953 until 1963. During this time the SRAF became the Royal Rhodesian Air Force, retaining this title for the air force of Southern Rhodesia when the federation was dissolved, with Northern Rhodesia (Zambia) and Nyasaland (Malawi) becoming independent. In 1965, a Unilateral Declaration of Independence was declared to avoid any grant of independence on terms unacceptable to the country's government. This led to United Nations sanctions, and in 1969, a republic was declared and the 'Royal' prefix dropped from the air force's title. By this time, the RhAF was operating a squadron of twelve Hawker Hunter FGA9 fighter-bombers, one of twelve de Havilland Vampire FB9 fighter-bombers, and another of twelve Canberra jet bombers. A squadron of twelve BAC Provost T52 armed-trainers handled COIN duties. Transport used four Douglas C-47s, and eight Sud Alouette III helicopters. Despite sanctions, a Beech Baron was obtained in 1967 for communications work. Guerrilla activity intensified during UDI, and the RhAF was heavily involved, remaining fully operational despite sanctions. Alouette III helicopters frequently ferried troops while overloaded, achieving this through making a 'running' or STOL take-off.

Rhodesia became independent in 1980, and after a spell as Zimbabwe-Rhodesia became Zimbabwe. The new state changed its allegiances, receiving from China more than twenty X'ian F-7 (MiG-21) fighter-bombers as well as FT-7 (MiG-21U) and FT-5 (MiG-17U) trainers. Air and ground crew for these aircraft were trained in China. BAe Hawk 60 armed-trainers were also introduced, while second-hand Alouette IIIs were obtained, with almost forty operated at one stage. Zimbabwe became involved in regional conflicts, sending forces to Mozambique to support the government against guerrillas, and later to the Democratic Republic of the Congo, where rebel forces claim to have shot down seven of the F-7s. Rapid expansion to eight squadrons was unsustainable, and in recent years serviceability has been poor, while attrition and poor maintenance has seen operational aircraft numbers fall sharply, not helped by the country's deteriorating economic situation in recent years.

Today, the ZiAF has 4,000 personnel. Average annual flying hours are around 100. There are just seven F-7s operational in one squadron, another has eleven K-8 armed-trainers, but while some sources show a squadron of twelve Hawker Hunters, these are believed to be in store. A reconnaissance squadron has fourteen Cessna 336 Super Skymasters. Transport has a composite squadron with eight CN-212s, an Il-76 and five BN2 Islanders, while ten C-47s are believed to be in store. A single helicopter squadron operates six Mi-35 attack helicopters, two VIP AS532 Cougars, eight AB412s and four Mi-8s. Training uses the K-8 force as well as a few surviving MB326s and eight SF-260s, with two Alouette IIIs.